AF763057

Innovation District Planning

This book aims to fill the knowledge gap on how to plan, develop and manage innovation districts that are competitive in terms of both productivity and quality of living, justifying the massive investment put into place and at the same time doing both in a delicate and harmonious way.

There is a need for smart urban land use that is wired with both hard infrastructures (e.g., telecommunication and transport) and soft infrastructures (e.g., diversity and tolerance). The reader learns this knowledge through conceptual expansions for key insights, frameworks for potential and performance assessment and best practices for global innovation districts. The authors begin innovation district planning with the role and effectiveness of planning a branding in the development of innovation districts. The next key topic of place making is recognised as a key strategy for supporting knowledge generation and innovation activities in the contemporary innovation districts. Another important topic is place quality where the reader learns to identify and classify indicators of place quality by studying global innovation districts best practices. The reader also expands their understanding on the classification of innovation districts based on their key characteristics through a methodological approach. The book concludes with district smartness studied through the socio-cultural role played by anchor universities in facilitating place making in innovation districts. Smart campuses, enabled by digital transformation opportunities in higher education, are seen as a miniature replica of smart cities and serve as living labs for smart technology.

The book serves as a repository for scholars, researchers, postgraduate and undergraduate students as it communicates the complex innovation district phenomenon in an easy-to-digest form by providing both the big picture view and specifics of each component of that view.

"This book on innovation district planning is a must needed 'how to' companion for the expanding drive to construct hard and soft urban infrastructures that attract, maintain, and support the rise of creative smart people and firms, and building tomorrow's economy."
Edward J. Blakely, *University of California Berkeley, USA*

"This timely book provides insights into different aspects of innovation districts, particularly the critical importance of collaborative relationships in the design and operation of innovation districts. The book focuses on the critical role that innovation districts can play in helping urban city-regions become dynamic innovation ecosystems."
David Wolfe, *University of Toronto, Canada*

"This book focuses on development of innovation districts from its origin as science and technology parks to its present form that integrates business climate with people climate, important for attracting and retaining talents and investments. It deals with this development in a convincing way and is a must read for academics, planners, and practitioners."
Bjørn T. Asheim, *University of Stavanger, Norway*

"Innovation—so much desired by urban policymakers and built environment practitioners, but so very hard to foster in many cities. This book fills a large gap that has existed in the literature on how exactly to go about designing innovation into cities. It brings interdisciplinary insights and a wealth of case studies to bear on the challenge of placemaking for urban economic innovation."
Nicholas Phelps, *University of Melbourne, Australia*

"The rise of innovation districts and the knowledge economy are reshaping the future of our cities, and this groundbreaking work provides critical insights about the geography of innovation and the development of place."
Matt Collins, *CEO, Planning Institute of Australia*

"Innovation districts can be a powerful tool for cities to achieve knowledge-based development. To succeed, we need to understand

thoroughly the various types of innovation districts and how they fit different contexts. This book provides the reader with valuable conceptual and practical tools to support planning successful innovation districts."
Antti Lonnqvist, *Tampere University, Finland*

"This book injects innovation district planning into placemaking to explore and advance new understanding and approaches. It integrates strong theoretical debates and empirical analysis into forging novel frameworks for best practices. It stands out for its usefulness and useability for scholars, students, and practitioners in pursuit of new urban planning."
Richard Hu, *University of Canberra, Australia*

Innovation District Planning

Concept, Framework, Practice

Tan Yigitcanlar
Surabhi Pancholi
Niusha Esmaeilpoorarabi
Rosemary Adu-McVie

CRC Press
Taylor & Francis Group
Boca Raton London New York

CRC Press is an imprint of the
Taylor & Francis Group, an **informa** business
A CHAPMAN & HALL BOOK

Designed cover image: Mountain View, California, USA, February 19, 2023: Colorful bikes on the territory of Google headquarters

First edition published 2024
by CRC Press
2385 NW Executive Center Drive, Suite 320, Boca Raton FL 33431

and by CRC Press
4 Park Square, Milton Park, Abingdon, Oxon, OX14 4RN

CRC Press is an imprint of Taylor & Francis Group, LLC

ISBN: 978-1-032-65742-4 (hbk)
ISBN: 978-1-032-65701-1 (pbk)
ISBN: 978-1-032-65743-1 (ebk)

DOI: 10.1201/9781032657431

Typeset in Minion
by MPS Limited, Dehradun

This book is dedicated to the following beloved, beautiful, and brilliant people who have shaped our lives:

Cahide, Susan, Ela, Selin (for Tan)
Brijbala, Saurabh Gaur, Ajay, Shrinika (for Surabhi)
Soraya, Mohsen, Amir Abbas, Touba (for Niusha)
Louis, Susan, Mhyce, Rowland (for Rosemary)

Contents

Foreword

TOWARDS MISSION-ORIENTED INNOVATION DISTRICTS

Innovation districts are rapidly becoming a novel, prominent member in the family of territorial innovation models. Pioneered by the Brooking Institute, a Washington, DC-based policy think-thank and following the initial success of 22@Barcelona, innovation districts are readily adopted by many cities as a global best practice for cities to become leaders in the knowledge economy, as districts are emerging and evolving in places like Oslo, Melbourne, Berlin, London, Eindhoven, Sydney, Medellin, Montreal, Brisbane, Seoul, Stockholm and Toronto. The concept brings together a range of ideas, practices and evidence about the conducive influence of geographical proximity on innovation. Borrowing insights from neo-Marshallian industrial districts, Porter's cluster theory, Regional Innovation Systems and Richard Florida's work on the Creative Class, it highlights the creative and innovative potential of spatial agglomeration and urban density in places where people share knowledge, skills and ideas as they work, meet and socialise together.

So far, the merit of innovation districts lies primarily in its quality as a translational device to mobilise diverse insights on the geography of innovation for urban policymakers and practitioners. Innovation districts feature primarily as a place-based urban development strategy geared to regenerate (often underperforming) downtown neighbourhoods into desirable locations for innovative and creative companies and workers and increase public spill-overs of knowledge-based anchor organisations (such as universities) in their immediate environment. According to the Global Institute of Innovation Districts, there are now more than 100 innovation districts emerging around the world. The same proponents conceive of innovation districts as 'the ultimate mash up of anchor institutions, companies, start-ups and ecosystem-building intermediaries in hyper-local geographies that leverage density, proximity and accessibility'. These districts actively advance a 'collaborate to compete' agenda to solve some of our world's most complex challenges. Drawn together by transit, powered by clean energy, wired for digital technology and fuelled by caffeine, innovation districts are ambitious in their efforts to strengthen local and regional competitiveness, grow new jobs with living wages, create new firms and strengthen inclusive growth and equity across the region.

Innovation districts thus evoke an imagery that is highly desirable for urban planners, policymakers and leaders where universities, businesses, entrepreneurs and creatives constitute a dynamic urban eco-system of innovation imbued with a vibrant inner-city life. Yet, in spite of its theoretical roots, it is worth reiterating that innovation districts have emerged from practice. In fact, it can be argued that the rise of innovation districts makes a typical case where policy has rushed ahead of many fundamental conceptual, theoretical and empirical questions in academia and witnessed an impressive but equally concerning account of global policy mobility. This trade-off is not unfamiliar for the field of Territorial Innovation Models. Nearly two decades ago, Morgan (2004) already warned against fashionable narratives on the geography of innovation and a tendency to collapse abstract concepts and models into simple narratives to render them digestible for politicians and policymakers. This should, however, not give grounds to abandon the profound engagement with policy and practice that characterises much of the scholarship around innovation and local development. Rather, it serves as a reminder to speak 'truth to power', especially in dialogues that cut across academic and societal relevance. Against this background, this book, by Yigitcanlar and colleagues, makes a welcome exception and apposite contribution in providing a rich empirical evidence based on the workings of innovation districts 'on the ground'.

So far, our understanding of the dynamics, governance and impact of innovation districts has remained at best piecemeal, eclectic and rather anecdotal. Breaking a strange academic silence on this topic but mindful of its value outside academia, the book is skillfully positioned at the intersection of academic literature and professional practice. It provides a first-of-its-kind inquiry into the realisation and implementation of a promising boundary concept—innovation districts—that can help to achieve 'real change' and direct the (re)generative potential of human creativity, ingenuity and problem solving to address persistent and complex challenges at the urban level. The book invites a deep dive into the experiences of planning for innovation districts based on an exploration of evidence and practices from mostly Australian cities. Here, at the frontier of an increasingly interconnected and urbanised planet, we may find inspiration for hopeful urbanism to tackle the persistent, complex and increasingly troublesome challenges that require better collaboration and co-creation between practice, theory and research to achieve creative solutions and innovation that are meaningful for our growing population of urbanites.

Especially the back end of the book asks and addresses pertinent questions about the role of innovation districts amidst immanent digital and green urban transitions. Hopefully this alludes to a growing recognition that innovation districts, while having established a significant reputation and legitimacy as vehicles and instruments of urban growth and competitiveness, can (and should) be deployed in a novel but still embryonic paradigm for innovation and innovation policy geared to sustainable transformation (Schot & Steinmueller, 2018). Turning to mission-oriented innovation districts (Fastenrath et al., 2023) requires however granular insight about what works and what does not work in growing innovation districts, about the conditions for equitable and effective governance of innovation districts, about apposite metrics to monitor and measure their performance and about the impacts of innovation districts for its residents, firms, entrepreneurs, workers and universities.

This book lays an important first but hopefully not last stone to build a solid academic foundation to analyse, theorise, scrutinise, plan and govern innovation districts in Australian cities and beyond.

Lars Coenen
Western Norway University of Applied Sciences

REFERENCES

Fastenrath, S., Tavassoli, S., Sharp, D., Raven, R., Coenen, L., Wilson, B., & Schraven, D. (2023). Mission-oriented innovation districts: Towards challenge-led, place-based urban innovation. *Journal of Cleaner Production*, 138079. 10.1016/j.jclepro.2023.138079.

Morgan, K. (2004). Sustainable regions: Governance, innovation and scale. *European Planning Studies*, 12(6), 871–889.

Schot, J., & Steinmueller, W. (2018). Three frames for innovation policy: R&D, systems of innovation and transformative change. *Research Policy*, 47(9), 1554–1567.

Preface

Global knowledge and innovation economy, since the beginning of globalised neoliberal economy in 1980s, has been forcing major cities of the world to seek to become more competitive in reaping the benefits of attracting global investment and talent. To acquire global competitiveness requires cities not only investing in hard infrastructures, e.g., telecommunication and transport, but also soft infrastructures, e.g., diversity and tolerance, and at the same time doing both in such a delicate and harmonious way. Particularly during the last couple of decades, cities have been increasingly devoting their efforts on that front in developing innovation districts as a new urban land use form that are wired with both infrastructure types. There is, however, limited understanding and perspectives on how to plan, develop and manage innovation districts that are competitive both in terms of productivity and quality of living that justifies the massive investment put into place.

Against this backdrop, this book, *Innovation District Planning: Concept, Framework, Practice*, aims to fill the knowledge gap that exists in the academic literature and professional practice. The book on innovation districts serves as a handbook for policymakers, planners, developers and practitioners and provides insightful conceptual expansions, frameworks for potential and performance assessment along with understanding typologies and how they work in practice. This book is a companion to our 2014 Routledge book *Knowledge and the City: Concepts, Applications and Trends of Knowledge-Based Urban Development* and 2019 Springer book *Geographies of Disruption: Place Making for Innovation in the Age of Knowledge Economy.*

This book delves into the conceptual foundations of innovation district planning, providing insights into the key principles and theories that underpin this emerging field. It explores the multidimensional nature of innovation districts, examining their economic, social and environmental dimensions and uncovering the complex interplay between various stakeholders. From place-making strategies and design principles to governance structures and stakeholder engagement, this book offers a road map for transforming urban areas into thriving innovation ecosystem hubs.

The central theme of this book revolves around the significance of collaboration and knowledge exchange among key stakeholders, namely researchers, practitioners and policymakers. Emphasising this collaborative approach is crucial for the effective design and implementation of innovation districts. By fostering cooperation and shared learning, innovation districts can drive economic growth, promote entrepreneurship and tackle pressing societal and environmental challenges.

By presenting evidence-based analysis, theoretical insights and case studies, this book sheds light on the pivotal role played by innovation districts in shaping urban economies and consolidating socio-spatial expansions. The interdisciplinary nature of the content allows readers to gain a nuanced understanding of the complex dynamics and multifaceted impacts of innovation districts. It is our hope that this book will contribute to the scholarly discourse on innovation districts and serve as a reference for future research and practical applications. By engaging with the concepts, frameworks and practices presented in this book, readers will be equipped to contribute to the development of vibrant and sustainable innovation ecosystems.

The book specifically delivers insights and adds value to the following three interconnected and important areas.

- *Conceptual expansions of innovation districts:* The theory behind business clusters is not new. Since the 1980s, introvert, secluded and mono-functional business clusters have emerged in the form of science or technology parks. However, what sets current innovation districts apart from traditional monolithic business clusters is the prominence of urbanism within these districts. These urban innovation districts give value to the quality of place, blurred boundaries, mixed-used development, walkability, heterogeneity, diversity and authenticity. The current book expands this new concept through unifying interdisciplinary knowledge. It contains a wide range of studies from the economy disciplines and social sciences through to urban planning and architectural and urban design.

- *Hard and soft infrastructure planning in innovation districts:* Innovation districts follow a multidimensional perspective including both hard and soft factors in determining urban competitiveness. Since creating liveable and attractive environments is a vital strategy for attracting talents and investments, hard factors mainly include dimensions such as income, property values, office space, accessibility, traffic, technical infrastructure and tax regimes. Soft factors, in contrast, contain social, environmental and natural amenities such as residential amenities, aesthetics and cultural amenities; tolerance for alternative lifestyles, ethnic diversity and cultural scenes; and the creation of meeting places for business and leisure purposes. Planners and policymakers are faced with addressing and combining all these attributes in their innovation districts as places to work, learn, live and play. This book accelerates the understanding of such a multidimensional approach.
- *Best practices of global innovation districts:* Best practices of innovation districts characterise the topic in a practical context. This book presents varied cases and contexts to explore how innovation districts can be shaped and survive in the real world. It includes cases with different physical typologies, industrial types and worker communities, as well as contexts with different economic, environmental and social potentials. From effective governance models and collaborative ecosystems to smart infrastructure and inclusive community engagement, this book delves into the key elements that have contributed to their success. Moreover, the exploration of various cases and diverse contexts provides a unique opportunity for comparative analysis of the subject; the book ensures the in-depth exploration of attributes that promote competitive advantages of a district in comparison to others.

This book is structured under five main parts that are elaborated briefly below.

Part 1—Innovation District: This part of the book sheds light on the new concept of clustering known as innovation districts. This part clarifies that these districts are not solely the agglomeration of firms, which provides financial benefits. They are the place of personal and social life, with a dynamic interaction between people, firms and routine life. This part explains the advantages of clustering, which leads to utilising more shared inputs, gathering thicker labour markets and creating further problem-solving systems. Additionally, this new form of clustering, the innovation district, is considered a response to not only the distinct nature of creative, smart or innovative work but also the changing preferences of society. This part pictures innovation districts as new configurations of place with blurred boundaries between work, learn, life and the community; work is no longer a separate activity during the day. These districts remix the notion of work, learn, recreation and living spaces to create a blurring of activities across all hours of the day. Innovation districts, hence, have become a new nexus between innovation and urbanism to be perfectly aligned with the dynamic of our era.

Part 2—Place Making: This part of the book explores the critical role of place making as a strategic approach to enhance the socioeconomic and environmental success of innovation districts. It provides a multidimensional conceptualisation of place making and establishes its multi-staged and multifaceted nature within the distinct context of innovation districts, drawing upon interdisciplinary exploration and relevant case studies. The role of place making in innovation-nurturing locations—characterised by diverse actors and stakeholders with varied interests—extends beyond economic success. The part contributes to the development of a unique character, fosters a sense of place and facilitates the generation and dissemination of knowledge and innovation. Place making encompasses not only spatio-physical aspects such as design and planning but also comprehensive considerations of the broader context, featured image or branding, functional aspects involving internal and external socioeconomic processes and flows, as well as the meanings attributed by stakeholders. The significance of spatio-temporal factors, including the unique local socio-cultural context, in shaping the distinct identity of these districts is specifically highlighted. This part advocates that a unidimensional or partial exploration of the concept of place making, without due consideration of its diverse aspects and stages, would result in a disintegrative approach.

Part 3—Place Quality: This part of the book explores the notion of place quality regarding innovation districts. Place quality characterises the intersection of key elements that people experience in a place. The part clarifies that place quality is an extremely complicated phenomenon containing objective and tangible physical factors of a place, to subjective and intangible feelings of wellbeing, fulfilment and satisfaction. Place quality is also closely related to the ensemble concept of liveability and sustainability which combine economic, environmental, social and personal dimensions. This part aims to respond to a rising question—how innovation districts can effectively shape such complex place quality as they are ever more responsible for not only empowering knowledge industries and workers but also contributing to the sustainability of cities and societies. Since innovation districts' level is the scale of real day-to-day life and has been connected to touchable places and actual people, it is difficult to measure it solely through quantitative measures. This part combines both qualitative and quantitative measures and methods to define and evaluate place quality in innovation districts.

Part 4—District Classification: This part of the book reveals that innovation clusters have evolved overtime in terms of their functions, features and spatial characteristics from 'science and technology park' model to 'innovation district' model. The latter model's key functional and spatial characteristics vary due to the local contextual factors. Furthermore, this part argues that investing in innovation districts is a risky business as there is no one-size-fits-all innovation district type due to their varying characteristics. However, the part confirms that innovation districts can be holistically classified through a multidimensional framework consisting of numerous attributes—i.e., dimensions, indicators and measures. The part reports on the process of framework development and validation by a panel of international experts through a Delphi survey. The result, a multidimensional innovation district classification framework, is the first of its kind. This part also presents the results of the real practical application of the validated framework on three eminent Australian innovation districts—i.e., Macquarie Park Innovation District (Sydney), Monash Technology Precinct (Melbourne) and Kelvin Grove Urban Village (Brisbane). The results confirmed that innovation district performances can be measured and typologies can be developed through a novel approach.

Part 5—District Smartness: This part argues that there are limited studies investigating university and innovation district symbiosis, particularly in the context of place making. The part aims to investigate the socio-cultural role played by anchor universities in facilitating place making in innovation districts. The results revealed four key roles universities play including the pivotal roles as facilitators of place making through societal integration in innovation districts and they act as platforms for collaboration and exchange of knowledge by adopting extroverted and collaborative approaches. Additionally, this part of the book explores how the smart campus is conceptualised and practiced as a smart district. The part reports on the process of conceptual framework development through the systematic literature review. This part identifies that society, economy, environment and governance are the key smart campus domains to classify existing research. Nevertheless, the part also reveals that there is little evidence of a comprehensive real-world application of the smart campus towards addressing all the four domains. This confirms the infancy of the current conceptualisation and practice of smart campus as a smart precinct. The findings reported in this part offer a firm basis for comprehensive smart campus conceptualisation and provide directions for future research and development of smart campuses and smart districts.

Authors

Tan Yigitcanlar is an eminent Australian researcher with international recognition and impact in the field of urban studies and planning. He is a Professor of Urban Studies and Planning at the School of Architecture and Built Environment, Queensland University of Technology, Brisbane, Australia. Along with this post, he carries out the following positions: Honorary Professor at the School of Technology, Federal University of Santa Catarina, Florianopolis, Brazil; Director of the Australia-Brazil Smart City Research and Practice Network; Lead of QUT Smart City Research Group; and Co-Director of QUT City 4.0 Lab. He is a member of the Australian Research Council College of Experts. He has been responsible for research, teaching, training and capacity building programs on the fields of urban studies and planning at esteemed Australian, Brazilian, Finnish, Japanese and Turkish universities. His research aims to address contemporary urban planning and development challenges that are economic, societal, spatial, governance or technology related in nature. The main foci of his research interests, within the broad field of urban studies and planning, are clustered around the following three interdisciplinary themes: smart technologies, communities, cities and urbanism; sustainable and resilient cities, communities and urban ecosystems; and knowledge-based development of cities and innovation districts. He has been providing research consultancy services to all tier governments (i.e., federal, state, local), as well as international corporations and non-governmental organisations in Australia and overseas. These services help government and industry form their key strategies, become more resilient and be better prepared for emerging disruptive conditions. He has disseminated his research findings extensively, including over 300 articles published in high-impact journals and 25 key reference books published by esteemed international publishing houses. His research outputs have been widely cited and have influenced urban policy, practice and research internationally. His research was cited over 21,000 times, resulting in an h-index of over 85 (Google Scholar). According to the 2023 Science-wide Author Databases of Standardised Citation Indicators, amongst the urban and regional planning scholars, he is ranked #1 in Australia and a top-10 ranked researcher worldwide. He was also recognised as an Australian Research Superstar in the Social Sciences Category at The Australian's 2020 Research Special Report, and he was named as the Australia's Social Sciences 'Research Field Leader' for the Urban Studies and Planning discipline in the 2024 edition of *The Australian Research Magazine.*

Surabhi Pancholi is a researcher with international recognition and impact in the field of architecture, urban design and urban planning, particularly specialised in place making. She is currently a Lecturer of Architecture at the School of Architecture and Built Environment, Deakin University, Geelong, Australia. As a firm believer in collaboration and interdisciplinary approach, her research works have gained international recognition on the topics of multidimensional place making, design principles, strategies and planning processes of urban knowledge and innovation clusters named variously as smart cities, knowledge cities, smart innovation districts, technology hubs, creative precincts and the like. She has a successful record of extensively publishing in high-ranked and esteemed international journals in the field of urban design and urban planning. She has developed a strong network at local and international levels by her collaborative projects and joint research publications, frequently presenting at international conferences and seminars. In her professional roles, as an architect and urban planner, she has worked on a range of prestigious commercial, residential, institutional and government projects nationally and internationally. She advocates the adoption of a multidisciplinary perspective and an integrated approach for making cities and clusters more people-oriented, economy-wise, environmentally sustainable and societally integrated.

Niusha Esmaeilpoorarabi is an Assistant Professor at the School of Architecture, Faculty of Fine Art, University of Tehran, Iran. She earned a PhD at Queensland University of Technology and her master's and bachelor's degrees at the University of Tehran, specialising in architectural design and urban studies. Niusha's research aims to address interdisciplinary themes from architecture, urban studies and social sciences fields. The main focus of her research includes knowledge-based development of cities, neighbourhoods and in particular innovation districts; her research also supports the sustainable growth of neighbourhoods and communities. Her works have been published in high-ranked journals and received attention from the industry. Since 2019, she has been collaborating with several organisations, including the Ministry of Science, Research and Technology, Iran Association of Science Parks and Innovation Organisations, to deliver workshops and training events promoting place making and community engagement in innovation districts. Additionally, she has been involved in a wide range of national and international projects in industry, from the planning process to designing phases. This also includes multiscale developments from housing projects to public urban spaces.

Rosemary Adu-McVie is a researcher with international recognition in the field of urban planning, specialising in the classification of innovation districts. She is currently a Lecturer of Property Studies at the Department of Survey and Lands, Papua New Guinea University of Technology, Lae, Papua New Guinea. She is a practitioner-turn-academic with about two decades of industrial experience in the fields of property management and urban valuations. Her research works focus on innovation districts and the like, covering the topics of the evolution of innovation districts, multidimensional classification of innovation districts, performance assessment of innovation districts and innovation district typology. She has successfully published in high-ranked and esteemed international journals in the field of urban planning and built environment. She has developed a strong network at local and international levels through her collaborative projects and research publications.

1

Innovation District

This part of the book sheds light on the new concept of clustering known as innovation districts. This part clarifies that these districts are not solely the agglomeration of firms, which provides financial benefits. They are the place of personal and social life with a dynamic interaction between people, firms and routine life. This part explains the advantages of clustering which leads to utilising more shared inputs, gathering thicker labour markets and creating further problem-solving systems. Additionally, this new form of clustering, the innovation district, is considered a response to not only the distinct nature of creative, smart or innovative work but also the changing preferences of society. This part pictures innovation districts as new configurations of place with blurred boundaries between work, learn, life and the community; work is no longer a separate activity during the day. These districts remix the notion of work, learn, recreation and living spaces to create a blurring of activities across all hours of the day. Innovation districts, hence, have become a new nexus between innovation and urbanism to be perfectly aligned with the dynamic of our era.

DOI: 10.1201/9781032657431-1

CHAPTER 1

Place Branding and Innovation Districts

1.1 INTRODUCTION

In the era of global knowledge and innovation economy, cities have been pursuing 'knowledge-based urban development' (KBUD) to gain a global competitive edge (Carrillo, 2004). Subsequently in recent years, a strong emphasis has been put on planning and branding mechanisms for place making for knowledge and innovation milieus (Hospers, 2008). Many of these milieus are labelled with brands such as innovative, creative, smart, intelligent, and knowledge-based (Komninos, 2002). For instance, various well-known cities, including Singapore, Barcelona, Copenhagen, Manchester, and Melbourne have pursued this route, most of the times with overt planning and place branding support (Landry, 2000). However, the role and effectiveness of planning and branding are questioned as developers largely base their marketing approach on the promotion of a desired set of values and image, but most of the time they do not necessarily take action to invest in the long-term achievement of this promise and the development plan (Stigel & Frimann, 2006). Regrettably, development practice tends to focus on marketing new properties rather than the creation of a unique identity and place (Morgan *et al.*, 2002). Consequently, this eventuates in the generation of unsuitable or unattractive environments for knowledge communities.

Unique identity and sense of place are important attractors in place making for knowledge communities (Merrilles *et al.*, 2013). In other words, conceived vision for knowledge communities should be based on a legitimising identity, the ideal image that developers want to promote and also deliver. However, there are recurrent mismatches between the promoted and delivered spaces. This issue has not been tackled

DOI: 10.1201/9781032657431-2

thoroughly in the literature so far. Thus, addressing the question of 'how knowledge and innovation milieus can be conceived to achieve an intended vision' can provide an understanding of the actual influence of place branding in guiding planning and design as a prior strategy more than a final marketing device. This also forms the rationale for the research reported in this chapter.

This study is confined to the conceived dimension of space, and the chapter aims to investigate the role of planning and branding in place making by assessing the effectiveness of planning and branding strategies in the development of knowledge and innovation milieus—such as 'innovation district'. Brisbane is chosen as a testbed for this investigation as the city has recently been branded as 'Australia's New World City', in an attempt to help in catching up with and joining the group of cities that reap the benefits of global knowledge economy. The case study investigation focuses on Brisbane's three major innovation district developments—i.e., Diamantina Knowledge Precinct (DKP), Kelvin Grove Urban Knowledge Village (KGUV), and Sippy Downs Knowledge Town (SDKT). These cases represent innovation districts at different development stages, from near completion to advanced planning, and they document cycles of subsequent visions translated into specific branding and planning propositions, which have then directly influenced the detailed urban design of these cases.

1.2 LITERATURE BACKGROUND

1.2.1 Knowledge-Based Urban Development

As Gabe *et al.* (2012, p. 1179) indicate, '[i]t would be an understatement to suggest that knowledge plays a key role in today's economy; for much of the developed world, it might be more accurate to assert that knowledge is today's economy'. Thus, knowledge generation has been a priority for global city administrations, and there is a growing consensus amongst scholars, planners, politicians, and industrialists in identifying KBUD as a panacea to the burgeoning economic problems (Knight, 1995; Kunzmann, 2009; Yigitcanlar, 2010, 2011; Huggins & Strakova, 2012; Lönnqvist *et al.*, 2014). Although, in the era of global knowledge economy, KBUD is a critical factor for economic success (Pratt, 2000; Sheppard, 2002), it is not solely an economic policy. For many, KBUD is a policy that targets building an urban setting to form perfect climates for business, people, and governance in an environmentally friendly atmosphere (Carrillo, 2006; Ergazakis *et al.*, 2006; Angelidou *et al.*, 2012).

In terms of KBUD, traditionally places from the most developed nations—e.g., Cambridge Science Park, Massachusetts Route 128, Silicon Valley, and Sophia Antipolis Technology Park—were the most dominant ones. However, globalisation has created a more opportunistic arena where nowadays, newly developed places—e.g., Dubai Knowledge Village, Hsinchu Science Park, Multimedia Super Corridor, Singapore One-North, and Songdo Science Village—can compete with the charm and appeal of older and more established places (Hanna & Rowley, 2008; Yigitcanlar & Bulu, 2015; Yigitcanlar & Sarimin, 2015). Consequently, today KBUD is also a popular trend for less developed cities to gain a competitive edge to catch up with more established ones (Tomaney & Bradley, 2007; Yigitcanlar *et al.*, 2008a; Yigitcanlar & Lönnqvist, 2013).

Creative Class Thesis (Florida, 2012)—even though heavily criticised (see Markusen, 1996)—brought a new perspective to KBUD. It suggests that places with high concentrations of talent, technology, and tolerance exhibit a higher level of economic development and foster an open, dynamic, personal, and professional urban environment. This new 'investment follows talent' perspective encouraged many city administrations to develop new KBUD policies to transform their cities, create new knowledge and innovation milieus, and provide home and lifestyle options suitable for the target talent groups—i.e., knowledge communities (Wedemeier, 2010; Florida, 2012; Yigitcanlar *et al.*, 2014). In recent years, innovation districts have become a popular form of urban knowledge and innovation milieus. According to Baum *et al.* (2009, p. 44), '[t]here is a significant policy discussion regarding the role of [these] clusters in strengthening local economies and significant energy has been expended discussing the many positive outcomes of such developments'.

1.2.2 Knowledge Community Precincts

Making a place that concentrates on knowledge generation has become a priority for many nations and cities that are competing in the global knowledge economy (Asheim, 2007; Malecki, 2007; Claessens *et al.*, 2010; Pancholi *et al.*, 2014). Knowledge generation profoundly depends on creativity and innovation, which are human-centred activities and dominantly urban phenomena. Hence, particular attention has been paid to building urban places with a certain scale, accessibility to and intensity of knowledge infrastructure, as well as vibrant urban life with a full mix of diversity and tolerance (Kunzmann, 2004; Florida, 2012).

Such places that house knowledge communities are referred interchangeably as 'science /technology/innovation park', 'knowledge hub', 'knowledge/innovation district', 'knowledge precinct', 'innovation district', and so on (Ku *et al.*, 2005; Hu, 2008; Ratinho & Henriques, 2010). Innovation district is the spatial nexus of KBUD, as it is a mixed-use urban setting that includes a critical mass of knowledge enterprises and advanced networked infrastructures, and developed with the aim of collecting the benefits of blurring the boundaries of living, shopping, recreation, and working facilities of a knowledge community (Yigitcanlar & Dur, 2013).

High dependence of knowledge generation on a large pool of talented labour power and consumption is critical to form a functional urban region that is suitable for innovation district development (Petruzzelli *et al.*, 2007; Florida, 2012). Nevertheless, once established due to the dynamism and intangibility of knowledge community—that is highly footloose—it is a bigger challenge to sustain knowledge generation process and retain these communities (Darchen & Tremblay, 2010; Tremblay & Darchen, 2010). A global city with evidence of world-class city formation—accommodating high-quality urban services and a diversified economic base including extensive supplier and distribution networks and specialised services—has a big advantage (Yigitcanlar, 2009). However, today the global knowledge economy also creates opportunities for newly developed cities to compete with more established ones in attracting talent and investment (McGranahan & Wojan, 2007).

In such a landscape, the most common strategies to attract and retain knowledge communities include concentrating on extensive global networks as intense mediums of exchange for innovation districts to flourish (Van den Berg *et al.*, 2005) and focusing on knowledge workers as primary sources of innovation districts that prefer inspiring cities with a thriving cultural life, an international orientation, and high levels of social and cultural diversity (Yigitcanlar *et al.*, 2007). Furthermore, literature highlights the following five common characteristics of successful innovation districts: Mixed-use environment, centrality, connectivity, learning and playing, and branding (Yigitcanlar *et al.*, 2008b). However, making place for knowledge communities is not an easy task and requires more than just bringing together industry, infrastructure, land uses, and promotions. In innovation districts, generation and flow of knowledge is one of the elements that determine its identity by driving the design of space, defining the use of the place and determining the perceived image through patterns of behaviour and social interaction. In these communities, production and distribution of knowledge is what constructs the 'meaning' that makes these spaces a 'place', and determines the elements of its identity (Castells, 2009). The branding of these new places is generally based on the conceived ideas for the new spaces and focused on attracting a specific target audience. Planning and place branding therefore are the initial triggers to attract the talented audience that will construct the desired knowledge community base.

1.2.3 Planning, Branding, and Place Making

Knowledge industries and innovation districts—as working and living milieus of knowledge workers—have become the new hope for cities, policymakers, and urban development agencies after the gradual diminishing of manufacturing industries from the face of developed nation cities (Kunzmann, 2009). In the global knowledge economy, as the competition for resources, investors, skilled workforce, and visitors becomes more intense, recent years have observed increased efforts to differentiate places through planning and branding (Kotler & Gertner, 2002; Bennett & Savani, 2003). Urban planning—as Kunzmann (2004, p. 385) refers to it, '[m]aking of place and mediating of space'—has been employed to the formation of places of knowledge generation more effectively since the last quarter of the 20th century. Strategic planning and master planning are techniques largely involved in the process for generating places of knowledge concentration. Whilst strategic planning offers a strategic vision for the development, master planning delivers a comprehensive guidance and instruction about the features of the project to achieve the vision (Carrillo *et al.*, 2014).

Despite much scepticism, place branding has been a growing practice all over the world (Pasquinelli & Teräs, 2013). Place-branding strategies have been inspired by the principles of marketing strategy in business (Kotler *et al.*, 1993; Anholt, 2008). It refers to the practice of applying appropriate marketing strategies to differentiate places in the competition, with regards to economic, socio-cultural, spatial, and political aspects (Kavaratzis, 2005; Kaplan *et al.*, 2010). Today, place branding is a common practice to market the history, quality of place, lifestyle, and culture for opportunity, prestige, or power in capital accumulation in a competitive environment (Evans, 2003; Zhang & Zhao, 2009). It helps in catching up with the other successful places (Zenker, 2011). Branding involves influencing people's ideas by forging emotional and psychological associations with a place (Eshuis & Edwards, 2013). Thus, planners/

designers see place branding as an effective method for 'spatial cognition' that is the knowledge people have of spatial unities such as places or simply the 'subjective knowing of a space' (Golledge & Stimpson, 1987; Veitch & Arkkelin, 1995). Studies into spatial cognition show that the image of a place is influenced in a positive manner by the extent to which the area is known in the outside world (Kalandides, 2011). This explains why places increasingly invest in a strategy of positive image formation through branding (Anholt, 2007; Zimmerbauer, 2011).

Through the process of creating a brand for innovation districts, planners, designers, and developers make powerful decisions in relation to the sensorial and symbolic elements that might shape the perceptions of that place, and determine the image of the groups of people who experience that place (Dixon & Durrheim, 2000; Rantisi & Leslie, 2006). Brakus *et al.* (2009) define brand experience as the subjective, internal consumer responses and behavioural responses evoked by brand-related stimuli that are part of a brand's design and identity, packaging, communications, and environments. The brand stimuli communicate the newly designed space to potential consumers—e.g., inhabitants, workers, and visitors—attracting them to use the space, and by so doing, add meaning and transform this space into a place. The brand identity then rises from this lived space; people move in, add meaning to the space, and create their own place. This, in turn, emanates more stimuli to the external communities, which perceive the space and create an image in their minds about that place according to its perceived identity (Zimmerbauer, 2011).

In the global knowledge economy, place branding has become a central basis for successful competition (Okano & Samson, 2010). Knowledge community as a target group is particularly popular in place branding as cities are in a tough competition for residents and workers in general, but for the ones with talen. Zenker (2009, p. 31) indicates that the main desire of knowledge workers is urbanity and diversity, and claims that 'meeting the special needs of these social groups with place marketing and branding will support the goal of attracting the creative class to a city'. Innovation districts have a great symbolic value as they help distinguish a particular area. At present, many cities worldwide apply innovative strategies, including forming new niche markets through the development of innovation districts, for transforming themselves successfully into knowledge cities (Yigitcanlar, 2014). Creating brands for these innovation districts comes almost naturally as a stage of their planning. Place-branding techniques—the practice of applying brand strategy and other marketing techniques and disciplines to the economic, socio-cultural, spatial, and political developments of a place (Kerr, 2006)—are frequently accommodated during the planning phase of these milieus (Gold & Ward, 1994; Lucarelli & Berg, 2011). For instance, 22@bcn Barcelona, Brainport Eindhoven, One-North Singapore, Orestad Copenhagen, Oulu Technopolis/Medipolis, and Taipei 101 have been using place branding as an effective strategic tool (Hospers, 2006; Pasquinelli & Teräs, 2013).

1.3 EMPIRICAL STUDY

This empirical study is structured around a comparative analysis of innovation districts to explore the effectiveness of branding and planning strategies in the successful development of knowledge and innovation milieus. For the comparative case study investigation, the leading three innovation district projects from Brisbane are selected: Diamantina Knowledge Precinct (DKP), Kelvin Grove Urban Knowledge Village (KGUV), and Sippy Downs Knowledge Town (SDKT). Distinctive characteristics of these Innovation districts are provided in Table 1.1.

Documentations on planning, branding, and promotion of innovation districts were gathered and analysed to profile the intended identity of the case studies. The driving principles of the innovation district projects were then isolated, and recurrent keywords and graphic elements were identified and classified in relation to the analytical framework illustrated in Figure 1.1. This framework is used to explore the correlation between the branding and planning strategies and changes through different stages of the project. It consists of four key elements: Place (investigated through conceived place); brand (investigated through the published promotional material); plan (investigated through indicators of character, layout and design); and external and internal (pre)conditions (investigated through economic, socio-cultural, spatial, and political factors). These elements are elaborated below:

- *'Place'* is the outcome of the development efforts and only conceived dimension is the subject of this study.
- *'Brand'* is discussed based on the promotional materials in order to assess how the intended identity of the innovation district is conceived and presented to the public. Promise, features and stimuli are selected as brand indicators where: 'Promise' consists of the vision for

TABLE 1.1 Salient Characteristics of the Case Studies

Case Study	Broad Brand Declination	Knowledge Brand and Scale	Anchor Institution	Locational Context	Area	Distance to Brisbane CBD	Occupancy Take-up Rate	Presence of a University Unit
DKP	Ecology	Knowledge precinct/ zone	Queensland Government, Commonwealth Scientific and Industrial Research Organization (CSIRO), and University of Queensland	Inner city Southern suburb	9 ha	3.8 km	~50%	Research units of the University of Queensland
KGUV	Creativity	Knowledge village/ district	Queensland University of Technology	Inner city Northern suburb	16 ha	3.7 km	~90%	A campus of the Queensland University of Technology
SDKT	Lifestyle	Knowledge town/suburb	University of the Sunshine Coast, and Innovation Centre Sunshine Coast	Regional coastal suburb	100 ha	100 km (and 10 km to Maroochydore city centre)	~20%	Main campus of the University of the Sunshine Coast

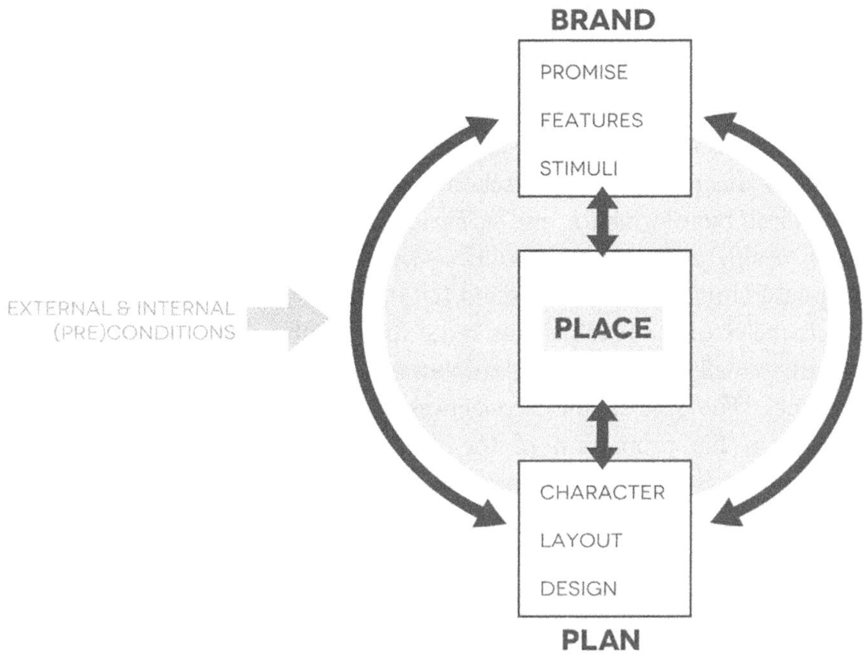

FIGURE 1.1 Analytical framework of place making.

innovation district articulated in the main message of the brand; 'Features' refer to the key elements reiterated in promotional material as fundamental components in defining the identity of development, and; 'Stimuli' include visual, rhetoric, and emotional elements used to engage the audience in the brand's vision.

- *'Plan'* refers to the actual development of the vision in a physical layout. This is assessed through the analysis of planning documents and guidelines. Character, layout, and design are selected as plan indicators where: 'Character' summarises the overall planning strategy to implement the brand's vision; 'Layout' includes urban form and land uses, evaluating the spatial distribution of the features promised in the brand's vision; and 'Design' refers to the actual design of the built environment—i.e., public and shared areas—and it is used to assess how the brand stimuli are translated into the built form.
- *'External and internal (pre)conditions'* include economic, socio-cultural, spatial, and political factors inside and outside of the innovation district that impact on its development, performance, and success levels.

1.3.1 Diamantina Knowledge Precinct

DKP is an inner-city innovation district developed on a brownfield site at the South of Brisbane CBD. In DKP, the idea of knowledge community was strongly promoted in the first stage of the project (2003–2006). Initially the vision for this precinct was loosely based on R&D with particular focus on biomedical, bioinformatics, and ICTs, to take advantage of the proximity to major health and research facilities—i.e., Princess Alexandra and Mater Hospitals, University of Queensland (UQ). In the second stage (2007–2008), the character of the precinct was more firmly defined and promoted as an environmentally and culturally responsible project envisioning sustainable outcomes. The idea of the Ecosciences Precinct was assumed as flagship element in the promotion of the project articulating the concept of a landmark development as promised in the preliminary proposal. During the third stage (2009–2014), the promotion of the precinct was realigned to the heritage and character of the site suggesting a vibrant and innovative urban innovation district. Although the presence of the historic Boggo Road jail was acknowledged in all stages of the project, it was only after the Ecosciences Precinct was under construction and the complementary mixed-used buildings were planned that this element became central in the promotional materials. Some constant features are promoted on DKP's communication material—proximity to CBD, accessibility through integrated public transport systems, mixed-use development, high-quality environment, and lifestyle. The concept of lifestyle is strongly promoted in the third stage of the development. The visual communication that illustrates the project through artists' impressions presents an environment characterised by an active community, based on pedestrian friendly streetscape with a clear presence of outdoor dining and coffee shops. Innovation district features are rendered through rhetoric that stresses the innovation and uniqueness of the precinct as a vibrant sustainable community.

From the planning point of view, DKP follows the typology of an urban innovation district with a layout based on a main street dominated by the anchor project—i.e., Ecosciences Precinct—and the historic jail. The master plan follows transit-oriented development principles promoting a pedestrian mixed-used environment strongly integrated with existing transport facilities. Built form is limited to nine stories in height and provisions are made to guarantee permeability through blocks. The promise of a vibrant community is translated in design guidelines aimed to promote a safe suburb—via crime prevention through environmental design—based on human scale areas and an interconnected open space system. The permeability of the site is enhanced by maintaining visual connections through the different areas of the innovation district and by setting views on landmark vistas outside the village—i.e., river and CBD. The legibility of the urban environment is also prescribed as a key factor in this development. The data collected about DKP makes it evident that the brand is constructed upon the vision of a sustainable community based on a research centre focused on environmental sciences and innovation. This characteristic has been a key factor in promoting the project to broader knowledge communities. This changes, however, when the target audience is potential buyers. The promotional message becomes more focused on the potential of a high-quality lifestyle in and around the innovation district. This is evident in the role of emotional stimuli—the heritage and history of the area. It is clearly observed that these qualities are not promoted in early stages of the development, but are increasingly presented in recently released promotional and advertising materials. Figure 1.2 illustrates the layout of the innovation district, and Table 1.2 lists the key findings.

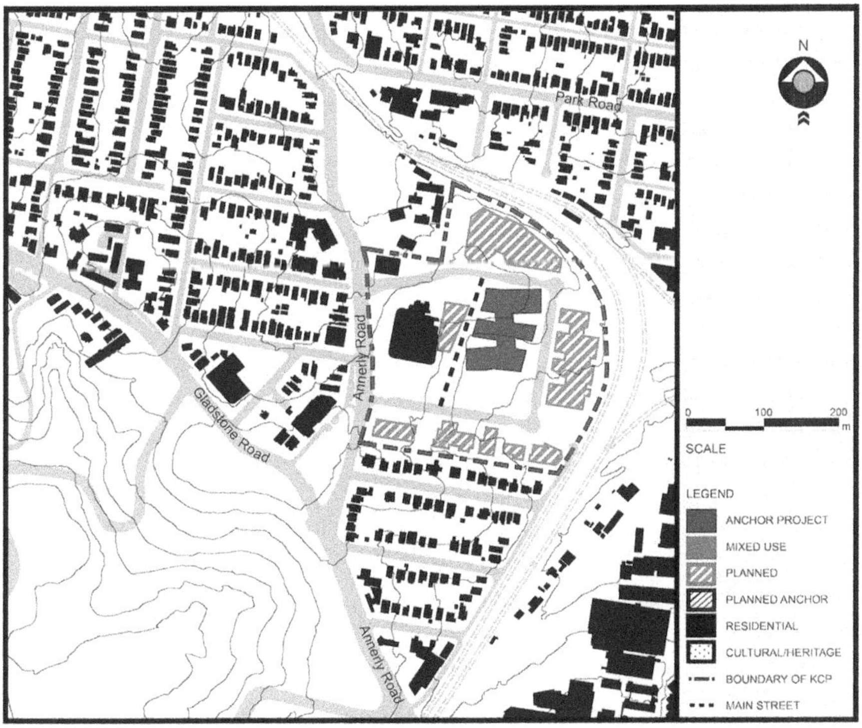

FIGURE 1.2 Layout of Diamantina Knowledge Precinct.

TABLE 1.2 Diamantina Knowledge Precinct Findings

	Stage I (2003–2006)	**Stage II (2007–2008)**	**Stage III (2009–2014)**
	QG (2003)	HRPPC (2007)	QG (2009)
Brand	• Former site of Boggo Road jail as opportunity to develop a unique precinct for knowledge industries and living community	• Importance of environmentally and culturally responsible development in order to achieve an environmentally and economically sustainable future	• DKP as unique and vibrant community based on historic heritage site
Promise	• Landmark project • Focus on medical research, ICT, computer modelling, and bioinformatics • Innovative business incubator	• Integration with the existing context • Mixed urban village lifestyle • Lifestyle opportunities	• World-class research facilities • Exciting and vibrant environment • Benefits for the local and broader community
Features	• Mixed-use development • Location close to the CBD • Accessibility through existing transport hub • Proximity to other major facilities and institutions	• Integration with the existing context • Mixed urban village lifestyle • Lifestyle opportunities • Ecosciences precinct • Affordable housing • Sustainable development • Mix of education and community facilities structured around open and public spaces	• Mixed-use development • Location close to the CBD
Stimuli	• Visual elements based on strong use of green • Vegetation and natural features are emphasised	• Rhetoric emphasises how the development will be innovative, sustainable and enable a specific unique community lifestyle	• Emphasis on the former Jail and the heritage legacy of the site • Vibrant colour scheme

	• Built form is presented as low impact, medium density blended in with the contest • Prominence given to existing transport hub		• Visual elements referring to technology, but with clear sustainable features • Images of vibrant environment with a clear focus on lifestyle and urban atmosphere
Plan	• Information is not available from the planning and branding documents	• Urban knowledge precinct based on the reuse of heritage structure and developed following as transit-oriented design (TOD) close to the central business district (CBD)	• Information is not available from the planning and branding documents
Character	• Information is not available from the planning and branding documents	• High-quality mixed-used development anchored by the Ecosciences precinct project • Sustainable and resource efficient development; passive design principles • Synergy between new development, heritage listed jail and existing neighbourhood	• Information is not available from the planning and branding documents
Layout	• Information is not available from the planning and branding documents	• Open space network as framework for the precinct • Urban form up to 9 stories in height • Highly active main street • Permeable urban form with linkages between public spaces through buildings	• Information is not available from the planning and branding documents

(*Continued*)

TABLE 1.2 *(Continued)* Diamantina Knowledge Precinct Findings

	Stage I (2003–2006)	**Stage II (2007–2008)**	**Stage III (2009–2014)**
Design	• Information is not available from the planning and branding documents	• Legible and safe pedestrian environment • Visual connections between elements within and outside the precinct • Human scale • Crime prevention through environmental design (CPTED) principles • Water-sensitive urban design (WSUD) • Enhancement of amenity of the area	• Information is not available from the planning and branding documents

1.3.2 Kelvin Grove Urban Village

KGUV is recognised as a successful innovation district building effort and a signature initiative in Brisbane (Charles, 2011). The development is characterised by a specific approach to achieve a real social mix providing different housing solutions and cater for different needs. The interest of Queensland University of Technology (QUT) to acquire and develop the site of disused army barracks close to existing teaching facilities was negotiated with Queensland Department of Housing. The general idea of the project was to promote an innovative joint venture between university and public and private sectors—i.e., triple-helix model. The preliminary brand was based on the idea of a vibrant village atmosphere where living, working, playing, and learning are integrated in a lively environment that is characterised by a constant presence of 'creativity'. The aim for uniqueness in the conceptualisation of this innovation district is strongly promoted in its planning documentations. Queenslander character and a design based on sustainable and subtropical principles were constantly present in the communication material. Two kinds of features were promoted: A specific lifestyle (including social housing option) based on the proximity to the CBD and other major urban facilities as well as strong integration with the surrounding neighbourhoods and accessibility to the area through major transport hubs. The promise was based on the peculiar character and history of the area. The use of public art and the involvement of artists in the development of the site were clearly communicated in promotional material and design guidelines. KGUV has a high level of sophistication in the detailing of the different components of public open spaces. Pavements, street furniture, landscaping, artwork, and the use of colour are all controlled elements that communicate the creative environment and its mixed Indigenous, European, and military heritages.

The built form is based on a main street connecting two landmark elements—the historic QUT campus at the top of the hill and the former Upper Parade Ground alongside Kelvin Grove Road. The latter was redesigned as the core of QUT Creative Industries Precinct comprising a system of interconnected squares. The built form was promoted as variegated and differentiated to achieve a sense of vibrancy and the promised village atmosphere. The legibility and permeability of the built form were promoted in the design guidelines for the site and prescriptions were made about making building entrances evident through awnings, landscaping, or arcades. While the creative component of the site is constant in the promotion of the development, this element is more evident in its design and promotions during the first (2001–2003) and second (2004–2007) stages of the project. During the third stage (2008–2009), the focus was placed on the community lifestyle and vibrancy. This coincides with the development of a major health research facility in the heart of the village that was promoted as a new element enriching the diversity of the area. In the fourth stage (2010–2014), the selling point was brought back to the creative nature of the site. Creativity then became so central that prescriptions were made for the provision of artworks on new buildings. The reuse of military facilities and the integration with university structures was later promoted as the main feature of KGUV. Figure 1.3 illustrates the layout of the innovation district, and Table 1.3 presents the key findings.

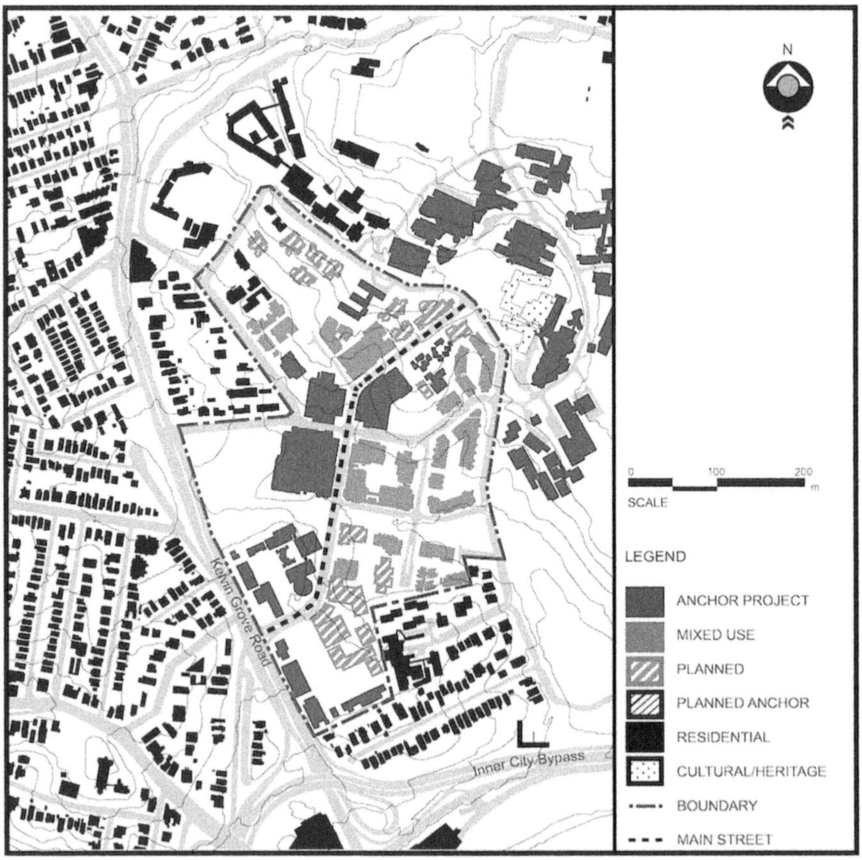

FIGURE 1.3 Layout of Kelvin Grove Urban Village.

TABLE 1.3 Kelvin Grove Urban Village Findings

	Stage I (2001–2003)	Stage II (2004–2007)	Stage III (2008–2009)	Stage IV (2010–2014)
	Queensland Government (2001, 2002)	QG (2004a, b, c), BCC (2005)	QG (2008a, b)	QG (2010), KGUV (2012)
Brand	• Dynamic village atmosphere joining together learning, working and leisure activities in a socially and culturally relevant site	• KGUV is an exciting first for Queensland and will lead the way for future sustainable developments	• Master planned community highly integrated with the rest of the city and offering unique village atmosphere	• Primacy of KGUV as first development of this kind in Australia based on learning, creativity, community, and sustainability
Promise	• Integrate urban environment blending University and village activities to achieve a national point of reference for creative Industries • New exciting living and learning opportunities in a sustainable development where creativity is at the centre • First creative industries precinct in Australia	• New urban environment that offers unique elements in the Brisbane context • Innovative and successful type of development	• A diverse city fringe community where learning is linked with enterprise, creativity industries with community • Connection with the CDB and the surroundings • Lifestyle community • Award-winning urban design	• KGUV in time will become a vibrant suburb with village atmosphere based on an inner-city community • The development is unique in terms of style, innovation, diversity and for its focus on people
Features	• High-quality facilities • Proximity to the CBD and accessibility of the area. • Lifestyle • Recognisable Queenslander architectural style	• Innovative research facilities such as the new health precinct • Vibrant and integrated urban village atmosphere • New development as business incubator	• Information is not available from the planning and branding documents	• Reuse of military heritage buildings • Integration with university facilities

(*Continued*)

TABLE 1.3 *(Continued)* Kelvin Grove Urban Village Findings

	Stage I (2001–2003)	Stage II (2004–2007)	Stage III (2008–2009)	Stage IV (2010–2014)
	• Use of subtropical design principles to achieve a sustainable development • Integration of art features to represent the history and relevance of the area • Themes of the development are creativity, indigenous as well as European heritage. Legacy of the former barracks • Housing and social mix • Mixed-use development	• New leading research and teaching facilities in Australia • State-of-the-art performance and teaching spaces		
Stimuli	• Rhetoric has strong emphasis on the creative element of the development • Text focuses on the heritage of the area • Documents reiterate the concepts like quality, high standards and innovation	• Rhetoric stresses innovation and primacy of the development • Visual communication is based on research-related images as well as modern and vibrant urban environment • Colour scheme based on grey, purple, Prussia blue, and green conveying an idea of sophistication	• Images present a vibrant community and a specific lifestyle where work, play, and research are integrated	• Rhetoric promotes innovation and uniqueness of the development • Strong images of a creative and dynamic environment

Plan	• Urban village on brown site aimed to integrate existing and new research/education facilities in a mixed-used development	• Integrated and master planned urban village within a mixed-use and highly urban environment	• Information is not available from the planning and branding documents	• Information is not available from the planning and branding documents
Character	• Distinctive character rendered in the quality of open and public spaces through you of art and visual stimuli	• Combination of different uses • Buildings to reflect subtropical and sustainable design principles • Retail at street level to enhance the main street image • Environment to be visually interesting	Information is not available from the planning and branding documents	Built form to educate public about sustainability and subtropical design principles
Layout	• Information is not available from the planning and branding documents	• Main street development • Village centre up to 7 stories • Mixed-use precinct 7 stories • Built form to respect heritage legacies as the Upper Parade ground • Built form to be differentiated • Integration with the surrounding suburbs • Provision of social housing	• Information is not available from the planning and branding documents	• Public real is the robust backbone of the development
Design	• Information is not available from the planning and branding documents	• Amenity and security • Accessibility and permeability • Attractive pedestrian environment with active frontage	• Information is not available from the planning and branding documents	• Buildings required to integrate public art, especially when in corner sites

(*Continued*)

TABLE 1.3 *(Continued)* Kelvin Grove Urban Village Findings

Stage I (2001–2003)	Stage II (2004–2007)	Stage III (2008–2009)	Stage IV (2010–2014)
	• Safe and supportive community • Design to provide a sense of community and highly legible urban environment • Recognisable feature to identify buildings entrances • Water-sensitive urban design (WSUD)		• Artworks to contribute to the general atmosphere of the area

1.3.3 Sippy Downs Knowledge Town

SDKT is a regional innovation district development on a greenfield site comprising the main campus of the University of the Sunshine Coast (USC), Innovation Centre Sunshine Coast, adjoining Sippy Downs town centre, and existing surrounding residential suburban areas. This case provides insight on the preliminary stages of design and branding allowing a reflection on the influence these two elements have on each other. The vision for the intervention is to provide the region with a high-quality urban environment with flexible and permeable spaces. The general idea takes advantage of the existing university structures and promises a new technology hub to diversify Sunshine Coast economy by attracting new activities. Sustainability, subtropical design and memorable environment are recurrent promotional elements. The unique selling feature is the location combined with the peculiar Sunshine Coast character and lifestyle. Local politicians have defined this development as 'a surfing community of highly skilled professionals' or as the 'Australian answer to Silicon Valley'.

At the first (2001–2006) and second (2007–2009) stages of the project, documentations about this innovation district are mainly institutional and aimed to promote the intervention to stakeholders rather than potential buyers. The message conveyed is mainly based on the sustainability of the project, its pedestrian friendly environment, and the opportunities of proximity to research institutions. Articles in local newspapers and online newsletters are often enriched by artist's impressions of a vibrant contemporary urban space. In some instances, images of USC high-tech buildings are presented with the contrasting presence of kangaroos in proximity of research facilities. The layout of the new town centre is based on a main street development. Perimeter blocks characterise urban form with 5–6 stories in the central area and 3–4 stories on access roads. The master plan prescribes buildings to have a character in line with the USC precinct and to adopt a tourist or beach style in line with the identity of this region. Mixed-use development and active frontage are other features promoted in the plan. In the stages of three (2010–2011) and four (2012–2014), the Sunshine Coast Regional Council particularly paid attention to support SDKT as a main project to stimulate the economy. However, the local administration clearly tries to avoid competition between the different urban areas on its territory specifying the character and role of each one. Although SDKT has a

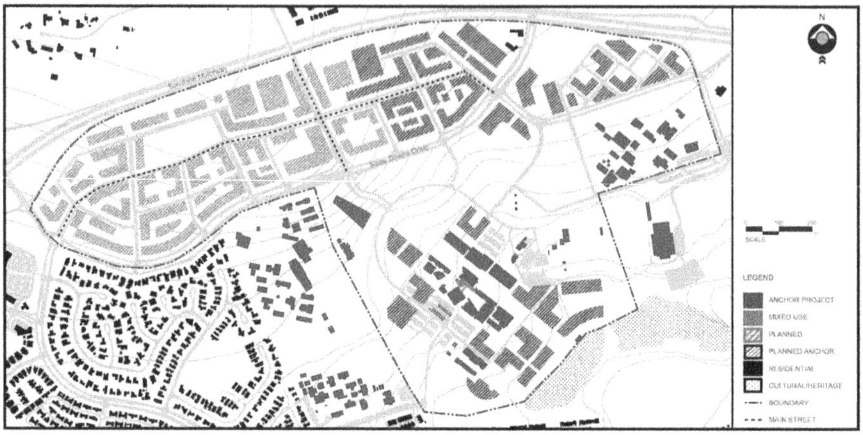

FIGURE 1.4 Layout of Sippy Downs Knowledge Town.

focus on knowledge generation and research, this is not strongly translated in the official policy documents other than referring them as opportunities. Figure 1.4 illustrates the layout of the innovation district, whereas Table 1.4 displays the key findings.

1.4 RESULTS AND DISCUSSION

1.4.1 Findings

The case studies represent three different conditions in terms of size, location, and development phases. KGUV is the most advanced and consolidated innovation district. Its built form is growing constantly and is closer to its completion than the other two. KGUV has a specific brand that in time has been revamped and restyled, but has maintained a clear focus. DKP is a smaller intervention and is in the phase where only the anchor project is completed. Its brand focuses on the cutting-edge sustainability and eco-sciences research, but this vision is not fully translated in the image or design of the space. SDKT is at its preliminary planning phase. This development is suggested as a new innovation district despite its major selling point being the broader lifestyle of the Sunshine Coast. While DKP and KGUV both start with a clear vision of the precinct as a unique site, and this is communicated in a specific and detailed way; SDKT promises a more generic approach. Interestingly in this case the stimuli are the more appealing and specific, while the concepts delivered are the most vague and

TABLE 1.4 Sippy Downs Knowledge Town Findings

	Stage I (2000–2006)	Stage II (2007–2009)	Stage III (2010–2011)	Stage IV (2012–2014)
	MSC (2000)	MSC (2007), SCRC (2008)	SCRC (2010)	SCRC (2012a, b)
Brand	• Development is required to protect and enhance the likely occurrence of memorable experiences	• Sippy Down centre offers opportunities for economic growth based on knowledge-based industries	• Surfing community of highly skilled professionals	• Sippy Down is an innovative development where research, education, learning, innovation, and investment are fully integrated
Promise	• Sustainable development that reflects the subtropical and coastal settings • Memorable town centre with a strong identity • Quality of the built environment and open spaces • Flexible and permeable spaces	• Connection and integration between the town centre, University of the Sunshine Coast, and technology hub • Fully integrated university town	• The new development will be a key factor in differentiating the Sunshine Coast economy • Sippy Downs presented as Australian answer to Silicon Valley	• The new development supports an appropriate mix of university-based business, employment, residential, and community activities
Features	• Subtropical design • Pedestrian friendly environment • Proximity of working places	• Information is not available from the planning and branding documents	• Information is not available from the planning and branding documents	• Information is not available from the planning and branding documents

(*Continued*)

TABLE 1.4 *(Continued)* Sippy Downs Knowledge Town Findings

	Stage I (2000–2006)	**Stage II (2007–2009)**	**Stage III (2010–2011)**	**Stage IV (2012–2014)**
Stimuli	• Suggestion of street naming related to research/ university context as well as relevant for the local cultural environment	• Rhetoric stresses the uniqueness of the first Australian planned university town	• Visual communication based on a contemporary urban space with high pedestrian activity • University of the Sunshine Coast presented as a modern environment where hi-tech buildings face open spaces populated by kangaroos	• The text promotes the articulation of the Sunshine Coast and its different centres, each one with its own character. Sippy Downs is not in competition with existing urban environments, but promotes new synergies
Plan	• New greenfield town centre based on pedestrian networks and public transport system	• Information is not available from the planning and branding documents	• Information is not available from the planning and branding documents	• Information is not available from the planning and branding documents
Character	• Variety of functions and facilities for the community • Integration with University and knowledge-based precincts	• Information is not available from the planning and branding documents	• Information is not available from the planning and branding documents	• Information is not available from the planning and branding documents
Layout	• Main street town centre • Buildings should reflect diversity, individual expression, but relate to the style and features of the	• Buildings with close relationship with the street	• Information is not available from the planning and branding documents	• Information is not available from the planning and branding documents

	University of Sunshine Coast • 5–6 stories in the town centre • 3–4 stories on access roads			
Design	• Active street frontage • Buildings to reflect the subtropical location, the proximity of the university. • Visually engaging buildings facades • Water-sensitive urban design (WSUD) • Subtropical design principles • Connectivity to the beach	• Information is not available from the planning and branding documents	• Information is not available from the planning and branding documents	• Information is not available from the planning and branding documents

stereotypical. In terms of features, only KGUV provides something unique; generally, all cases deliver Yigitcanlar et al.'s (2008b) findings—i.e., promoting a mixed-used environment, proximity to premium activities, connectivity and networking, a work-live-learn-play settings, place branding. In each case, the brand appeals to potential buyers and in different stages the message shifts from a more specific one to a more generic one based on lifestyle and possibilities. The specific site is generally discussed in terms of location and marginally for its actual character; although this is sold as unique often this is communicated with typical images of vibrant public spaces. The design of the three cases generally aligns with preconceived templates—a main street with commercial activities, interconnected public spaces, pedestrian environments. All cases aim to provide sustainable subtropical built environments; only KGUV suggests a unique approach stressing the importance of developing public art within the innovation district. Table 1.5 summarises the findings and reveals the alignment between the intended brands and how they are translated with planning to the development.

TABLE 1.5 Case Study Comparison

		Branding			**Planning**		
		Promise	**Features**	**Stimuli**	**Character**	**Layout**	**Design**
BRUV							
Stage I	2003–2006	@	#	#	$	$	$
Stage II	2007–2008	#	#	#	#	#	#
Stage III	2009–2014	@	#	@	$	$	$
KGUV							
Stage I	2001–2003	@	@	@	@	$	$
Stage II	2004–2007	#	@	@	#	#	#
Stage III	2008–2009	@	$	#	$	$	$
Stage IV	2010–2014	#	@	#	@	#	@
SDKT							
Stage I	2000–2006	#	#	@	#	#	#
Stage II	2007–2009	#	$	@	$	#	$
Stage III	2010–2011	#	$	@	$	$	$
Stage IV	2012–2014	#	$	@	$	$	$

@—Unique project specific strategy; #—Generic stereotypical approach; $—Information not available

1.4.2 Insights

This research reveals the following invaluable insights in Brisbane's journey in place making for knowledge and innovation milieus and communities:

- *Size:* Even being generally accepted that size only matters in financing and maintaining the development, smaller areas seem to be of advantage in establishing cohesiveness in the community, as exemplified by KGUV and DKP. SDKT is such a large and ambitious project—Australian answer to Silicon Valley—and it aims to take the innovation district to a 'knowledge town' level. With its large size, SDKT is likely to have voluminous challenges ahead, especially when considering the limited financial public sector support behind this development.
- *Industries:* The type and nature of knowledge-intensive businesses/industries and R&D activities undertaken in innovation districts are quite important for the long-term success. For example, if the knowledge-intensive activities were based on an endogenous asset, the success chance would likely be higher (Romer, 1994). KGUV invests on biosciences and creative industries based on the strengths of the university and its stakeholders. Likewise, DKP invests on eco-sciences specialising in fields that benefit and supported by the State of Queensland. The business and industrial specialisation of SDKT is not as clear as the other two. Even though investing on subtropical design and lifestyle options for knowledge workers sounds good to the ear, it is not clear how SDKT intends to attract knowledge communities when employment opportunities are generally limited in the region.
- *Anchors:* Having an anchor institution is critical for the performance and attractiveness of a innovation district. Fortunately, there exist one or more anchor institutions in all of the investigated innovation districts. These anchors are as follows. For DKP: Queensland Government, Commonwealth Scientific and Industrial Research Organization (CSIRO), and University of Queensland; for KGUV: Queensland University of Technology; and for SDKT: University of the Sunshine Coast and Innovation Centre Sunshine Coast.

- *Support:* Political support is an essential driver as well as an external precondition for innovation district formation. All three innovation districts were initiated during the Queensland Government 'Smart State Strategy' initiative, which has now been abolished (QG, 2005). With no strong political drive behind innovation districts, KBUD in the Brisbane region might be at risk. Considering the completion and take-up rates, SDKT is the one that seems to be the most disadvantaged by this. On the other hand, the current global financial crisis and the end of the mining boom have brought attention to the need for diversification of Australian economy, this might mean bringing back the policy of moving towards a knowledge economy.
- *Partnership:* Triple-helix model partnership plays an important role in the formation of knowledge communities. In this regard, all three innovation districts have received support from public-private-academia partnerships by being clustered around the major universities in the greater Brisbane area. However, it must be considered that USC's budget as a regional university is much lower in comparison to the sandstone Group of Eight University member UQ, Australian Technology Network of Universities member QUT, or even Griffith University that is included in Australia's top ten universities in research.
- *Community:* It is true that developing a innovation district requires large investment; however, the real cost associated with it is to maintain their attractiveness and competitive edge. This generally requires an onsite management that is supported by on-site stakeholders. Healthy social interactions and strong identity are essential to achieve the goals of forming a knowledge community worth living in. KGUV is a step ahead in forming such active community due to the efforts of its on-site management in comparison to the other two innovation districts—although KGUV is still quite far from reaching to an ideal level of cohesive community formation. DKP and SDKT are still in their infancy in terms of community formation. The changing of focus on the branding promotions observed both in KGUV and DKP can be seen as a response to these dynamics of maintaining attractiveness and competitive edge. This might be what SDKT need to do in order to get off the ground.

- *Planning:* Unlike brands, planning is static and lacks much dynamism. This is a serious contradiction as it signals that planning efforts might not follow the branding or vice versa. Planning and branding need to work as complementary activities where the branding offers the dynamism that the planning cannot offer, and therefore, planning policies could be prepared with this in mind, hence providing possible evolutionary adjustments. As highlighted through the analysis, in KGUV, the vision and spirit of two plans remain the same apart from some additional new activities and urban design features. In DKP, from the preliminary to final plan only the architectural design was changed. In SDKT, approved master plan has not been fully implemented yet. In all efforts have been made so that the design matches the features and the cultural aspects of these places, furthermore, the design principles suggested in the planning documents seem to align with the branding strategy. KGUV and DKP have the inner-city design character, where SDKT has the beach-style relaxed tourism site design characteristics. All cases follow the same layout principles, featuring a main street, square, and central urban design features. However, they all present a major disadvantage as being an enclosed development. Although they are not gated communities and there is access, flow and 'communication' with surrounding areas, the connection is artificial and not authentically integrated. The edges are blocked with busy roads or train tracks, which makes it hard to walk in and out, and marks them less permeable, despite their promises.
- *Design:* Investigated innovation districts seem to be accommodated with good design principles—including transit-oriented development, New Urbanism, crime prevention through environmental design, and water-sensitive urban design. KGUV's compact urban village development concept has been recognised with a national design excellence award. USC library building has won numerous design accolades including the Australian building of the year. DKP employs award-winning architectural style and sustainable design principles to create a highly sustainable and collaborative environment for the community. The Ecosciences precinct in DKP has been recognised by national-, state-, and regional-level awards including the awards for commercial architecture at national and

regional level and Brisbane regional building of the year (2011). These achievements demonstrate the design success and the degree of attractiveness that it entails for these innovation districts in terms of their architectural features.

- *Brand:* In all investigated innovation districts brand is not static; it evolves over time and is adapted to the needs of the market, audience, and development phases. For instance, in KGUV, it was clear that at the beginning of the project, 'knowledge' was the main message, then it is shifted to 'urban lifestyle', and once the innovation district is established, the branding message is returned to the original vision and focused on communicating 'knowledge'. In DKP, it started as 'knowledge' and then due to the expansion of the project it became 'urban lifestyle'. However, in SDKT, due to infancy of the project, 'beach-style (Bohemian) lifestyle' so far is the unchanged brand. This constant evolution of branding and promotional strategies can be understood as a response to the dynamics of maintaining attractiveness and competitive edge.
- *Image:* Establishment of the 'image' varies in all cases, since image is a social construction that depends on the acknowledgement of the places and respective brands by each of their audiences. Because of the age of each analysed precinct, it is needless to say that KGUV and DKP have better established images, while SDKT is still a vision embedded in the brand, not yet achieved an actual consistent image. Promoting a similar desired future and environment including authenticity and uniqueness is a question for the cases. All three cases' policy documents say they are 'unique'; however, the features they promote are pretty much the same—vibrant community, workable environment, safe, sustainability, promoting a lifestyle. The differentiation lays mainly on each specific flagship development and unique location—KGUV's Creative Industries Precinct and inner-city location, DKP's Ecosciences Precinct and inner-city location, and SDKT's Innovation Centre Sunshine Coast and close to the beach location. KGUV's rebranding indicates to a certain degree failure in the previous brand's appeal to the market. DKP was launched as an innovation district without living facilities and as it is now evolving into a true innovation district, its promotional efforts are following and so might its future perceived image. The branding

and promotions of SDKT's are still in initial stages, it might prove an interesting case as the way this brand is managed might be one of the internal factors that define success of this development despite the scale of envisioned project.

- *Challenges:* In the journey from space to place—through brand, legibility, image, and identity—all three innovation districts stand at different points. KGUV, being the most mature precinct, is a good example of using place branding at the planning stage to form a knowledge community with strong support from university and government. However, at the initial stage a stronger support from the private sector was missing. Such involvement is essential in becoming a globally recognised innovation district. DKP has a strong government backing that helps its establishment as a innovation district. Although initially it is planned to include surrounding areas offering wider residential and lifestyle facilities, in its current formation as a innovation district it fails to provide a home for knowledge communities. Thus, the main challenge for DKP is to evolve into a healthy innovation district by providing onsite living, shopping, cybering and recreation activities. SDKT, on the other hand, is an ideal innovation district conception located on a prime coastal position. Due to the state and local governments' economic development programs, limited economic activities in the region might forbid or significantly delay greater knowledge community gain.
- *Drawbacks:* There are some interesting inconsistencies between the promoted brand (intended image) and actual place identity and possible perceived image of the place in Brisbane's innovation districts. For example, KGUV branding promises a vibrant local community, but nowadays Kelvin Grove hosts a large undergraduate student population that has not reside at the precinct more than a few years or that does not live in the area at all and simply travels to and from KGUV. Due to the limited industry take up and initiatives, future employment opportunities for these young knowledge workers in the same environment where they studied are very slim. This means KGUV does not benefit adequately from the talent it has fostered. DKP's brand aims at the formation of a knowledge community, but provides limited non-work (residential) land uses. This limits the community

formation, despite the branding efforts promoting a lively environment that makes no necessary reference to the existing scientific work/precinct. SDKT has not yet been developed enough to be able to host a knowledge community. Additionally, even all innovation districts aim ambitiously to become 'a place to visit' by people outside their communities, unfortunately, none have managed to achieve that yet.

1.5 CONCLUSION

Literature findings suggest that just providing space and facilities for knowledge generation and innovation does not suffice to become competitive in attracting talent and investment. The quality of the environment as well as the specific character and appeal of new locales are central elements in the competition between urban centres in general and knowledge communities. Place branding, hence, has become a popular practice adopted by many cities around the world in the context of intensified urban competition (Zhang & Zhao, 2009). Location is indeed a critical issue in relation to innovation district developments. Although there are rare successful examples of regional or rural innovation districts—such as the Desert Knowledge Precinct of Australia—the innovation district is mainly an urban phenomenon that is reshaping the inner core of global cities. Best practice examples around the world take advantage of infrastructure, services, and amenities that inner-city locations provide.

Empirical study in assessing the effectiveness of knowledge and innovation milieus in the case of Brisbane generates invaluable insights. It indicates that Brisbane's signature innovation district initiatives demonstrate various levels of potential for success in terms of place making for knowledge and innovation milieus and communities. The analysis also reveals the challenges and barriers Brisbane's innovation districts are facing and suggests as much as good planning, branding strategies, and practice, required external and internal (pre)conditions also need to be met—e.g., leadership, political support, financial incentives, and other offerings of the city/region the innovation district is located in. This is to say, in Brisbane, a stronger investment—both in policy and financial terms—to generate these conditions is urgently needed, if these innovation districts to flourish and provide a promising pathway for other successful initiatives to come.

Planning and design methods as well as worldwide availability of material and globalisation of social values have generated homogenous design outcomes, creating places that lack aesthetic distinctiveness, identity, and sense of community. However, planning, design, and branding are just one side of what builds a place or community image and identity. As Paasi (2002) argues, cities are not capable of making decisions and achieve societal goals, cities do not operate, but social actors do. This is to say it is the daily community life that happens inside the designed environment hold the power to determine the dynamics of its identity and uniqueness, and the pace of change and evolution of a place. Our future research, therefore, will focus on the perceived and lived dimensions of place with an aim to provide a complete picture of identity evolution and perceived image of innovation districts (Yigitcanlar *et al.*, 2016).

ACKNOWLEDGEMENTS

This chapter, with permission from the copyright holder, is a reproduced version of the following journal article: Yigitcanlar, T., Guaralda, M., Taboada, M., & Pancholi, S. (2016). Place making for knowledge generation and innovation: Planning and branding Brisbane's knowledge community precincts. *Journal of Urban Technology, 23*(1), 115–146.

REFERENCES

Angelidou, M., Gountaras, N., & Tarani, P. (2012). Engaging digital services for the creation of urban knowledge ecosystems. *International Journal of Knowledge-Based Development*, 3(4), 331–350. DOI:10.1504/IJKBD.2012.051547

Anholt, S. (2007). *Competitive identity*. London: Palgrave Macmillan.

Anholt, S. (2008). Place branding. *Place Branding and Public Diplomacy*, 4(1), 1–6. DOI:10.1057/pb.2008.2

Asheim, B. (2007). Differentiated knowledge bases and varieties of regional innovation systems. *Innovation*, 20(3), 223–241. DOI:10.5172/impp.2007.20.3.223

Baum, S., O'Connor, K., & Yigitcanlar, T. (2009). The implications of creative industries for regional outcomes. *International Journal of Foresight and Innovation Policy*, 5(1-3), 44–64. DOI:10.1504/IJFIP.2009.023506

Brisbane City Council (BCC). (2005). *Kelvin Grove Urban Village local plan: Planning document*. Brisbane: BCC.

Bennett, R., & Savani, S. (2003). The rebranding of city places. *International Public Management Review*, 4(2), 70–87. DOI:10.1080/1471903032000109837

Brakus, J. J., Schmitt, B. H., & Zarantonello, L. (2009). Brand experience. *Journal of Marketing*, 73(3), 56–68. DOI:10.1509/jmkg.73.3.52
Carrillo, J. (2004). Capital cities. *Journal of Knowledge Management*, 8(5), 28–46.
Carrillo, J. (Ed.). (2006). *Knowledge cities*. New York: Elsevier. DOI:10.1108/13673270410552268
Carrillo, J., Yigitcanlar, T., Garcia, B., & Lönnqvist, A. (2014). *Knowledge and the city*. New York: Routledge. DOI:10.4324/9781315818459
Castells, M. (2009). *Power of identity*. London: Wiley-Blackwell.
Charles, D. (2011). The Role of Universities in Building Knowledge Cities in Australia. *Built Environment*, 37(3), 281–298. https://doi.org/10.2148/benv.37.3.281
Claessens, S., Dell'Ariccia, G., Igan, D., & Laeven, L. (2010). Cross-country experiences and policy implications from the global financial crisis. *Economic Policy*, 25(62), 267–293. DOI:10.1111/j.1468-0327.2010.00239.x
Darchen, S., & Tremblay, D. (2010). What attracts and retains knowledge workers/students? *Cities*, 27(1), 225–233. DOI:10.1016/j.cities.2009.10.006
Dixon, J., & Durrheim, K. (2000). Displacing place-identity. *British Journal of Social Psychology*, 39(1), 27–44. DOI:10.1348/014466600164358
Ergazakis, K., Metaxiotis, K., & Psarras, J. (2006). Knowledge cities. *Journal of Information and Knowledge Management Systems*, 36(1), 67–81. DOI:10.1108/13673270610650956
Eshuis, J., & Edwards, A. (2013). Branding the city. *Urban Studies*, 50(5), 1066–1082. DOI:10.1177/0042098012455118
Evans, G. (2003). Hard-branding the cultural city. *International Journal of Urban and Regional Research*, 27(2), 417–440. DOI:10.1111/1468-2427.00447
Florida, R. (2012). *The rise of the creative class*. New York: Basic Books. DOI:10.1177/0042098011417689
Gabe, T., Abel, J., Ross, A., & Stolarick, K. (2012). Knowledge in cities. *Urban Studies*, 49(1), 1179–1200.
Gold, J., & Ward, S. (1994). *Place promotion*. Chichester: John Wiley & Sons. DOI:10.1057/palgrave.pb.6000076
Golledge, R. G., & Stimpson, R. J. (1987). *Analytical behavior geography*. Beckenham: Croom Helm.
Hanna, S., & Rowley, J. (2008). An analysis of terminology use in place branding. *Place Branding and Public Diplomacy*, 4(1), 61–75.
Hospers, G. J. (2006). Borders, bridges and branding. *European Planning Studies*, 14(8), 1015–1033. DOI:10.1080/09654310600857878
Hospers, G. J. (2008). Governance in innovative cities and the importance of branding. Innovation: Management. *Policy & Practice*, 10(2/3), 224–234. DOI:10.5172/impp.453.10.2.224
Humphreys Reynolds Perkins Planning Consultant. (2007). *Boggo Road Urban Village structure plan: Planning document*. Brisbane: HRPPC. DOI:10.1080/09654310701746518

Hu, T. (2008). Interaction among high-tech talent and impact on innovation performance. *European Planning Studies*, 16(2), 163–187.

Huggins, R., & Strakova, L. (2012). Knowledge-based economic development in emerging regions. *Regional Studies*, 46(1), 961–975. DOI:10.1080/00343404.2010.510238

Jauhiainen, J., Kapyla, M., Salonius, H., & Yigitcanlar, T. (2014). Knowledge that matters. *European Planning Studies*, 22(10), 2011–2029. DOI:10.1080/09654313.2013.871125

Kalandides, A. (2011). The problem with spatial identity: revisiting the "sense of place". *Journal of Place Management and Development*, 4, 28–39. https://doi.org/10.1108/17538331111117142

Kaplan, M. D., Yurt, O., Guneri, B., & Kurtulus, K. (2010). Branding places. *European Journal of Marketing*, 44(9/10), 1286–1304. DOI:10.1108/03090561011056648

Kavaratzis, M. (2005). Place branding. *The Marketing Review*, 5(4), 329–342. DOI:10.1362/146934705774538226

Kerr, G. (2006). From destination brand to location brand. *Journal of Brand Management*, 13(4/5), 276–283. DOI:10.1057/palgrave.bm.2540246

Kelvin Grove Urban Village (KGUV). (2012). *Kelvin Grove Urban Village: Master plan*. Brisbane: KGUV.

Knight, R. (1995). Knowledge-based development. *Urban Studies*, 32(2), 225–260. DOI:10.1080/00420989550012949

Komninos, N. (2002). *Intelligent cities*. London: Spon Press. DOI:10.4324/9780203304973

Kotler, P., Haider, D., & Rein, I. (1993). *Marketing places*. New York: Free Press.

Kotler, P., & Gertner, D. (2002). Country as brand, product and beyond. *Journal of Brand Management*, 9(4/5), 249–261. DOI:10.1057/palgrave.bm.2540114

Ku, Y., Liau, S., & Hsing, W. (2005). The high-tech milieu and innovation-oriented development. *Technovation*, 25(4), 145–153. DOI:10.1016/S0166-4972(03)00030-4

Kunzmann, K. (2004). Culture, creativity and spatial planning. *Town Planning Review*, 75(4), 383–404. DOI:10.3828/tpr.75.4.h7875058h67272g5

Kunzmann, K. (2009). The strategic dimensions of knowledge industries in urban development. *DisP*, 45(177), 40–47. DOI:10.1080/02513625.2009.10556932

Landry, C. (2000). *The creative city*. Upper Saddle: Prentice Hall.

Lönnqvist, A., Käpylä, J., Salonius, H., & Yigitcanlar, T. (2014). Knowledge That Matters: Identifying Regional Knowledge Assets of the Tampere Region. *European Planning Studies*, 22(10), 2011–2029. https://doi.org/10.1080/09654313.2013.814621

Lucarelli, A., & Berg, P. (2011). City branding. *Journal of Place Management and Development*, 4(1), 9–27. DOI:10.1108/17538331111117035

Malecki, E. J. (2007). Cities and regions competing in the global economy. *Environment and Planning C: Government and Policy*, 25(1), 638–654. DOI:10.1068/c0666r

Markusen, A. (1996). Sticky placesin slipperyspace: a typol ogy of industrial districts. *Economic Geography*, 72(3), 293–313.

McGranahan, D. A., & Wojan, T. R. (2007). Recasting the creative class to examine growth processes in rural and urban counties. *Regional Studies*, 41(1), 197–216.

Merrilles, C., Miller, D., & Herrington, C. (2013). City branding. *Journal of Business Research*, 66(1), 37–44. DOI:10.1016/j.jbusres.2012.05.014

Morgan, N., Pritchard, A., & Piggott, R. (2002). The creation of a powerful niche destination brand. *Journal of Brand Management*, 9(4/5), 335–354. DOI:10.1057/palgrave.bm.2540105

Maroochy Shire Council. (2000). *Development of Sippy Downs Town Center: Planning document*. Maroochydore: MSC.

Maroochy Shire Council. (2007). *Local growth management strategy: Planning document*. Maroochydore: MSC.

Okano, H., & Samson, D. (2010). Cultural urban branding and creative cities. *Cities*, 27(1), S10–S15. DOI:10.1016/j.cities.2009.09.003

Paasi, A. (2002). Place and region: regional worlds and words. *Progress in Human Geography*, 26(6), 802–811. https://doi.org/10.1191/0309132502ph404pr

Pancholi, S., Yigitcanlar, T., & Guaralda, M. (2014). Urban knowledge and innovation spaces. *Asia Pacific Journal of Innovation and Entrepreneurship*, 8(1), 15–38. DOI:10.1108/APJIE-11-2013-0075

Pasquinelli, C., & Teräs, J. (2013). Branding knowledge-intensive regions. *European Planning Studies*, 21(10), 1611–1629. DOI:10.1080/09654313.2013.776254

Petruzzelli, A. M., Albino, V., & Carbonara, N. (2007). Technology districts. *Journal of Knowledge Management*, 11(5), 98–114. DOI:10.1108/13673270710826223

Pratt, A. C. (2000). New media, the new economy and new spaces. *Geoforum*, 31(4), 425–436. DOI:10.1016/S0016-7185(00)00017-8

Queensland Government. (2001). *Kelvin Grove Urban Village: Village news brochure*. Brisbane: Queensland Government.

Queensland Government. (2002). *Kelvin Grove Urban Village: Urban village brochure*. Brisbane: Queensland Government.

Queensland Government. (2003). *Investing in science: Research, education and innovation brochure*. Brisbane: Queensland Government.

Queensland Government. (2004a). *Kelvin Grove Urban Village: Development opportunities brochure*. Brisbane: Queensland Government.

Queensland Government. (2004b). *Kelvin Grove Urban Village: Design guidelines planning document*. Brisbane: Queensland Government.

Queensland Government. (2004c). *Kelvin Grove Urban Village: Creative community brochure*. Brisbane: Queensland Government.

Queensland Government. (2005). *Smart Queensland: Smart state strategy 2005-2015: Policy document*. Brisbane: Queensland Government.

Queensland Government. (2008a). *Kelvin Grove Urban Village: Clever community brochure*. Brisbane: Queensland Government.

Queensland Government. (2008b). *Kelvin Grove Urban Village: Development opportunities brochure*. Brisbane: Queensland Government.
Queensland Government. (2009). *Boggo Road Urban Village at Dutton Park: Project update brochure*. Brisbane: Queensland Government.
Queensland Government. (2010). *Kelvin Grove Urban Village: Design guidelines planning document*. Brisbane: Queensland Government.
Rantisi, N. M., & Leslie, D. (2006). Branding the design metropole. *Area*, 38(4), 364–376. DOI:10.1111/j.1475-4762.2006.00703.x
Ratinho, T., & Henriques, E. (2010). The role of science parks and business incubators. *Technovation*, 30(1), 278–290. DOI:10.1016/j.technovation.2009.07.008
Romer, P. M. (1994). The origins of endogenous growth. *Journal of Economic Perspectives*, 8(1), 3–22. DOI:10.1257/jep.8.1.3
Sunshine Coast Regional Council. (2008). *Sunshine Coast growth management: Planning document*. Caloundra: Sunshine Coast Regional Council.
Sunshine Coast Regional Council. (2010). *Media Releases: Apr 29, May 11 and 21, Nov 17*. Caloundra: Sunshine Coast Regional Council.
Sunshine Coast Regional Council. (2012a). *Draft Sunshine Coast planning scheme: Planning document*. Caloundra: Sunshine Coast Regional Council.
SCRC. (2012b). *Media Releases: Jan 13, Jun 22 and 28, 2012*. Caloundra: SCRC.
Sheppard, E. (2002). The spaces and times of globalization. *Economic Geography*, 78(3), 307–330. DOI:10.2307/4140811
Stigel, J., & Frimann, S. (2006). City branding. *Nordicom Review*, 27(2), 243–266. DOI:10.2478/v10114-011-0015-x
Tomaney, J., & Bradley, D. (2007). The economic role of mobile professional and creative workers. *Town Planning Review*, 78(1), 511–529. DOI:10.3828/tpr.2007.12
Tremblay, D. G., & Darchen, S. (2010). The creative class thesis and the mobility patterns of knowledge workers. *International Journal of Knowledge-Based Development*, 1(3), 176–203. DOI:10.1504/IJKBD.2010.033430
Van den Berg, L., Pol, P., Van Winden, W., & Woets, P. (2005). *European cities in the knowledge economy*. London: Ashgate.
Veitch, J. A., & Arkkelin, H. A. (1995). *Environmental psychology*. Englewood Cliffs: Prentice Hall.
Wedemeier, J. (2010). The impact of the creative sector on growth in German regions. *European Planning Studies*, 18(1), 505–520. DOI:10.1080/09654310903485632
Yigitcanlar, T., Baum, S., & Horton, S. (2007). Attracting and retaining knowledge workers in knowledge cities. *Journal of Knowledge Management*, 11(5), 6–17. DOI:10.1108/13673270710773577
Yigitcanlar, T., O'Connor, K., & Westerman, C. (2008a). The making of knowledge cities. *Cities*, 25(2), 63–72. DOI:10.1016/j.cities.2007.10.002
Yigitcanlar, T., Velibeyoglu, K., & Martinez-Fernandez, C. (2008b). Rising knowledge cities. *Journal of Knowledge Management*, 12(5), 8–20. DOI:10.1108/13673270810892195

Yigitcanlar, T. (2009). Planning for knowledge-based development. *Journal of Knowledge Management*, 13(5), 228–242. DOI:10.1108/13673270910984534

Yigitcanlar, T. (2010). Making space and place for the knowledge economy. *European Planning Studies*, 18(11), 1769–1786. DOI:10.1080/09654313.2010.523422

Yigitcanlar, T. (2011). Position paper: Redefining knowledge-based urban development. *International Journal of Knowledge-Based Development*, 2(4), 340–356. DOI:10.1504/IJKBD.2011.044006

Yigitcanlar, T., & Dur, F. (2013). Making space and place for knowledge communities. *Australasian Journal of Regional Studies*, 19(1), 36–63.

Yigitcanlar, T., & Lönnqvist, A. (2013). Benchmarking knowledge-based urban development performance. *Cities*, 31(1), 357–369. DOI:10.1016/j.cities.2012.09.001

Yigitcanlar, T. (2014). Position paper: Benchmarking the performance of global and emerging knowledge cities. *Expert Systems with Applications*, 41(12), 5549–5559. DOI:10.1016/j.eswa.2014.02.036

Yigitcanlar, T., Lönnqvist, A., & Salonius, H. (2014). Analysis of a city-region from the knowledge perspective. *VINE*, 44(3), 445–466. DOI:10.1108/VINE-02-2013-0005

Yigitcanlar, T., & Bulu, M. (2015). Dubaization of Istanbul. *Environment and Planning A*, 47(1), 89–107. DOI:10.1068/a130248p

Yigitcanlar, T., & Sarimin, M. (2015). Multimedia super corridor, Malaysia. *VINE*, 45(1), 126–147. DOI:10.1108/VINE-03-2013-0013

Yigitcanlar, T., Guaralda, M., Taboada, M., & Pancholi, S. (2016). Place making for knowledge generation and innovation: planning and branding Brisbane's knowledge community precincts. *Journal of Urban Technology*, 23(1), 115–146. DOI:10.1080/10630732.2015.1090198

Zenker, S. (2009). Who's your target? *J. Place Management and Development*, 2(1), 23–32.

Zenker, S. (2011). How to catch a city? *J. Place Management and Development*, 4(1), 40–52. DOI:10.1108/17538331111115280

Zhang, L., & Zhao, S. (2009). City branding and the Olympic effect. *Cities*, 26(1), 245–254. DOI:10.1016/j.cities.2009.03.003

Zimmerbauer, K. (2011). From image to identity. *European Planning Studies*, 19(2), 243–260. DOI:10.1080/09654313.2011.5419

CHAPTER 2

Public Space Design and Innovation Districts

2.1 INTRODUCTION

Public spaces are at the heart of democratic living (Carr *et al.*, 1992). They are the main stage of urban life and facilitate encounters, exchange of experiences, and foster a tolerant urban society through the exposure to different people and their traditions (Worpole & Greenhalgh, 1996). These spaces are constantly charged with symbols and meanings that enhance the identity of a society (Zukin, 1995). Different disciplines have different definition of what is a public space, highlighting for example land tenure, accessibility, or social interaction. This research focuses on public spaces and their important role in the creation of local identity or place making (Amin & Graham, 1997; Massey, 2005). Policymakers more and more acknowledge the importance of place globally (Yigitcanlar *et al.*, 2007; Van Winden, 2010).

The concept of place has been defined in different ways (Arefi, 2004), but it is generally recognised that place making is a process depending on people's experience of a locale. Design plays a central role in characterising and defining this experience (Buttimer, 1980). Places do not just exist; they are the product of a society and embed different level of symbolism (Cresswell, 2004). An understanding of the socio-cultural context of a locale is considered one of the factors to achieve effective place making through design (Massey, 2005). Scholars, such as Carmona *et al.* (2003), support the idea that urban design can be implemented as a process of facilitating a sense of place for a specific context. Urban design

DOI: 10.1201/9781032657431-3

emphasises the importance of people and place (Adams & Tiesdell, 2013), but it is also a process affected by globalised forces, market, and regulatory factors (Kabachanik, 2010). For instance, Carmona (2003) asserts that urban design can be considered to involve direct and indirect design factors that shape the nature of a place through establishing policy, making investment decisions, and managing space. Public spaces, thus, include both physical and the social dimensions (Duff, 2010). Whilst Shaftoe (2008) indicates that production of successful public spaces involves an integrated approach with a proper interplay of design and management within the broader context of urban policies. Worpole (2005) observes that design alone cannot solve the problems being placed into the deep social and economic nature of circumstances that undermine a vibrant community and public space culture. Hence, it is necessary to take an integrated approach to understand place making in contemporary public spaces, especially in contexts that are specifically designed with the intention of providing a unique experience to a particular social group (Watson & Bentley, 2007).

This study considers place making in public spaces within the context of knowledge-based urban development (KBUD). During the last few decades, KBUD has been at the centre of policymaking in many global or world cities that seek to attain long-term competitiveness (Yigitcanlar, 2009; Lonnqvist *et al.*, 2014; Yigitcanlar & Bulu, 2015). Globally, the acceptance of KBUD as the panacea for all the social, economic, environmental, and spatial challenges being faced by contemporary development (Knight, 1995; Kunzmann, 2009; Yigitcanlar, 2010; Yigitcanlar & Sarimin, 2015) has manifested into the consequent emergence of new typology of knowledge milieus in the form of innovation districts (Asheim, 2007). A few global successful examples include, but are not limited to Silicon Valley, Silicon Alley in New York, Silicon Roundabout in London, Ørestad in Copenhagen, Brainport in Eindhoven, One-north in Singapore, Australian Technology Park in Sydney, Parkville Knowledge Precinct in Melbourne, and Kelvin Grove Urban Village (KGUV) in Brisbane (Carrillo *et al.*, 2014).

One of the major objectives of innovation districts is to attract and retain skilled human capital in the form of knowledge workers by providing them with quality of life in ultimately unique places (Yigitcanlar *et al.*, 2007; Florida, 2012). Place making plays a major role in retaining talent, as considerations about the quality and character of place are essential components of KBUD policies (Pratt, 2002; Sheppard, 2002).

Nevertheless, accommodating and absorbing the flows and processes of globalisation—i.e., diversity of population—and retaining these people is a significant challenge for cities. Knowledge workers, an emergent class, have special needs and requirements (Zenker, 2009). More importantly, with the acceptance of open innovation model as the key to success in global economy, these traditionally secluded innovation districts are transfiguring themselves as more extrovert and accessible locations (Chesbrough, 2003; Kodama & Shibata, 2015). Additionally, one of the major objectives of KBUD, in sync with the global climate change awareness, is sustainable enviro-urban development (Yigitcanlar & Lonnqvist, 2013; Cooke, 2015; Yigitcanlar *et al.*, 2015). These needs and requirements are impacting the formation of new typologies of settlements and public spaces within innovation districts.

This gives rise to the following question: What are the role and characteristics of public spaces to facilitate place making in contemporary innovation districts? To provide answers to this question, a case study investigation is undertaken in one of the most consolidated innovation districts of Brisbane—the capital city of Queensland in Australia. Australia is one of the rapidly growing knowledge economies globally (Yigitcanlar *et al.*, 2008a) and Brisbane is one of the emerging knowledge centres in the country (Yigitcanlar & Velibeyoglu, 2008). Several Australian cities have been witnessing a similar trajectory of development, where KBUD have been implemented as key components of strategic developments. Other renowned examples of this process are Melbourne and Sydney (Yigitcanlar, 2010).

The overall aim of the chapter is to contribute to the existing understanding of place making in the context of emerging innovation districts' public spaces. To do so, we seek to develop an understanding of the major attributes that have been attached to the public spaces in these new emerging typologies aiming to achieve a thriving KBUD. The chapter, first, reviews the literature about the special conditions of KBUD, and the profile of knowledge workers, which sets the broader context for the case study. The chapter identifies the attributes that make innovation districts viable, vibrant, and capable of attracting and retaining global knowledge workers. KGUV is then evaluated, taking into consideration which design strategies have been implemented to develop a unique sense of place in this innovation district. The study attempts to understand in what ways different attributes have been implemented in the design of this contemporary innovation district.

2.2 LITERATURE BACKGROUND

An innovation district is considered the 'spatial nexus' for the generation and dissemination of knowledge in KBUD (Yigitcanlar *et al.*, 2008b). Yigitcanlar & Lonnqvist (2013) summarise KBUD as a policy that targets of building a 'place' to form perfect climates not only for business, but also for people, place, governance, and their integration. Thus, these four developmental perspectives establish the four main pillars of KBUD—i.e., economy, society, environment, and governance (Yigitcanlar & Sarimin, 2011; Yigitcanlar, 2014). Brisbane is one of the emerging knowledge cities of Australia (Yigitcanlar, 2008a); policies such as the 'Smart State Strategy' developed by the Queensland government, or the 'Smart City Strategy' for Brisbane had a considerable impact over the spatial restructuring of this city. These projects concentrated on the social and urban development policy areas to establish the city in a global knowledge economy by producing an investment-friendly business climate as well as enhancing the endogenous skill set of the residents (Wiltshire, 2003). They shifted the economic base from traditional industries to knowledge sectors (Mort & Roan, 2003); thus, transitioning the modest local economic development policy to the more ambitious KBUD perspective.

The pentagon prism model that forms the base of the Smart City Strategy deals with five aspects: Creative, administrative, business, natural and built environments, with connectivity playing a crucial role in strengthening local and global linkages (Yigitcanlar & Velibeyoglu, 2008). These innovation districts, generally clustered together, are spatially characterised by the presence of blurred boundaries between the spaces of live-work-play-learn-cyber (Yigitcanlar, 2010). The strong foundation of Brisbane in a number of knowledge industry sectors such as biotechnology and biosciences, aviation and aerospace, and information technology has fuelled the development of innovation districts in and around the city (Yigitcanlar & Velibeyoglu, 2008), such as KGUV, Da Vinci Airport Precinct, information technology sector in Milton, CBD and Fortitude Valley (also hosting the entertainment sector), Diamantina Knowledge Precinct (formerly known as Boggo Road Urban Village) and Knowledge-based Research and Business Precinct and Dutton Park, on the Southside Nathan with Griffith University campus, Mt Gravatt Research Park and Brisbane Technology Park, Griffith University on the Gold Coast, on the north side University of the Sunshine Coast Knowledge Town, and so on. A few of these locations are in the developmental stage and others are already fully fledged and functional.

Creative Class Thesis of Florida (2002), which is based on the idea of the 3 Ts—i.e., technology, talent, and tolerance—is at the centre of these types of development. Florida's thesis asserts the central importance of knowledge workers in the contribution to global knowledge economy (Baum *et al.*, 2009). With about 21% of Brisbane's population being young people, special attention is being given by policymakers to knowledge workers as is evident in the recent formulation of the Youth Strategy 2014–2019 aiming at making youth engaged, empowered, valued, and celebrated. These modern age workers possess a few characteristics that make them distinct (Zenker, 2009). They are well informed and participative, seek a better quality of life, culturally active and artistically expressive, displaying diversity and tolerance, and having high competency in human social interactions (Carrillo, 2004).

The objectives attached with the public spaces in innovation districts, therefore, are getting reconfigured with their changing preferences and progressive sense getting attached to place making in innovation districts with globalisation (Pancholi *et al.*, 2014). It needs to be noted that progressive sense of place is a concept based on the wider world integration of a locale in the local and global contexts; this process enriches itself by the merging of all flows and processes (Massey, 1991; Lippard, 1997). Hence, the definition of place making in contemporary public spaces, that we are referring to here, is based on the idea of developing 'permanence' from the flow of 'processes' (Cresswell, 2004). We, hence, define a place as a: (a) spatio-temporal entity dependent on uniqueness of location and (b) process dependent on flows.

At large, Carr *et al.* (1992) list the following characteristics of good public spaces: They should be meaningful, democratic (in accessibility, expression, and rights), responsive, comforting, engaging (passively and actively), and provide a sense of discovery. Historically, innovation district developments were constructed as secluded campuses with an introvert character as required for research-oriented environment. In many of the older version of innovation districts, i.e., science and industrial parks, the use of public spaces was majorly limited to relaxation and engagement of internal users. With growing extroversion of such spaces, Yigitcanlar *et al.* (2008b) summarise the holistic objective of modern innovation districts as not only purely generation, but also transfer and transmission of knowledge for socioeconomic growth of cities and their regions (Yigitcanlar *et al.*, 2008b); therefore, attaching a new purpose to public spaces of modern innovation districts along with the above-mentioned characteristics outlined by Carr *et al.* (1992).

Contemporary public spaces in innovation districts are not only required to be serving as entertaining, relaxing, comforting, and engaging points, but also as a place that acts as the first point of connection between the knowledge space and the city, allowing the maximum knowledge dissemination within its boundaries and as well as outside the boundaries. This leads us to two major objectives for contemporary public spaces to facilitate place making in innovation districts: (a) developing a sense of meaning and attachment, i.e., by attracting and retaining knowledge workers within innovation districts by providing them a sense of permanence in a unique 'place' that is developed by being contextual, connected, sustainable, and vibrant internally; and (b) facilitating the external flow and processes of knowledge, i.e., by being responsive towards the broader context, connected by an active relationship of give-and-take between the innovation district and the city, the region and also at the global levels keeping it viable, vibrant, resilient, and adaptable. These characteristics are further explored in the case of diverse and globalised KBUD community of KGUV.

2.3 METHODOLOGY

The methodology of the study is twofold. First, we have undertaken a concise review of the literature to set the 'broader context', as above. Then, we undertake 'site observations' to reveal insights from the case of KGUV. For the case investigation, the study applied a simple but effective analytical framework that is dependent on two major attributes derived from the aforesaid definition of place making. These are as follows: (a) 'Character', i.e., arising from culture, scale, climate, and environment; and (b) 'Connectedness', i.e., harder layers, such as spatial and visual elements, or softer ones, such as socio-cultural milieu. The specific character serves as the context for the creation of a sense of place; connectedness acts as the catalyst in developing and maintaining a specific identity. In the case examined, this is achieved by being context-sensitive and human-scaled. This approach provides a sense of familiarity to the newly arrived population that can relate to the local culture, heritage, and natural context. On the other hand, connectedness with the outer world is necessary for the dissemination of knowledge. Connectedness integrates the innovation district with its surroundings and wider context, thus, contributing not only to knowledge dissemination, but also in regional development.

KGUV is a sustainable urban community developed as an integrated and master-planned urban village within a mixed-use and highly urban environment (Brisbane City Plan, 2000). This Queensland Government's flagship urban renewal project is developed as a joint initiative by Queensland University of Technology (QUT) and Queensland Government Department of Housing, only two kilometres from Brisbane's CBD. One of the major prestigious awards won by this development is the Planning Institute of Australia's 'National Design Excellence Award' in 2004 (Yigitcanlar & Dur, 2013). Acknowledged as a landmark project of KBUD in Brisbane and a design that aims at creating strong sense of place and community in a safe environment, the case gives an opportunity to explore the characteristics of public spaces in the context of KBUD (Foth & Adkins, 2006; Foth *et al.*, 2008; Yigitcanlar *et al.*, 2008b; Klaebe *et al.*, 2009).

KGUV case study is analysed at two levels: (a) development of sense of meaning and permanence is investigated within the innovation district and externally and (b) progressive place to facilitate external and internal flows is being analysed at the innovation district and city or global levels, considering the give-and-take process at a broader scale. Each of these major attributes is further divided into sub-attributes. Site-specific character is analysed under the sub-attributes of context integration and people-orientedness, while character at a broader scale is analysed under creative image and the role of wider climate and global image on built character. On the contrary, connectedness of public spaces is analysed internally and externally at spatial, visual, and socio-cultural levels.

2.4 RESULTS

KGUV is a well-established inner-city suburb and an example of an award-winning knowledge community. KGUV is a result of advanced planning processes that have produced a master plan document defining specific codes for the design of individual buildings (Figures 2.1 and 2.3). Master plan and the development are based on the triple-helix model—i.e., a joint venture between university and public and private sectors—and provided public spaces with unique character. This intervention is centred on creative industries and biosciences; QUT is the most recognisable presence in the precinct, which is promoted for its creative and vibrant village atmosphere providing a highly urban and well-integrated live-work-play-learn environment. KGUV applies some recognisable principles in its design that are

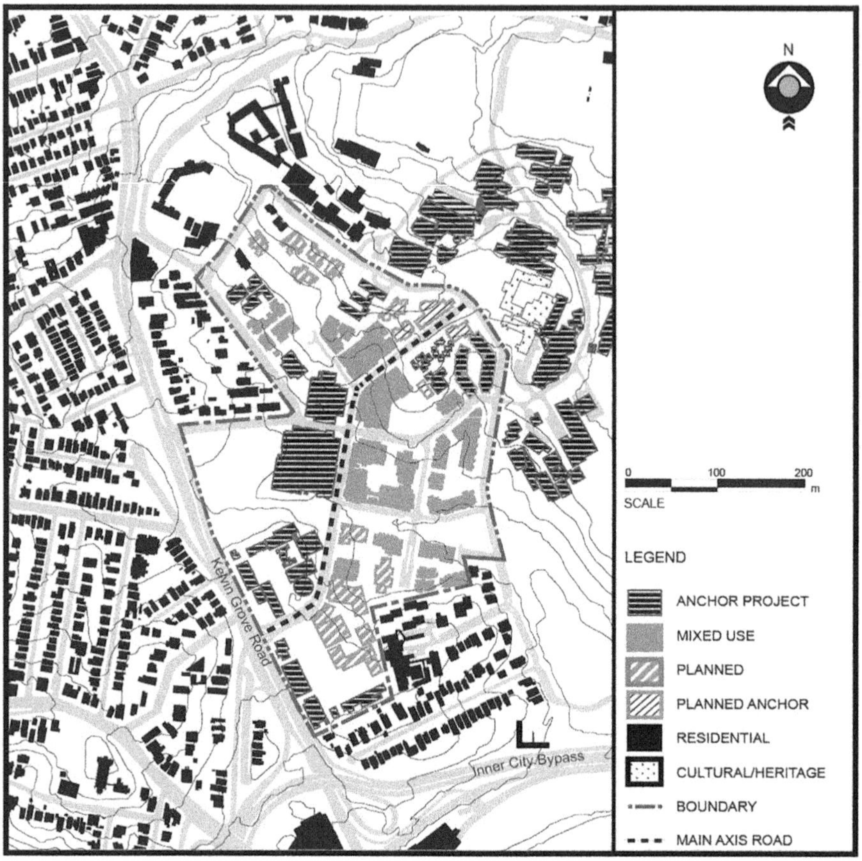

FIGURE 2.1 Figure ground map of KGUV.

commonly suggested as key factor in the establishment of an innovation district (Yigitcanlar *et al.*, 2008b; Yigitcanlar & Dur, 2013).

The salient design characteristics are listed and briefly elaborated below:

- *Mixed land uses:* Innovation districts are characterised by the presence of mixed uses that contribute to their vitality and economic diversity (Charles, 2011). This is evident in KGUV (Figure 2.1).
- *Contemporary high-tech environment:* Spaces that generate knowledge should reflect the innovation and creativity by their contemporary high-tech environment. The use of modern materials in built form, integration of high-tech elements in public spaces, and strong digital connections are few of such initiatives as evident in the case considered.

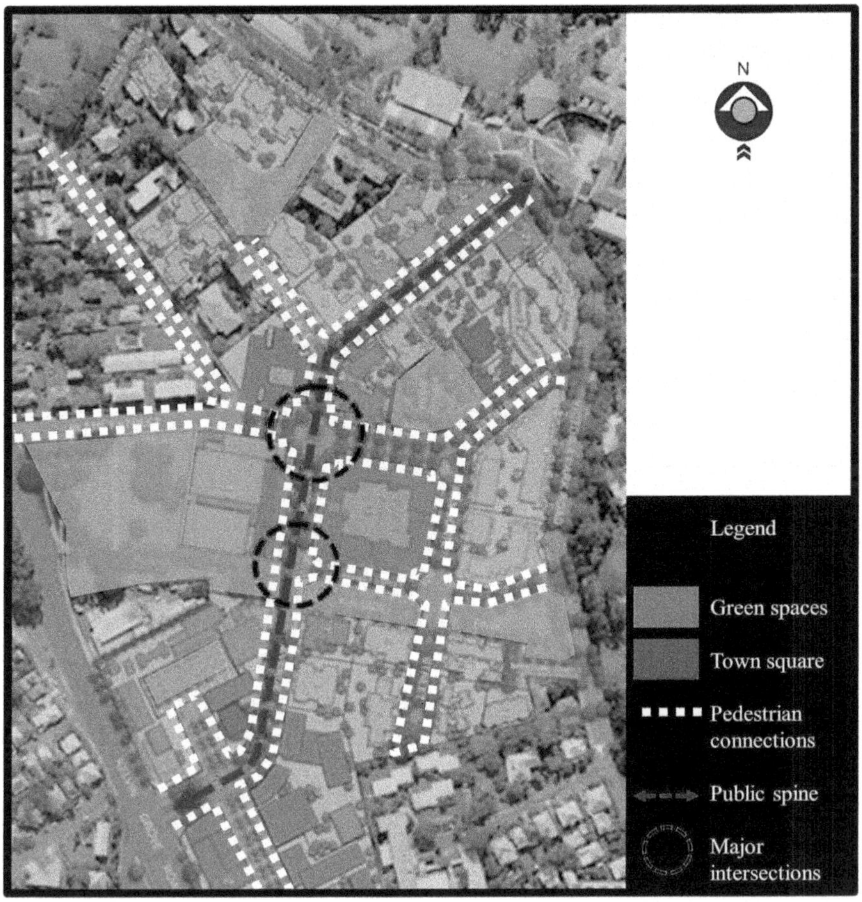

FIGURE 2.2 Interconnectivity of public spaces (derived from Brisbane City Plan, 2000).

- *Anchor project dominancy:* Innovation districts are characterised by the existence of one main building and complex as an anchor project that provides character and flavour to the area, and acts as a base support for all ancillary activities. In KGUV, QUT is the main anchor project (Figure 2.1).
- *Presence of an axis:* In Brisbane's innovation districts, one main street leads to the anchor project and acts as major axis or public spine. This boulevard acts as the major spine of KGUV with the entire main commercial, institutional, and public activities distributed around it, e.g., Musk Avenue in KGUV (Figure 2.2).

FIGURE 2.3 Location and vicinity of the urban village.

- *Compact development:* Contemporary innovation districts are mostly developed as compact development. With a concentration on variety of housing typologies catering to the diverse population and ranging from medium to high density, KGUV is also developed to have a dense and compact built form. The residential development is mostly in the form of multi-unit dwellings and apartments.
- *Central public and open spaces:* The public and open spaces, including the town centre, are given central locations in KGUV (Figure 2.2).

The analysis of planning documents (Brisbane City Plan, 2000) and site observations support the understanding on how place-making strategies have been integrated in the overall design of KGUV (Figure 2.3). The findings of these analyses, including major public space features, are presented below under a categorical classification and summarised in Table 2.1.

2.4.1 Character

A. Site-specific scale:

a. Context integration:

- *Integration of culture:* The area has a rich military and indigenous history. Army barracks are retained as heritage and revitalised. To reflect the indigenous 'Turrbal' community's strong relationship with the land, parks and public arts

TABLE 2.1 Summary of Major Public Space Features in KGUV

Place Level	Derived Attributes	Attributes Addressed	Description
SITE-SPECIFIC	Character	Context integration	Integration of historical military context, indigenous Turrubal community and native residents in built form, park and street names.
			Integration of nature by retaining native old trees in the McCaskie Park.
		People-orientedness	Human scale, buildings up to seven stories high and use of human scale elements at front entrances.
			Pedestrian-friendly streetscape by high level of detailing in pavements, street furniture, use of artworks, and landscaping.
SCALE	Connectedness	Spatial connectivity	Interconnected open space network by continuous pathways.
			Seamless integration of main project with built environment.
		Visual and socio-cultural connectivity	Concept of 'eyes-on-street' by houses overlooking the public spaces.
			Accessibility of public to all display areas of the university (i.e., QUT).

(Continued)

TABLE 2.1 *(Continued)* Summary of Major Public Space Features in KGUV

Place Level	Derived Attributes	Attributes Addressed	Description
BROADER	Character	Creative image	Transparent façades of the lower floors, where creative arts are displayed engage the outsiders.
			Use of public art reflects the creative context of creative industries.
		Built character	High-tech built environment based on knowledge and urban lifestyle that gives a global innovation district character.
			Implementation of climate-responsive subtropical design principles.
SCALE	Connectedness	Spatial connectivity	Considerable integration with surrounding neighbourhoods.
			Well connected by transport to other parts of city, but a less permeable boundary.
		Visual and socio-cultural connectivity	Remarkable CBD and mountain panorama from the residences and offices.
			Local markets attract visitors from other parts of the city; however, the site lacks large-scale national and international cultural events.

are named accordingly. Besides, some of the streets are named after former eminent residents of the city to recognise their contributions to the society (Klaebe, 2006).

- *Integration of nature:* The old native trees are conserved, including the ones in McCaskie, Grey Gums and Kulgun Parks, and large active and passive green spaces are developed (Figure 2.4).

b. People-orientedness:

- *Human-scale:* Height of the buildings is restricted to seven stories. Urban form is planned to directly relate to human scale with a control on building heights and the use of inviting elements like awnings, podiums, and sitting areas as part of the front entrances on pathways.
- *Pedestrian environment:* Considered as a key innovation district feature extensive pedestrian areas and networks are planned. The provision of casual seating, landscaping, and integration of arts on the main axis make the environment conducive for pedestrian activities (Figures 2.4 and 2.5).

FIGURE 2.4 McCaskie Park.

FIGURE 2.5 Mixed-use development and public spaces.

B. Broader scale:

a. Creative image:

- *Engaging lower floors:* Buildings relate directly to public space through the planning of commercial activities and use of transparent façades, mainly in anchor projects as well as other public buildings. An example is the presence of an art gallery with a transparent façade at the ground floor of a creative industries precinct. This acts as showcase of creativity-related activities that take place inside the buildings and sends an extrovert image and open character of contemporary innovation districts, and engages outsiders visually by catching their attention (Figure 2.6).
- *Use of public arts:* By the integration of public arts in the form of sculptures and creative façades at visually prominent locations, the context of creativity is further strengthened. This is helpful in attaching a creative and innovative image to the innovation district (Figures 2.6 and 2.7). It is to be noted that both factors not only produce an image and identity in the minds of people, but also act as contact points for knowledge transfer disseminating the fresh creative ideas externally to the rest of the city.

FIGURE 2.6 Façade design and compact development.

FIGURE 2.7 Pedestrian-oriented human-scale active public spaces.

b. Built character:

- *Climate-responsiveness:* Considering the broader climatic context, sub-tropical design principles are followed throughout. Special attention has been given to develop the built form that responds to the tropical climate and includes features that recall traditional elements of Queensland architecture, such as latticework, awnings, and screens (Figure 2.7).

- *Global image:* Despite climate-sensitive design principles, the use of modern construction material in façades reflects the progressive context of 'innovation'. A high-tech built environment relating to knowledge and urban lifestyle gives a global contemporary ambiance to the innovation district.

2.4.2 Connectedness

A. Site-specific scale:

a. Spatial level:

- *Interconnected public spaces:* The main axis of KGUV sets up a system of connections thus increasing the permeability within the innovation district. All the green and public spaces are connected through continuous pedestrian pathways (Figures 2.1 and 2.8).
- *Integration of anchor project:* The anchor project, i.e., QUT's campus, is seamlessly integrated with the rest of the innovation district and has boundaries that are blurred (Figure 2.3).

b. Visual and socio-cultural level:

- *Private-public visual connections:* Provision of private balconies comprising 30–75% of the length of the building façade and 30–80% of length for windows at frontage that face the main street ensures passive safety by 'eyes-on-the-street' environment, and creates visual connectivity of residents with public domain (Figures 2.7 and 2.8).
- *Internal events:* KGUV community has access to all display areas and organised events of the university—i.e., QUT. Additionally, the projects such as 'Sharing Stories' provide an opportunity for residents to share their experiences informally, and act in bringing the community together. Furthermore, QUT and La Boite theatre contributes to cultural liveliness of the innovation district (Figure 2.9).

FIGURE 2.8 Central axis, anchor project, and use of climate-sensitive architectural elements.

FIGURE 2.9 QUT and La Boite Theatre.

B. Broader scale:

a. Spatial connectivity:

- *Physical connections*: KGUV is well connected with the city and its surroundings. Not only vehicular, but also direct pedestrian and cycle routes ensure connectivity and accessibility with the CBD. Brisbane airport is also just an 11-minute drive away and connected through a direct rail line. This ensures the innovation district's physical linkages with the rest of the world.
- *Spatial integration:* The transition between the innovation district and its neighbourhoods is quite smooth (Minnery *et al.*, 2009), and the development is well integrated with the surroundings (Figures 2.1 and 2.3).

b. Visual and socio-cultural connectivity:

- *Vistas:* Brisbane City Plan (2000) sets a requirement for all developers to ensure the opening of vistas of KGUV towards the CBD and mountain views. The plan also dictates the surrounding residential areas and other types of buildings to follow this rule as well.
- *Events and markets:* Local arts, crafts, and food markets are held on Saturdays and act as a major factor in bringing people within the innovation district and surrounding suburbs together. Apart from the public events organised inside the university campus, which are accessible to all, the area still lacks the organisation of a large-scale event that connects the innovation district nationally and globally.

Permeability is a major attribute in facilitating the progressive sense of place making, creating a sense of spatio-temporal permanence for knowledge workers in public spaces in innovation districts (Pancholi *et al.*, 2014). This stands evident in the case of KGUV from the above discussion. Permeability here can be seen emerging as a key defining attribute not only in spatial, but also in social, economic, and environmental terms. The defining characteristics of a permeable innovation district typology for KGUV can be summarised as follows:

- *Spatial permeability:* Spatially, this is seen in KGUV by the presence of: (a) mixed-land uses: the blurring of boundaries between the living, working and playing is one of the defining qualities of these places; (b) permeable urban form: physical and visual connectivity is promoted throughout the innovation district by the interconnections between public spaces and opening up of internal vistas; (c) permeable built forms: the façades of public and anchor project buildings are made transparent towards public domains to allow maximum visibility of activities going inside to engage people outside. The windows, entrances, and footpath uses (supported by awnings, weather-shades, etc.) at lower floors enable active street frontage; (d) external vistas: the innovation district should visually connect to the outer world by opening up the vistas towards the CBD and surroundings. The spatial permeability at the broader level seems to be lacking at the level of connections with the city because of the less permeable outer layer.
- *Social permeability:* The social permeability in KGUV is manifested by developing: (a) character: careful considerations are taken for the development of these places in a way that they respond to the original context as seen in KGUV, i.e., the revitalisation of old military barracks and recognition of history of Turrbal community by integrating the aboriginal arts in public spaces and naming the parks and streets after them. This way, they communicate the local arts, culture, and history to the migrant population and develop a sense of place by letting them attach meaning to the place and (b) connectedness: the display of local music, arts, and crafts as a part of the weekly market in KGUV is an attempt to communicate and share some of the local knowledge assets with knowledge workers and the public. Such markets also act as a platform for people from various cultures to display their arts, culture, and music, thus giving all the social groups an equitable chance to express them. The use of ICT for providing wireless communication within the precinct also enhances the internal connectedness; (c) integration of diversity: emphasis has been made on the provision of a variety of housing typologies responding to the needs of people from all social and cultural layers. However, again in this case, the social and cultural activities are not yet seen to be effectively sufficient. The enhancement is necessary to make the social layer permeable with the city and region attracting visitors from outside and disseminating the knowledge.

- *Economic permeability:* The economic permeability is seen at two levels in KGUV: (a) economic ownership: an innovation district, unlike a traditional precinct, is based on the triple helix model, thus reflecting the permeability in its economic model. It is a joint collaboration between the private, public, and economic sectors; and (b) affordability: the permeable character is reflected in the presence of economic options for all people in such places. In KGUV, special attention is being paid to develop activities catering to all economic groups.
- *Environmental permeability:* Special care has been taken to blend the architecture with the context by keeping it to a scale that matches its surroundings. In case of KGUV, the low impact, medium density housing blended with the existing context is an example of connectivity with environment. Built environment needs to keep itself permeable to the wildlife and nature around. Moreover, the provision of sustainable designs stresses the integration of nature with human-made designs. Thus, a considerable extent of connectivity with between natural and man-made environment is seen in KGUV.

In sum, KGUV is succeeding as permeable typologies driven by some common underlying principles shaping it. This facilitates the flow within the confines of the space and from the space to the overall veins of the city, i.e., Brisbane. The accessibility and permeability of the area allow interactions between knowledge workers, city dwellers, and occasional visitors, contributing to the vibrancy of the urban village. The interaction between different social groups and the urban fabric confer to this innovation district a specific character, and in doing so attaining the objective of place making for talented knowledge workers and users.

2.5 CONCLUSION

The research reported in this chapter aimed to address the research question focuses on investigating the role and characteristics of public spaces to facilitate place making in contemporary innovation districts. The literature findings have shown that the design of public domain within innovation districts is a central element in providing the vibrancy and diversity requested by the creative class of knowledge workers (e.g., Yigitcanlar & Dur, 2013; Pancholi *et al.*, 2014). The competitiveness between global cities has generated an attention to local contexts; the specificity of locale, a unique sense of place, can provide that character and experience that young knowledge workers are looking for.

In the specific case of KGUV, some constants in the design of built environment and of public spaces systems have been identified. The approach taken in this innovation district from Brisbane is to focus on character and connectedness aspects in both site-specific and broader scales as literature indicates importance of both unique character and high connectivity and these aspects (e.g., Southworth & Owens, 1993). These aspects have influenced the planning principles in KGUV, aiming to ensure a clear and legible identify for the precinct based on its creativity, heritage, and its specific geographical context, for example with the introduction of subtropical design principles in the development and design of the innovation district.

In terms of the key public space design and place-making strategies that the master plan is implementing, the use of anchor projects, main street development, and especially a layout based on connectivity and permeability have been highlighted as KGUV's main features. These features are also present in several other innovation districts in southeast Queensland, for example, Diamantina Knowledge Precinct and Knowledge-based Research and Business Precinct, and the proposed Knowledge Town of Sippy Downs at the Sunshine Coast.

This is to say a local typology for KBUD is clearly emerging slowly but surely in Brisbane, which is also considering the local subtropical character that allows open space living and the traditional outdoor living of the Australian culture. In terms of the planning approach, all innovation districts from Brisbane have a proposed master plan that details specific codes to achieve a precise control on the quality of the built environment. Whilst the design of these developments has been recognised as successful through the award of prizes and certification, the success from the social point of view is yet to be addressed and requires a more extensive and detailed investigation. This constitutes our future research direction (Pancholi *et al.*, 2015).

ACKNOWLEDGEMENTS

This chapter, with permission from the copyright holder, is a reproduced version of the following journal article: Pancholi, S., Yigitcanlar, T., & Guaralda, M., (2015). Public space design of knowledge and innovation spaces: Learnings from Kelvin Grove Urban Village, Brisbane. Journal of Open Innovation: Technology, Market, and Complexity, 1(1), 13.

REFERENCES

Adams, D., & Tiesdell, S. (2013). *Shaping places: urban planning, design and development*. New York: Routledge.

Amin, A., & Graham, S. (1997). The ordinary city. *Transactions of the Institute of British Geographers*, 22(1), 411–429. DOI:10.2307/623200

Arefi, M. (2004). The pedagogy of the American city: revisiting the concepts of place, non-place, and placelessness. *Urban Design International*, 9(1), 103–117. DOI:10.1057/palgrave.udi.9000092

Asheim, B. (2007). Differentiated knowledge bases and varieties of regional innovation systems. *Innovation: The European Journal of Social Sciences*, 20(3), 223–241. DOI:10.1080/13511610701581102

Baum, S., O'Connor, K., & Yigitcanlar, T. (2009). The implications of creative industries for regional outcomes. *International Journal of Foresight and Innovation Policy*, 5(1), 44–64. DOI:10.1504/IJFIP.2009.021906

Brisbane City Plan. (2000). *Kelvin Grove Urban Village local plan*. Brisbane: Brisbane City Council.

Buttimer, A., & Seamon, D. (1980). *The human experience of space and place*. London: Croom Helm. DOI:10.4324/9780203404346

Carmona, M., Heath, T., Oc, T., Tiesdell, S., & Carmona, M. (2003). *Urban Spaces-Public Places: The Dimensions of Urban Design*. Oxford: Architectural Press. https://doi.org/10.4324/9780080515427

Carmona, M., Heath, T., Oc, T., & Tiesdell, S. (2010). *Public places urban spaces: the dimensions of urban design*. London: Architectural Press. DOI:10.4324/9780080455718

Carr, S., Francis, M., Rivlin, L., & Stone, A. (1992). *Public space*. Cambridge: Cambridge University Press.

Carrillo, F. (2004). Capital cities: a taxonomy of capital accounts for knowledge cities. *Journal of Knowledge Management*, 8(5), 28–46. DOI:10.1108/13673270410560143

Carrillo, F., Yigitcanlar, T., Garcia, B., & Lonnqvist, A. (2014). *Knowledge and the city: concepts, applications and trends of knowledge-based urban development*. New York: Routledge. DOI:10.4324/9780203071329

Charles, D. (2011). The role of universities in building knowledge cities in Australia. *Built Environment*, 37(3), 281–298. DOI:10.2148/benv.37.3.281

Chesbrough, H. (2003). The era of open innovation. *MIT Sloan Management Review*, 44(3), 35–41. DOI:10.1002/smj.4250140309

Cresswell, T. (2004). *Place: a short introduction*. London: Wiley.

Cooke, P. (2015). Green governance and green clusters: regional & national policies for the climate change challenge of Central & Eastern Europe. *Journal of Open Innovation: Technology, Market and Complexity*, 1(1), 1–17. DOI:10.1186/s40852-015-0006-2

Duff, C. (2010). On the role of affect and practice in the production of place. *Environment and Planning D*, 28(5), 881–895. DOI:10.1068/d10808

Florida, R. (2002). The economic geography of talent. *Annals of the Association of American Geographers*, 92(1), 743–755. DOI:10.1111/1467-8306.00315
Florida, R. (2012). *The rise of the creative class: revisited*. New York: Basic Books.
Foth, M., & Adkins, B. (2006). A research design to build effective partnerships between city planners, developers, government & urban neighbourhood communities. *The Journal of Community Informatics*, 2(2), 116–133. DOI:10.15353/joci.v2i2.30
Foth, M., Klaebe, H., & Hearn, G. (2008). The role of new media & digital narratives in urban planning and community development. *Body, Space & Technology*, 7(2), 1–18. DOI:10.16995/bst.87
Kabachanik, P. (2010). England or Uruguay? The persistence of place and the myth of the placeless Gypsy. *Area*, 42(2), 198–207. DOI:10.1111/j.1475-4762.2009.00893.x
Klaebe, H. (2006). *Sharing stories: a social history of Kelvin Grove Urban Village*. Sydney: Focus Publishing.
Klaebe, H., Adkins, B., Foth, M., & Hearn, G. (2009). Embedding an ecology notion in the social production of urban space. In *Handbook of Research on Urban Informatics: The Practice and Promise of the Real-Time City*, edited by Foth, M., 179–194. Hershey: IGI Global. DOI:10.4018/978-1-60566-152-0.ch011
Knight, R. (1995). Knowledge-based development: policy and planning implications for cities. *Urban Studies*, 32(2), 225–260. DOI:10.1080/00420989550013096
Kodama, F., & Shibata, T. (2015). Demand articulation in the open-innovation paradigm. *Journal of Open Innovation: Technology, Market and Complexity*, 1(2), 1–21. DOI:10.1186/s40852-015-0009-2
Kunzmann, K. (2009). The strategic dimensions of knowledge industries in urban development. *disP-The Planning Review*, 45(177), 40–47. DOI:10.1080/02513625.2009.10556958
Lippard, L. (1997). *The lure of the local: senses of place in a multicultural society*. New York: The New Press. DOI:10.1080/09654313.2013.778065
Lonnqvist, A., Kapyla, J., Salonius, H., & Yigitcanlar, T. (2014). Knowledge that matters: identifying regional knowledge assets of Tampere Region. *European Planning Studies*, 22(10), 2011–2029.
Massey, D. (1991). A global sense of place. *Marxism Today*, 35(6), 24–29. DOI:10.1016/0277-5395(91)90004-J
Massey, D. (2005). *For space*. London: Sage. DOI:10.1080/02697450903329884
Minnery, J., Knight, J., Byrne, J., & Spencer, J. (2009). Bounding neighbourhoods: how do residents do it? *Planning, Practice & Research*, 24(4), 471–493.
Mort, G., & Roan, A. (2003). Smart state: Queensland in the knowledge economy. *Queensland Review*, 10(1), 859–870. DOI:10.1017/S1321816600007365
Pancholi, S., Yigitcanlar, T., & Guaralda, M. (2014). Urban knowledge and innovation spaces: concepts, conditions and contexts. *Asia Pacific Journal of Innovation and Entrepreneurship*, 8(1), 15–38. DOI:10.1108/APJIE-07-2013-0046

Pancholi, S., Yigitcanlar, T., & Guaralda, M., (2015). Public space design of knowledge and innovation spaces: learnings from Kelvin Grove Urban Village, Brisbane. *Journal of Open Innovation: Technology, Market, and Complexity*, 1(1), 13. DOI:10.1186/s40852-015-0015-7

Pratt, A. (2002). Hot jobs in cool places: the material culture of new media product spaces: the case of south of the market, San Francisco. *Information, Communication and Society*, 5(1), 20–27. DOI:10.1080/13691180275357372

Shaftoe, H. (2008). *Convivial urban spaces: creating effective public spaces.* London: Earthscan. https://doi.org/10.4324/9781849770873

Sheppard, E. (2002). The spaces and times of globalization: place, scale, networks, and positionality. *Economic Geography*, 78(3), 307–330. DOI:10.2307/4140812

Southworth, M., & Owens, P. (1993). The evolving metropolis: studies of community, neighborhood, and street form at the urban edge. *Journal of the American Planning Association*, 59(3), 271–287. DOI:10.1080/01944369308975634

Van Winden, W. (2010). Knowledge and the European city. *Economic and Social Geography*, 101(1),100–106 DOI:10.1111/j.1467-9663.2009.00550.x

Watson, G., & Bentley, I. (2007). *Identity by design*. New York: Routledge.

Wiltshire, K. (2003). Queensland Smart State: positioning Queensland. *Queensland Review*, 10(1),1–10. DOI:10.1017/S1321816600002907

Worpole, K., & Greenhalgh, L. (1996). *The freedom of the city*. London: Demos.

Worpole, K. (2005). *No particular place to go? Children, young people and public space*. London: Groundwork.

Yigitcanlar, T., Baum, S., & Horton, Stephen. (2007). Attracting and retaining knowledge workers in knowledge cities. *Journal of Knowledge Management*, 11(5), 6–17. DOI:10.1108/13673270710832361

Yigitcanlar, T., & Velibeyoglu, K. (2008). Knowledge-based urban development: the local economic development path of Brisbane, *Australia. Local Economy* 23(3), 195–207. DOI:10.1080/02690940802055458

Yigitcanlar, T., O'Connor, K., & Westerman, C. (2008a). The making of knowledge cities: Melbourne's knowledge-based urban development experience. *Cities*, 25(2), 63–72. DOI:10.1016/j.cities.2007.11.002

Yigitcanlar, T., Velibeyoglu, K., & Martinez-Fernandez, C. (2008b). Rising knowledge cities: the role of knowledge precincts. *Journal of Knowledge Management*, 12(5), 8–20. DOI:10.1108/13673270810901657

Yigitcanlar, T. (2009). Planning for knowledge-based development: global perspectives. *Journal of Knowledge Management*, 13(5), 228–242. DOI:10.1108/13673270910986709

Yigitcanlar, T. (2010). Making space & place for the knowledge economy: knowledge-based development of Australian cities. *European Planning Studies*, 18(11), 1769–1786. DOI:10.1080/09654313.2010.522057

Yigitcanlar, T., & Sarimin, M. (2011). The role of universities in building prosperous knowledge cities: the Malaysian experience. *Built Environment*, 37(3), 260–280. DOI:10.2148/benv.37.3.260

Yigitcanlar, T., & Dur, F. (2013). Making space and place for knowledge communities: lessons for Australian practice. *Australian Journal of Regional Studies*, 19(1), 36–63. https://www.questia.com/library/journal/1G1-340620374/making-space-and-place-for-knowledge-communities-lessons
Yigitcanlar, T., & Lonnqvist, A. (2013). Benchmarking knowledge-based urban development performance: results from the international comparison of Helsinki. *Cities*, 31(1), 357–369. DOI:10.1016/j.cities.2012.05.002
Yigitcanlar, T. (2014). Position chapter: benchmarking the performance of global and emerging knowledge cities. *Expert Systems with Applications*, 41(12), 5549–5559. DOI:10.1016/j.eswa.2014.02.042
Yigitcanlar, T., & Bulu, M. (2015). Dubaization of Istanbul: insights from the knowledge-based urban development journey of an emerging local economy. *Environment and Planning A*, 47(1), 89–107. DOI:10.1068/a130209p
Yigitcanlar, T., & Sarimin, M. (2015). Multimedia Super Corridor, Malaysia: knowledge-based urban development lessons from an emerging economy. *VINE: The Journal of Information and Knowledge Management*, 45(1), 126–147. DOI:10.1108/VINE-07-2014-0033
Yigitcanlar, T., Inkinen, T., & Makkonen, T. (2015). Does size matter? Knowledge-based development of second-order city-regions in Finland. *disP-The Planning Review*, 51(3), 48–63. DOI:10.1080/02513625.2015.1075076
Zenker, S. (2009). Who's your target? The creative class as a target group for place branding. *Journal of Place Management and Development*, 12(5), 8–20. DOI:10.1108/13632540911001424
Zukin, S. (1995). *The cultures of cities*. Oxford: Basil Blackwell.

2

Place Making

This part of the book explores the critical role of place making as a strategic approach to enhance the socioeconomic and environmental success of innovation districts. It provides a multidimensional conceptualisation of place making and establishes its multi-staged and multifaceted nature within the distinct context of innovation districts, drawing upon interdisciplinary exploration and relevant case studies. The role of place making in innovation-nurturing locations—characterised by diverse actors and stakeholders with varied interests—extends beyond economic success. This part contributes to the development of a unique character, fosters a sense of place and facilitates the generation and dissemination of knowledge and innovation. Place making encompasses not only spatio-physical aspects such as design and planning but also comprehensive considerations of the broader context, featured image or branding, functional aspects involving internal and external socioeconomic processes and flows, as well as the meanings attributed by stakeholders. The significance of spatio-temporal factors, including the unique local socio-cultural context, in shaping the distinct identity of these districts is specifically highlighted. This part advocates that a unidimensional or partial exploration of the concept of place making, without due consideration of its diverse aspects and stages, would result in a disintegrative approach.

DOI: 10.1201/9781032657431-4

CHAPTER 3

Place-Making Attributes of Innovation Districts

3.1 INTRODUCTION

Place making is increasingly manifesting itself as a popular strategy adopted by the competitive cities for retaining talent force to sustain creativity and knowledge generation in their blooming spatial assemblages of innovation and knowledge-intensive activities and functions (Yigitcanlar, 2009; Edvardsson *et al.*, 2016). These locations—better known in the literature as knowledge precincts, knowledge locations, and innovation clusters or districts—are referred to as innovation districts in this chapter. Incorporating multifaceted developmental motives, innovation districts are seen as the growth engines aimed at holistic and sustainable growth in terms of economy, society, governance, and environment (Yigitcanlar *et al.*, 2008a; Van Winden *et al.*, 2013; Carrillo *et al.*, 2014). However, these globalised knowledge-nurturing locations—in addition to the retainment of talent force—are facing number of economic, social, political, and environmental challenges (Florida, 2005; Baum *et al.*, 2009; Peck, 2010). Interestingly, despite acknowledgement of place making as a strategic solution and its multidimensional definition—as accentuated by recent studies—knowledge regarding the applicable approaches for its practical integration in innovation districts remains an understudied subject. This research gap widens further considering the mentioned challenges unique to innovation districts (Yigitcanlar *et al.*, 2008b, 2017a; Pancholi *et al.*, 2014; Carvalho and Winden, 2017).

DOI: 10.1201/9781032657431-5

This study aims to explore planning and policy approaches for place making as a strategic solution for complex challenges of contemporary innovation districts. The research question, thus, is: 'How can place making be integrated as a strategy to combat the challenges of innovation districts?' This paper places case studies of three leading Australian innovation districts—Macquarie Park (MQP), Melbourne Employment Cluster (MEC), and Dutton Park Knowledge Precinct (DPKP)—located in Sydney, Melbourne, and Brisbane, under the microscope. An interview-based qualitative method is adopted as the methodological approach, and findings are presented in line with a multidimensional conceptual framework of place making. Conclusions are drawn based on identifying the common as well as unique patterns related to place making through an empirical spatio-temporal investigation across the three cases. In doing so, the study scrutinises the refined role of place making and elucidate key attributes of the framework. The paper does not focus on comparing the three cities or evaluating their success; rather it aims to derive general implications on the grounds of research findings. The paper contributes to the existing knowledge by pointing out the theoretical and practical implications for planners and policy-makers looking for innovative approaches to integrate place making as a strategy to achieve desired denouements from innovation districts —the spatial nexus of knowledge-based urban development (KBUD).

3.2 LITERATURE BACKGROUND

3.2.1 Place Making

To comprehend the 'contingent, multi-layered, complex and contested' nature of place making as asserted by Arefi (2014, p. 4), it is necessary to consider related theoretical assertions. Place has remained a topic of long-held debate amongst scholars from crosscutting disciplines—such as sociology, cultural geography, political geography, human geography, economic geography, psychology, spatial science, planning, and design. Considering place as a location in spatial science, stature of place is juxtaposed within two major schools of thought. One stream of discussions strongly accentuates the 'placelessness' of knowledge with the 'death of distances' and world getting flat due to rapid globalisation (Harvey, 1989; Massey, 1991; Markusen, 1996). A parallel thought stream contends how the claims over diminished stature of location are overemphasised and place—as a location—still remains as relevant as it was in shaping up of innovation and creativity and the significant role of physical environment in it (Ophir and Shapin, 1991; Livingstone, 2003; Zhao, 2005).

Not limiting the extent of meaning of 'place' in a static sense like spatial science, theoreticians from interrelated disciplines draw attention towards place as a product of social, economic, and political processes over a period of time extending far beyond the confines of location (Lefebvre, 1991; Massey, 1991; Merrifield, 1993). Pioneered by the classic work of Lefebvre (1991), in consecutive years, a group of renowned theoreticians has postulated the need to understand place as coherence between its conceived, perceived, and lived forms (Cresswell, 2004). Montgomery (1998) proposes a well-defined framework that identifies place as form, activity, and image. In addition, with people, the role of internal and external factors such as market, regulatory force or globalised forces also cannot be ignored in the shaping up of a place. The solution for problems of a place can be solved not only by design, but also by apprehending it as interplay of management placed in urban policy context and the wider social and economic processes that they are embedded into (Jones, 2009; Carmona *et al.*, 2010; Adams and Tiesdell, 2013). In his recent works, Arefi (2014) asserts that places are not only physically built, but also socially mobilised and politically contested. Place making, therefore, needs to be considered multidimensionally not only as a physical product but as socioeconomic, political, and organisational processes, people's perceptions, and experience of a locale as well as the context that shapes it (Pancholi *et al.*, 2015a; Makkonen and Weidenfeld, 2016).

3.2.2 Contemporary Innovation Districts

Knowledge flows related to innovative activities are a core component of innovation districts. As stated by Witt (2016, p. 3), 'innovations mirror new cognitive concepts, which emerge when inventors, researchers, developers, or entrepreneurs (re-)combine—often in team work—already existing cognitive concepts into something that can be given a novel meaning'. Rapidly gaining importance of innovation along with bringing together urban development in evolutionary economic geography with regional resilience in urban planning and Markusen's (1996) industrial districts concept, in recent years, generated a new model of space formation for innovation in cities—so-called innovation districts (Clark *et al.*, 2010).

Globally, the effect of shifting from manufacturing-based economy towards knowledge economy can be seen as manifested as a few significant transformations in the contemporary cityscape (Yigitcanlar and Lonnqvist, 2013; Yigitcanlar and Lee, 2014; Stehlin, 2016). The most evident spatial transformation is growth of spontaneously developed or deliberately planned locations or innovation districts. In these locations innovation and knowledge-intensive activities aggregate together and socially, which has led to the growth of knowledge workers as a new class of society (Landry, 2000; Yigitcanlar *et al.*, 2007; Elia *et al.*, 2016). Depending on the context and conditions—in addition to their contribution as spatial nexus of KBUD, these knowledge-producing locations also contribute in: (a) knowledge economy stimulation, (b) strengthening of the local innovation ecosystem, (c) creative revitalisation of inner-city areas, and (d) global competitiveness of their cities and regions (Scott, 2006; Van Winden *et al.*, 2013; Reynolds and Uygun, 2017).

The phenomenon of spatial agglomeration of economic activities is not new and remains a widely discussed concept in literature (Jacobs, 1969; Glaeser, 1999). Scholars like Saxenian (1994), Porter (2000), and Storper and Venables (2004) favoured the role of geographical proximity in enabling 'face-to-face' interactions that create 'buzz'—necessary for cross-fertilisation of ideas and innovation. Recent works have asserted the role of 'relational proximity', i.e., proximity in terms of society, organisation, and institution, to develop a social ecosystem of learning—vital for the new knowledge generation and, henceforth, the economic success of innovation districts (Boschma, 2005; Berkes, 2009; Bathelt *et al.*, 2013; Phillips and Yeung, 2016). With the rise of network society and a new 'open innovation' model, the economic environment of contemporary innovation districts has advanced towards being more collaborative and people-oriented rather than the earlier competitive and merely economy-oriented model (Chesbrough, 2003; Tallman *et al.*, 2004; Yun, 2015; Kodama and Shibata, 2015; Yun *et al.*, 2016a, 2016b, 2017).

Katz and Wagner (2014) emphasise the necessity of an integrated relationship between people, firms, and place for simultaneous development of economic, networking, and physical assets of innovation districts. With the growth of cultural economy, spatial ensembles characterised by the intersection of technology, culture, and place mushroomed up—better known as 'new media spaces', 'new economy spaces', 'creative industry clusters', or 'creative clusters' followed (Hutton, 2004; Evans, 2009). Spatially, therefore, transfiguration or 'urban turn'—as referred by Van Winden *et al.* (2013)—of innovation districts from their earlier introvert, secluded and mono-functional counterparts to modern physically extrovert and functionally mixed-use versions distinguishes the contemporary agglomerations from their older counterparts (Yigitcanlar and Sarimin, 2011; Pancholi *et al.*, 2017).

Socio-organisationally, triple-helix model of governance, i.e., a partnership between public, private, and academia, has become their modus operandi (Lonnqvist *et al.*, 2014; Leydesdorff and Ivanova, 2016.). The need for a synergistic and holistic balance of economy, society, environment, and governance for successful KBUD has been emphatically accentuated in renowned works (Yigitcanlar and Dur, 2013; Carrillo *et al.*, 2014). Further to their economic contribution, the specialised role of innovation districts in providing social equity and inclusion for a democratic society achieved through strong social and human capitals has been propounded by a number of scholars (Fernandez-Maldonado and Romein, 2010; Pancholi *et al.*, 2017). Besides, the redefined role of governance, acknowledged role of public participation, as the factor of social change, and recognised role of smart environments for smart communities, that encourage innovation, complement the process (Henton and Held, 2013; Jung, 2013; Komninos, 2016; Yigitcanlar *et al.*, 2017b).

3.2.3 Challenges for Contemporary Innovation Districts

To apprehend the specialised role of place making of innovation districts, it is also inevitable to understand the major challenges that contemporary innovation districts are facing. Recent research studies have delineated that despite the theoretical acknowledgement regarding the significance of multi-faceted factors in the success of innovation district—as explored above—there exist certain challenges in the process (Esmaeilpoorarabi *et al.*, 2016; Cooke, 2017). Firstly, the knowledge workers generally prefer the vibrant and diverse downtown areas to the secluded and monotonous environment of innovation districts—making it a challenge for the companies to retain the talent force (Castells, 2000; Friedmann, 2007; Yigitcanlar *et al.*, 2016). Consequently, innovation districts face an exodus of businesses, as shown by Graham and Guy (2002) in their study on the Silicon Valley. As a result of this exodus and added economic value, issues such as what the gentrification of areas entails—that drives the existing users out to the cheaper areas, pushing land prices to soar high (Peck, 2010). Recent works also emphasise other social pitfalls such as partial integration of culture—largely limited to the integration of physical assets such as heritage structures—and the development of 'mono-culture' in the city, marginalising the groups other than knowledge workers in the society

(Stehlin, 2016). Economically, studies have displayed a lack of collaboration, despite the clustering and agglomeration of economic activities. Though displaying strong international connections, a number of concurrent innovation districts fail to establish strong knowledge-exchange relations at local level (Garnsey and Heffernan, 2005). The role of place making, therefore, extends from merely developing a creative physical environment for attracting talent force to finding solutions for these enlisted challenges.

3.3 EMPIRICAL INVESTIGATION

This research applied a case study method for the empirical investigation. The method was considered appropriate for this research because it allows defining the topic more broadly (i.e., to identify the issues around successful integration of place making as a strategy to combat the challenges of innovation districts) by considering contextual issues in each case and relying on multiple sources of evidence (Yin, 2011). The two most common approaches of case study research include an inductive approach based on the Grounded Theory (Glaser & Strauss, 1967) and the deductive/testing approach (Yin, 2003). The basic difference between the approaches is that while the Grounded Theory relies on data to generate new theories (there is no initial preconceived framework of concepts and hypotheses), the other approach develops a theory at the beginning of the research and focuses on testing and validating the theory in case settings. Another approach proposed by Eisenhardt (1989) lies in-between these two approaches—i.e., aligns with the Grounded Theory approach—that is inductive, but there are elements that follow a more planned approach. Given the rich literature on the role of incentives on innovation, this research follows the deductive approach to identify the issues around successful integration of place making as a strategy to combat the challenges of innovation districts.

To carry out the empirical investigation in three cases from Australia, a semi-structured interview-based qualitative analysis approach has been adopted. Data from primary and secondary sources—i.e., policy and plan documentations obtained from government organisations, planning and design firms, developers, research institutes, and on-site tenant firms are also utilised by integrating them with the interview findings. Additionally, sources such as field observations, photographs, physical plans, and maps also contributed to the analysis as primary data sources. A questionnaire—hinged on the conceptual framework—is moulded in each case in accordance with the issues identified with the help of data from secondary sources. To gain a comprehensive understanding regarding place making, a wide range of key stakeholders related to projects were interviewed—in total, 44. The interviewees are selected to form five major groups—government executives, planners and designers, networking groups, firms and institutions, and knowledge workers (Table 3.1). For each group, interviewees were selected as experts at key positions associated with the case. Interviews commenced in the second half of 2015, each lasted about 45–60 minutes, were digitally recorded, and then transcribed into text manually. An inductive approach of content analysis—informed by the phenomenographic methodology—was adopted to analyse findings.

To carry out the analysis, a multidimensional conceptual framework for place making in innovation districts has been adopted that integrates the diverse interdisciplinary perspectives together (Pancholi *et al.*, 2015a). It rests on the theoretical paradigm propounding the multi-faceted nature of place-making manifested as coherence between conceived, perceived, and lived spaces (Lefebvre, 1991; Montgomery, 1998; Castells, 2000). The five dimensions of place making here are context, feature, function, form and image (Figure 3.1). Any innovation district is an outcome of the broader context that surrounds it—that forms the first and foremost dimension. This refers to spatial, political, institutional, economic, and socio-cultural context—each to be considered spatio-temporally, i.e., in terms of its history, current dynamics, and related future plans in place. In order to succeed, it is necessary for an innovation district to display unique selling factors that give it a competitive edge over other innovation districts. Henceforth, in the conceived layer, 'feature' as a second dimension includes the attributes defining marketability of innovation district. Another dimension in a conceived layer is 'form' that refers to the spatial and physical factors pertaining to innovation districts. Groups 1 and 2 mostly contributed to assessing feature and form. Pertinent to the 'lived space' layer, 'function' forms the third dimension, integrating all the processes happening in the innovation district to be assessed for the functional factors required for knowledge generation and dissemination. Data from Groups 3, 4, and 5 were considered for analysing function. In the 'perceived' layer, 'image' forms the last dimension, which analyses the joint perceptions of all the groups of stakeholders and the meanings associated to them as pertains to the innovation district.

TABLE 3.1 List of Interviewees

Group	ID	Position	Relevance
Group 1: Government executives	Interviewee#1	State government executive	Leading team member of the regional plan
	Interviewee#2	Local council executive	Involved in local economic development plans of the site
	Interviewee#3	Local council manager	Involved in strategic plans effecting the site
	Interviewee#4	Local council executive	Key involvement in planning and execution
	Interviewee#5	Local council manager	Key involvement in local economic development plans
	Interviewee#6	Executive director	Key involvement in planning and functioning
	Interviewee#7	Department head	Leading team member of the precinct plan
Group 2: Planners and designers	Interviewee#8	Senior urban designer	Involved in master planning of the site
	Interviewee#9	Senior urban planner	Involved in master planning of the site
	Interviewee#10	Chief architect	Involved in master planning of the university
	Interviewee#11	Senior urban planner	Key involvement in master planning of the precinct
	Interviewee#12	Senior urban designer	Key involvement in master planning of the campus
	Interviewee#13	Urban designer	Key involvement in on-site infrastructure projects
	Interviewee#14	Senior urban planner	Key involvement in social and urban planning of site
	Interviewee#15	Senior urban designer	Key involvement in urban design of site
	Interviewee#16	Senior architect	Involvement as a design expert
	Interviewee#17	Design manager	Involvement as a design expert
Group 3: Networking groups	Interviewee#18	Director	Head of a formal local networking group
	Interviewee#19	Manager	Coordinator of a formal local networking group
	Interviewee#20	Manager	Key role in formal group
	Interviewee#21	Community team leader	Key member of leading team
	Interviewee#22	Chief executive officer	Involvement by leading a major on-site networking group
	Interviewee#23	Committee member	Key member of an active community group

Group 4: Firms and institutions	Interviewee#24	Chief executive officer	Active participant of a local networking group
	Interviewee#25	Manager	Executive of an on-site anchor business
	Interviewee#26	Director	Executive of an on-site business
	Interviewee#27	Director	Key position in an on-site institution
	Interviewee#28	Director	Key position as collaborator in an on-site institution
	Interviewee#29	Director	Leading an on-site business
	Interviewee#30	Executive manager	Executive position in an on-site anchor institution
	Interviewee#31	Senior manager	Key position in an on-site anchor institution
	Interviewee#32	Division director	Leading a key anchor institution
Group 5: Knowledge workers	Interviewee#33	Chief technology officer	All participants in this group are interviewed for their perceptions as users of the site
	Interviewee#34	Director	
	Interviewee#35	Manager	
	Interviewee#36	Senior researcher	
	Interviewee#37	Executive position	
	Interviewee#38	Lead associate	
	Interviewee#39	Technology expert	
	Interviewee#40	Research officer	
	Interviewee#41	Technical officer	
	Interviewee#42	Principal scientist	
	Interviewee#43	Senior researcher	
	Interviewee#44	Senior researcher	

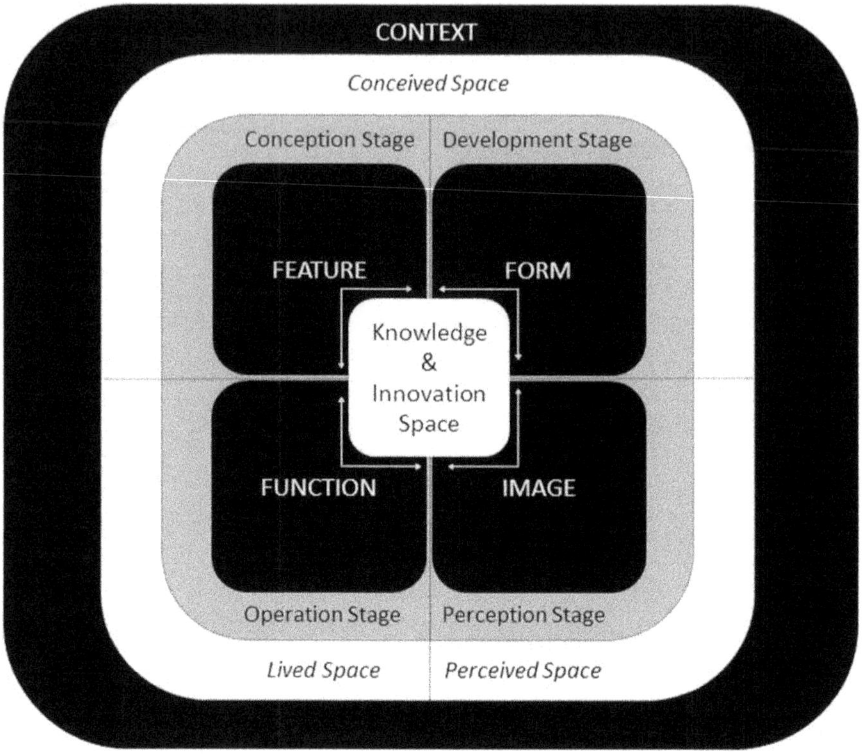

FIGURE 3.1 Conceptual framework of place making (Pancholi *et al.*, 2017, p. 77).

The three cases selected for investigation are MQP (Sydney), MEC (Melbourne), and DPKP (Brisbane)—these cases are the top innovation districts in their cities, with an established or emerging global reputation (Figure 3.2). The diversity of context in each case—in terms of age, proximity to city, leading sectors, and spatial design, i.e., whether planned or spontaneous—makes the study further robust and interesting. While two of the innovation districts, i.e., MQP in Sydney and DPKP in Brisbane, have been deliberately planned on the lines of the internationally successful cases, MEC in Melbourne evolved as a spontaneous agglomeration.

3.3.1 Context

Spatio-environmentally, the location of MQP is noteworthy as it forms an integral part of the global economic corridor of Sydney, being only 12 km from the centre of the city. Melbourne's MEC has a significant location 20 km from Melbourne CBD towards its southeast, displaying

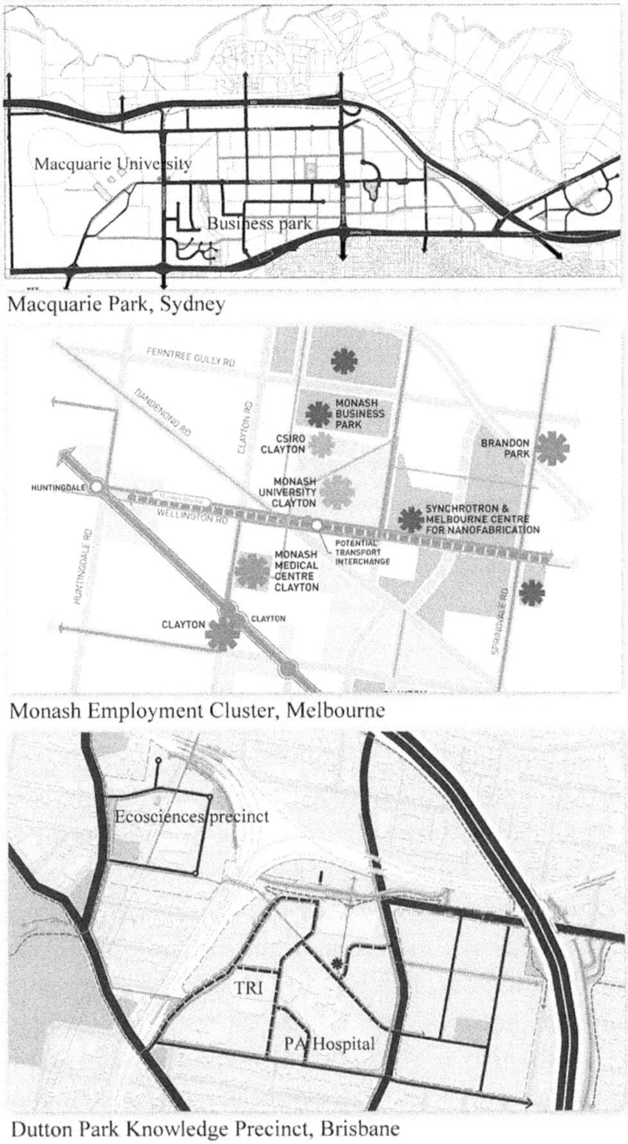

FIGURE 3.2 Layout of investigated innovation districts.

strong connectivity with key transport corridors crossing through it—Monash Freeway, and Clayton, Blackburn, and Dandenong Roads. Being the second biggest innovation district agglomeration in Australia, it is acknowledged to play a significant role in the decentralisation of urban core. Similarly, Brisbane's DPKP boasts a strategic location being only 3.5 km south of the city. Socio-culturally, existing cultural and ethnic diversity acts as another success factor of the innovation district in the form of a local asset. Sydney, Melbourne, and Brisbane boast a culturally diverse community and are all known for their art, sports, and culture, making them a lucrative choice for the young talent force representing different nationalities and origin. Greater Sydney has an average of 20%—with the case study area having 48%—of its population as overseas, born reflecting the multiculturally rich and tolerant society. Melbourne displays a representation from about 200 nationalities within its boundaries, as authenticated by Interviewee#14, *One of the advantages that Australia has is that we are quite multiculturally diverse.*

Politically, support for innovation is a critical factor in determining the development of the innovation district and creation of a reliable local and international identity. Interviewee#30 expresses, *Without government none of this would have happened because government is the only one that will invest where there is no return, when you are still doing the discovery.* This support was in the form of: (a) policy and planning and (b) economic support. More importantly, as evident from interviews, continued high-level prioritisation in policies is necessary for sustaining success as is evident by the cases of Melbourne and Sydney. Conversely, in DPKP, shifted prioritisation with the change in government resulted in slowing down the developmental plans. Group 4 in Brisbane's case discussed the necessity for extending governmental support from seed funding to operational funding. Institutionally, the existence of a strong institutional framework at the state and city level facilitates a favourable context. While high-level universities and research institutions provide the knowledge base, specifically dedicated bodies and networking organisations act as a catalyst—initially, in the process of design and development with coordination and networking in later stages. Metropolitan Planning Authority in Melbourne is a specifically dedicated organisation focusing on planning of such sites that function by bringing stakeholders together. The existence of a dedicated body for urban design at the local level further complements this. Interviewee#8 believes the potential for MEC's future development rests in this: *The fact that we have an in-house urban design and sustainability group has some influence.*

3.3.2 Feature

Existence of strong anchor and sub-anchors strengthen the profile of innovation district as is evident in all three cases with renowned universities and other well-acclaimed institutions acting as anchors. In Sydney,

Macquarie University (one of the world's top-250 universities) is the key anchor, with many other global players from diverse fields and Australia's top 100 companies, such as Johnson and Johnson, Microsoft, Sony, Optus, Cochlear, and Foxtel. In Melbourne, a number of world-class institutions and sub-clusters such as Monash University (one of the world's top-100 universities), Commonwealth Scientific and Industrial Research Organisation, Synchrotron, Melbourne Centre for Nanofabrication, Monash Health Translation Precinct, Monash Business Park and Monash Technology Precinct and so on gives MEC as an area a distinguished address that attracts others. Similarly, in Brisbane, collocation of regional health providers like Princess Alexandra Hospital, Mater Hospital, and others as well as renowned institutions, such as the University of Queensland (one of the world's top-50 universities), Queensland University of Technology (one of the world's top-250 universities), Griffith University (one of the world's top-350 universities), Ecosciences Precinct, Translational Research Institute, Pharmacy Australia Centre of Excellence, Biopharmaceutical Australia, and Commonwealth Scientific and Industrial Research Organisation make it the most-renowned innovation district at a regional level. Additionally, interviews demonstrate that specialisation as sector-based product or process specialisation and availability of world-class researchers serve as a unique selling point. In MEC and DPKP, globally acclaimed specialisation of institutions in the field of biopharmaceuticals contributes as a key feature. Pointing towards the role of world-class researchers as a prime feature, Interviewee#30 elucidates, *The way companies come here is through a leading researcher. Because they know that she will get these projects that will mean that more people will use their technology.*

Effective marketing strategies are required to make an innovation district lucrative both to the firms and people. Unlike MQP with a well-defined integrated identity, Melbourne's innovation district was seen as an agglomeration of individual sub-clusters rather than having a whole identity of its own—until recently when in 2014 Plan Melbourne designated the overall area as MEC. Similarly, in DPKP, the need for an integrated identity due to the lack of a well-defined name of an innovation district is apparent from the interviews. Other uniqueness defining factors, such as a specific focus on societal and place-making aspects, contribute in indirect branding of an innovation district. In MQP, this refers to the core focus in agenda on developing a live-work-learn-play environment. In DPKP, heritage integration by redevelopment of Boggo Road Gaol makes the project unique. Summarising the feature as a place-making element, Interviewee#14 asserts, *It comes back to that unique selling point, that unique advantage that your precinct has over anybody else.*

Highlighting the need to provide additional support to small-scale businesses in the success of innovation districts, Interviewee#22 conveys, *Start-ups are only as good as their access to cash and their next experiment.* In MEC, the local council—with the support of state, national government, and key on-site anchor tenants—has developed the Monash Enterprise Centre and Eastern Innovation Business Centre to support SMEs and innovative businesses. In MQP, plans are under way for the establishment of an innovation district with the university's support. However, DPKP is still lagging behind in providing such amenities, except for the individual initiatives by anchor institutes that organise regular training sessions and host companies. Cost-effectiveness emerges as another key factor from interviews that affect the viability of location. As seen in the case of Melbourne, due to competitive land and rent prices, it proves cost-effective for companies to locate themselves in MEC in comparison to CBD, with an added advantage of decent block sizes. As a CEO, Interviewee#24 highlights, *Firstly you have got the space for industry to expand and grow but also the rent is also much lower than in the city as well.* Interviewee#5 from Sydney exclaims, *The biggest thing for companies was it's the cheapest place.* Conversely, in DPKP, soaring land prices have been an impediment in attraction of companies, owing to the site's proximity to the city. This has also resulted in a push in market demand towards residential development. In addition, lack of affordability of houses is one of the major pitfalls in developing a live-work-learn-play environment. As Interviewee#14 points out, *Again it is a bit of a problem in this area because the land is so expensive.*

3.3.3 Form

Both Melbourne and Sydney's cases have been historically developed as low-scale business park settings and strictly zoned developments on big blocks of land. As is evident in MEC wherein, to ensure a sense of quality and greenery, development followed the 'garden city' character under the local council's strict site planning guidelines and land use zoning—catering to the companies' preference of consolidating their segregated offices at one location. While it gave a unique character to the area and facilitated the favourable conditions for its development as an innovation district, it also conversely led to non-walkable environment and segregation. Large, isolated corporate campuses behind a boom gate—having their own cafes and restaurants—exist as a current norm that Interviewee#13 in Sydney—an urban designer—denominates as *A private internalised world not a public rich environment.* Interviews pointed toward the lack of common green areas, vibrant public realm, and walkability as major issues. Interviewee#33 expresses, *Buildings are so close to each other without enough spaces to generate activities between them. I would love to have more green areas and larger open spaces between buildings.* In recent proposals, therefore, there has been a growing acknowledgement to develop more connected, high-density, and people-oriented environments, as revealed from the interviews with Groups 2 and 5. Interviewee#13 emphasises, *We need public realm that's going to encourage I suppose the high-end workers to leave the city and not see it as a major disadvantage. You got to deal with it to get people out of the buildings and engaging with each other.* Additionally, due to the absence of public art, heritage, or innovative element, the innovation districts lack uniqueness of character necessary for developing a sense of place. Initiatives like climate-sensitive, creative, collaboration-oriented designing and heritage integration strengthen the innovative character of innovation district, as apparent in DPKP. Buildings like Ecosciences Precinct and Translational Research Institute are awarded at a national level as buildings 'without walls' for their collaborative environment through the use of shared spaces and resources as well as sustainability. Interviewee #17 considers the redevelopment of the historical jail as *a beneficial prospective move*, which not only aims to make the innovation district more vibrant, creative, and people oriented, but heritage integration also adds to its innovative character.

An integrated development as a site is a critical aspect to ensure proper exchange of knowledge and people. In DPKP, interviews highlight the lack of order and flow. The site—being punctuated by the passenger and freight lines—gets divided into two halves. Interviewee#15 emphasises, *What missing out in the precinct is a highlighting order* and Interviewee #17 states the *need to connect two parts of the precinct.* In MQP, the current planning proposal, i.e., Herring Road Urban Activation Precinct Proposal, is demonstrative of initiatives to make the precinct more people-oriented and enhance its vibrancy. These include, but are not limited to developing: (a) land-use re-zoning to mixed land-uses; (b) higher density urban community with more height allowance; (c) pedestrian-friendly environment by providing better connectivity and fine-grained roads; and (d) landscaping, amenities, community facilities, green spaces, and places to meet. Connect Macquarie, i.e., a formal group with representation from key stakeholders aimed at transport solutions, contributes in strengthening the people-oriented character further. Herring Road Urban Activation Precinct Proposal illustrates that to integrate the diverse knowledge community and society in an innovation district, a genuine and fine intermix of diversity in housing options is necessary. The plans for redevelopment of Ivanhoe Estate—an on-site social housing estate—will enhance the current 259 existing social dwellings into a mixed-housing estate neighbourhood with at least 556 social housing dwellings. Integrating more social housing will also contribute to combating social issues such as gentrification and marginalisation.

3.3.4 Function

Analysing function as a place-making dimension, participants from Groups 1, 2, 3, and 4 across the three cases equally acknowledged the enhancement in importance of networking and its critical role in the sustainable success of innovation district. Interviewee#20 synopsises, *Final comment on this is, the whole concept of open innovation is much stronger as compared to keeping everything secret. Companies are recognising they need new ideas and to work in collaboration rather than just relying on themselves.* Melbourne's MEC emerges as the leading example out of three displaying the strong role of virtual networks existing as its networking assets. Here, groups like Victorian Platform Technologies Network—a state-level centralised networking group—and Monash Technology Research Platform—an innovation district-level subset platform of Victorian Platform Technologies Network active in Monash—play a proactive role in MEC as an ideal collaboration for government, research organisations, universities, industry, researchers, and innovators. Managers from Melbourne commended the role of these groups in social networking expedited within groups of managers and others leading to fruitful collaboration. Interviewee#27 formulates, *So I think in addition to the physical structure and the spaces and so forth, you cannot underestimate the social networks and structures that are important for success.*

Another key aspect that defines place making through function is how vital the location is. This depends largely on the vibrancy of activities, diverse functions or uses that exist in the innovation district, as well as the usability of its public amenities. Analysis shows that the role, exchange, and interaction of universities with the rest of the innovation district slightly vary from case to case, depending on their simultaneity of establishment relative to the innovation districts. While already established ones like the University of Queensland in the case of Brisbane tend to have a higher influence as an anchor, the process has been time consuming for those that have been established simultaneously, like Macquarie University in Sydney and Monash University in Melbourne. With their opening up in the last few years, they have already started reflecting a significantly positive augmentation in their impact by lubricating the networking within the innovation district. They act as a common platform for firm, institution, government, and society as is evident by the statement of Interviewee#28, *Some of the companies don't have a natural way. It's hard whereas we can introduce them naturally in a very relaxed way. And they can follow up by doing business together.* The cases of Melbourne and Sydney have been initially developed as business parks and hence functions were deliberately zoned to prioritise the development of corporate and commercial-based functions. This lack of proper mixed-use functions and public amenities can negatively impact the vitality of the innovation district safety and achieving a 24/7 vibrancy. Consequently, the current plans are addressing this as an issue by introducing more mixed uses. Explaining the future plans, Interviewee#15 says, *The proposal is to find ways to bring activity into the precinct both through retail and leisure, food, obviously. Try to make it more porous, try to open it up a bit, try to get people to come in and spend time in these spaces—like a third space.*

3.3.5 Image

This section focuses on capturing the perceptions of group of stakeholders that investigates attraction, image sustenance, and repulsion factors related to three innovation districts. In MQP and MEC, while users appreciated the quality of environment and openness of site, but physical factors such as lack of amenities, casual spaces, walkability, 24/7 active precinct model and economically housing affordability are common concerns raised by knowledge workers and firms in all three cases. Interviewee#36 elucidates, 'I like the peace and quiet but at the same time I did not like that there is nothing else there'. In addition to these, factor such as commuting distance from the city emerges as an issue for the companies to retain the knowledge workers who prefer CBD to remote locations in innovation districts as evident from interviews of Group 4 in MQP and MEC. However, local community in MQP displayed great sense of attachment and pride in belonging to the area owing to its accessibility, amenities, and other assets. The cases display a quadruple helix model of governance and planning, i.e., an equal participation from public, private, academia and community through consultation from initial stages to final stages of planning and functioning. In all the three cases, ample efforts from policymakers and planners, i.e., Groups 1 and 2, are evident to make the planning processes democratic. Analysing firm and institution's perceptions, Interviewees from Group 4 show high degree of satisfaction regarding the role of government in three cases. Interviewee#25 from Melbourne expresses, *The government provides implicit support.*

Considering its initial stages, Brisbane's DPKP revealed that even a people-oriented proposal can face conflicts due to lack of clarity in its communication to community. These conflicts in DPKP were later converted into coordination by the active role of planners in community consultation. MQP is a strong case representing the efficiency of communication and public engagement as evident by interviews with the key community group. To make the communication effective, initiatives here were not limited to online dissemination of information but extended to hand-delivered letters with clear information translated into multiple languages. Interviewee#21 recounts, *Staff on the site was very understanding, very compassionate, very patient with people, I do not think we could have asked for much more in terms of how the government has handled itself,* Pointing towards the significance of effective communication, he sums up, *We did the research, with such teams like Macquarie University and we know having an answer just had a caustic effect on people's well-being.* Coordination within different hierarchy of government is critical too. In Sydney and Melbourne, while interviews with Group 1 display substantial coordination, certain issues at planning stage such as land use conversion in innovation district from zoned to mixed-use emerges as a point of conflict between local and state level. Interviewees from Group 1 associated at local level displayed some reluctance towards the conversion of existing land-use model owing to their qualms about the effect of change on current commercial success of both innovation districts.

3.4 DISCUSSION AND CONCLUSION

As stated by Witt (2016, p. 4), *in a historical perspective, successful innovations have been the main drivers of economic growth creating wealth and raising the standard of living of the masses. Innovations are, therefore, often considered in a wholesale manner a desideratum in politics, in the public, and, not least, in innovation-related research.* The literature points out innovation districts; hence, as the nexus of innovation activities and KBUD, which have caught the attention of urban administrators and policymakers. In theory, place-making practices provide an opportunity to shape innovation districts as innovation hubs to deliver desired outcomes concerning economic prosperity and beyond (Pancholi *et al.*, 2015b). However, in practice it is not clear how place making can be integrated as a strategy to combat the challenges of innovation districts. This paper, hence, aimed to address this issue though an investigation of the Australian context. The key findings of this qualitative empirical investigation are discussed below.

Summarising the empirical investigation, each of the cases—on account of their respective strengths and weaknesses—represents the role of different dimensions in the overall process of place making in innovation districts (Table 3.2). MQP in Sydney is an illustrative example of place making through design and societal integration in contemporary innovation districts. Investigation of MEC in Melbourne strongly demonstrates the role of virtual networks—formal and informal—as well as institutions and governance in place making. DPKP is noteworthy for exploring the community-related conflicts and critical role of individual actors in shaping a knowledge location and, additionally, exhibits how heritage integration and innovative designs bring character to innovation districts.

While the investigation on three major cities of Australia explores individual spatio-temporal integration of place making, few significant theoretical implications that stand common in terms of definition, characteristics, and role of place making in innovation districts are identified. These are as follows:

- Any location is shaped by the complex interplay between its context, socioeconomic and political processes, physical and spatial factors and relationship between actors. Place making, thus, is a multifaceted concept that needs to be seen in the conceived, perceived and lived aspects of space being manifested as five pivotal dimensions, i.e., form, feature, function, image, and context. Innovation districts evolve over time.
- The role of place making corroborates itself across the various stages of innovation districts from pre-establishment conditions, planning, functioning to perceptions of its users.

TABLE 3.2 Attributes, Components, and Implications

Attributes	Components	MQP	MEC	DPKP	Implications
Context					
Conditions					
Spatio-temporal	Site location	+	+	+	Building upon local assets
	Development history	+	+	+	
Socio-economic	Socio-cultural diversity	+	n/a	+	Sustained funding
	Sectoral strength	n/a	+	+	
Support					
Political	Policy and planning	+	+	+	Continued prioritisation in policies
	Economic	+	+	n/a	
Institutional	Organisational	–	+	–	Dedicated institutionalisation
	Knowledge assets	+	+	+	
Feature					
Profile					
Marketing	Strategic branding	+	–	–	Effective marketing
	Indirect branding	+	+	+	
Identity	Anchor Image	–	+	+	Anchor and specialisation as asset
	Specialisation	n/a	+	+	
Competitiveness					
Resource	Innovation support	+	+	+	Cost-effective resources for SMEs
	Business support	+	+	–	
Viability	Rent and land prices	+	+	–	Ancillary amenities for knowledge workers
	Housing affordability	+	n/a	–	

Form					
Integration					
Permeability	Urban-spatial integration	+	–	+	Horizontal and vertical permeability
	Land use mix	+	–	–	
Order	Sub-cluster level flow	+	+	–	Special local bodies for design
	Campus-site relationship	–	+	–	
Character					
Uniqueness	Design innovativeness	+	+	+	Unique identity
	Enviro-cultural sensitivity	+	+	+	
People-orientedness	Public space character	+	–	–	Developmental incentives
	Housing diversity	+	n/a	–	
Function					
Connectedness					
Formal networks	Anchor-tenant collaboration	+	+	+	Hierarchical networking groups
	External formal networking	n/a	+	n/a	
Social networks	Knowledge worker level	+	+	+	Focus on social networking
	Community-integration	+	n/a	+	
Vitality					
Vibrancy	Role of anchor	+	+	+	Timely planning of mixed functions
	Use of space and amenities	+	–	–	
Diversity	Size and sector of firm	+	+	–	Planning for diverse size and type of firms
	On-site functions	+	–	–	

(*Continued*)

TABLE 3.2 *(Continued)* Attributes, Components, and Implications

Attributes	**Components**	**MQP**	**MEC**	**DPKP**	**Implications**
Image					
Sense of ownership					
Transparency	Planning stage	+	+	+	Effective communication
	Implementation stage	+	+	+	
Inclusiveness	Pre-planning stage	+	+	–	Democratic participation
	Decision-making stage	+	+	+	
Identity					
Sense of trust (of stakeholders)	Stakeholder coordination	+	+	+	Coordinated governance
	Government coordination	–	–	n/a	
Sense of attachment (of users)	Knowledge worker satisfaction	+	+	–	Collaborative body and leadership
	Community pride and belonging	+	n/a	+	

Note: (+) = Strong, (-) = Weak, (n/a) = Not available.

- Considering the current challenges of innovation districts—the role of place making in innovation districts, hence, should not be considered limited only to the retention of talent force by creating a desired physical environment. But it needs to be extended to integrate the solutions that cater to the social, economic, organisational, and cultural needs of knowledge communities, society, and spaces.
- This refined and extended multidimensional role of place making is to ensure:
  - An efficient governance environment that facilitates supportive conditions for the growth and sustenance of innovation district;
  - Developing and strengthening a unique identity derived from identifying and building upon local assets;
  - Economic sustainability of innovation district by facilitating support and an efficiently managed vision;
  - Sustained knowledge and innovation generation, nurturing, exchange, and dissemination in a well-networked and vibrant functional space;
  - A sense of attachment and place by creating a connected, permeable, creative, character-rich and people-oriented spatial environment;
  - A sense of trust and ownership by fortifying a transparent, well-informed, democratic society displaying equal participation of all stakeholders, and most importantly;
  - Constructive societal integration for sustainable growth of a cohesive and resilient community as well as to disseminate knowledge and reap the benefits of KBUD.

The role of planners and policymakers in place making for innovation districts shifts from direct intervener to facilitator of conducive conditions. Consequently, for practically integrating place making as a strategy for the definition of planning policies, aimed at combating the unique challenges of innovation districts, major implications—as relative to each dimension—derived here based on investigation are as follows:

A. *Context-wise—for ensuring support and favourable conditions:*

a. *Continued prioritisation:* Irrespective of change in government and henceforth political agendas, it is crucial to prioritise innovation districts in policies for the sustenance of growth and success. This prioritisation specifically refers to policies such as strong recognition as key project; multidimensional vision for their planning and physical development; and investment in innovation districts for innovation and knowledge growth. More importantly, specific consideration of place making in relation with innovation districts is also inevitable.

b. *Sustained funding:* The funding support by government needs to extend from seed funding to operational funding for sustainable growth and success.

c. *Dedicated institutionalisation:* Specific government authorities dedicated to innovation district's development, design, management, and coordination—between stakeholders and different hierarchy levels in government—should be institutionalised to make the processes related to them efficient, fast, coordinated, and organised.

d. *Building upon local assets:* Place-making strategies need to identify and build upon the local economic and socio-cultural assets.

B. *Feature-wise—for enhancing competitiveness and viability of project:*

a. *Effective marketing:* Branding strategies make them unique and competitive at local and international level. A brand that gives innovation district an integrated identity can be defined by: A well-defined name and logo for the area; integration of socio-cultural aspects; and specific focus on place making.

b. *Anchor and specialisation:* During conception, it is to be considered that the global and local image of institutions existing as anchors and their talented researchers contribute as a strong appeal nationally and internationally. Specialisation in particular sectors, processes, or at individual level is also critical for the success.

c. *Cost-effective resources for SMEs:* Investing and planning for cost-effective resources required by SMEs and start-ups that provide them innovation support—in the form of physical infrastructure such as presence of accelerators, a range of collaborative working environments such as hot desks, break out zones, work stations, laboratories and private office spaces with 24/7 accessibility and shared facilities—as well as business support—through flexible leasing arrangements, access to finance, training, and exposure to information—make innovation districts economically more inclusive and innovative.

d. *Ancillary amenities:* Provision of formal and informal amenities required by knowledge workers act as a key attractive feature. These include a common accessible management, physical and digital infrastructure, and other casual amenities.

C. *Form-wise—for ensuring permeability and uniqueness:*

a. *Horizontal and vertical permeability:* For promoting 24/7 activities and diversity, mixed-use zoning is vital. An integrated and permeable design, i.e., having connection at various levels such as between: Buildings and their surroundings; sub-clusters; lower floors and outside of buildings; and adjacent campuses.

b. *Local bodies:* Facilitating local bodies or departments specifically dedicated for designing can contribute positively in the process of place making.

c. *Unique identity:* By integration of heritage and cultural elements as well as integrating sustainability and innovativeness in design features, the innovation district can achieve the unique character necessary for establishing an identity for itself.

d. *Developmental incentives:* Developmental incentives can be used as planning tool to achieve the desired people-oriented physical character of innovation district, such as ensuring housing diversity, on-site permeability, and fine-grained mixed-use development. Introducing diversity in housing not only caters to the diverse socio-cultural needs and income groups to make resilient, cohesive, and vibrant community, but initiatives such as integration of social housing also act as a strong place-making factor at social level.

D. *Function-wise—for improving connectedness and vitality:*

a. *Hierarchical networking groups:* Centralised networking organisations with subsets at innovation district-level that strengthen formal networks in diverse sectors need to be supported and invested into. They contribute by being a common platform bridging the interface through common events and meetings; providing open access to core facilities, common service and expertise; skill development through workshops; bringing diverse sectors together; leading to fruitful collaborations; and social networking. The success of their functioning depends on the openness of university that act as lubricator, ability of leadership, support and investment from government, accessibility to researchers, innovators, small-scale businesses, and private and public sector as well as their scale of reach.

b. *Social networking:* Social or informal networks that develop in lieu of the formal networking sessions and training prove to be highly helpful for strengthening knowledge exchange and dissemination and developing a sense of attachment and trust.

c. *Timely planning of diversity of functions:* For planned innovation districts, process of designation or rezoning the land area to mixed uses needs to be implemented in earlier stages for its fruitful integration, i.e., to ensure functional use of spaces from commencement itself and discouragement for market demand to push towards dominant sector. This diversity in functions and amenities is vital for making the contemporary innovation districts vital.

d. *Diverse size and type of firms:* This diversity needs to be extended to the size and type of firms, too. While the innovation depends on an exchange between diverse sectors such as biopharmaceuticals, technology, engineering, and creative sectors, it also depends on the co-existence of small-scale, medium-scale, and large-scale firms.

E. *Image-wise—for developing a sense of ownership and identity in range of stakeholders:*

a. *Effective communication:* Through an efficient plan for proper communication of information, team of architects and planners can garner support by avoiding conflicts and developing a sense of trust. Three main requisites for effective communication are ensuring it is transparent, personalised, and explicit.

b. *Democratic participation:* Knowledge communities should pioneer the development as a quadruple helix model with an equal participation from public, private, academia, and community and effective stakeholder engagement in decision-making process, plan implementation, and functioning of an innovation district.
c. *Coordinated governance:* Equally important is the coordination between various tiers and departments of government for the successful and timely accomplishment of vision.
d. *Collaborative body and leadership:* A common organisational setup with efficient leadership and a team of experts in designing, developing, functioning, and coordinating acts as a lubricator by overseeing the overall process and serving as a common platform for various stakeholders for a quick and fair decision-making process.

The investigation on the Australian innovation district experience reveals the necessity of an integrated, efficient and multidimensional place-making approach—aimed at not only physical outcomes, but also a holistic set of objectives including economic, societal and management aspects. This holistic approach will help in the achievement of a milieu containing the qualities of: (a) sustainable and collaborative business environment; (b) supportive governance and organisational environment; (c) connected and characterful spatial environment; and (d) vibrant people environment comprising cohesive communities with a strong sense of place (Pancholi *et al.*, 2019).

ACKNOWLEDGEMENTS

This chapter, with permission from the copyright holder, is a reproduced version of the following journal article: Pancholi, S., Yigitcanlar, T., & Guaralda, M. (2019). Place making for innovation and knowledge-intensive activities: The Australian experience. Technological Forecasting and Social Change, 146, 616–625.

REFERENCES

Adams, D., & Tiesdell, S. (2013). *Shaping places*. New York: Routledge.

Arefi, M. (2014). *Deconstructing placemaking*. New York: Routledge.

Bathelt, H., Feldman, M., & Kogler, D. F. (2013). *Beyond territory*. London: Routledge.

Berkes, F. (2009). Evolution of co-management. *Journal of Environmental Management*, 90(5), 1692–1702. 10.1016/j.jenvman.2008.12.001

Boschma, R. (2005). Proximity and innovation. *Regional Studies*, 39(1), 61–74. 10.1080/0034340052000320887

Baum, S., O'Connor, K., & Yigitcanlar, T. (2009). The implications of creative industries for regional outcomes. *International Journal of Foresight and Innovation Policy*, 5(1/2/3), 44–64.

Carmona, M., Heath, T., Oc, T., & Tiesdell, S. (2010). *Public places urban spaces*. London: Architectural Press.

Carrillo, J., Yigitcanlar, T., Garcia, B., & Lonnqvist, A. (2014). *Knowledge and the city*. New York: Routledge.

Carvalho, L., & Winden, W. V. (2017). Planned knowledge locations in cities. *International Journal of Knowledge-Based Development*, 8(1), 47–67. 10.1504/IJKBD.2017.10003130

Castells, M. (2000). *The rise of network society*. Malden: Blackwell.

Chesbrough, H. (2003). The era of open innovation. *MIT Sloan Management Review*, 44(3), 35–41.

Clark, J., Huang, H. I., & Walsh, J. P. (2010). A typology of innovation districts. *Cambridge Journal of Regions, Economy and Society*, 3(1), 121–137.

Cooke, P. (2017). Complex spaces. *Journal of Open Innovation*, 3(1), 9. 10.1186/s40852-017-0075-9

Cresswell, T. (2004). *Place: a short introduction*. London: Wiley.

Edvardsson, I., Yigitcanlar, T., & Pancholi, S. (2016). Knowledge cities research and practice under the microscope. *Knowledge Management Research and Practice*, 14(4), 537–564. 10.1057/s41275-016-0043-5

Elia, G., Petti, C., & Sarcina, A. (2016). Promoting communities of innovation. *International Journal of Knowledge-Based Development*, 7(3), 207–224. 10.1504/IJKBD.2016.078083

Esmaeilpoorarabi, N., Yigitcanlar, T., & Guaralda, M. (2016). Towards an urban quality framework. *International Journal of Knowledge-Based Development*, 7(3), 290–312. 10.1504/IJKBD.2016.081970

Eisenhardt, K. M. (1989). Building theories from case study research. *Academy of Management Review*, 14(4), 532–550. 10.5465/amr.1989.4308385

Evans, G. (2009). Creative cities, creative spaces and urban policy. *Urban Studies*, 46(5-6), 1003–1040. 10.1177/0042098009103850

Fernandez-Maldonado, A., & Romein, A. (2010). The role of organisational capacity and knowledge-based development. *International Journal of Knowledge-Based Development*, 1(1–2), 79–96. 10.1504/IJKBD.2010.034057

Florida, R. (2005). *Cities and the creative class*. New York: Routledge.
Friedmann, J. (2007). The wealth of cities. *Development and Change*, 38(6), 987–998.
Garnsey, E., & Heffernan, P. (2005). High-technology clustering through spin-out and attraction. *Regional Studies*, 39(8), 1127–1144. 10.1080/00343400500387889
Glaeser, E. (1999). Learning in cities. *Journal of Urban Economics*, 46(2), 254–277.
Glaser, B. G., & Strauss, A. L. (1967). *The discovery of grounded theory*. Chicago: Aldine.
Graham, S., & Guy, S. (2002). Digital space meets urban place. *City*, 6(3), 369–382.
Harvey, D. (1989). *The condition of postmodernity*. Oxford: Blackwell.
Henton, D., & Held, K. (2013). The dynamics of Silicon Valley. *Social Science Information*, 52, 539–557. 10.1177/0539018413476132
Hutton, T. (2004). The new economy of inner cities. *Cities*, 21(2), 89–108.
Jacobs, J. (1969). *The economy of cities*. London: Edward Arnold.
Jones, P. (2009). Putting architecture in its social place. *Urban Studies*, 46(12), 2519–2536.
Jung, N. (2013). Relational governance and the formation of a new economic space. *International Journal of Urban and Regional Research*, 37(4), 1233–1253.
Katz, B., & Wagner, J. (2014). *The rise of innovation districts*. Washington: Brookings.
Kodama, F., & Shibata, T. (2015). Demand articulation in the open-innovation paradigm. *Journal of Open Innovation*, 1(1), 2. 10.1186/s40852-015-0005-0
Komninos, N. (2016). Smart environments and smart growth. *International Journal of Knowledge-Based Development*, 7(3), 240–263. 10.1504/IJKBD.2016.079051
Landry, C. (2000). *The creative city*. London: Routledge.
Lefebvre, H. (1991). *The production of space*. Malden: Blackwell.
Livingstone, D. (2003). *Putting science in its place*. Chicago: University of Chicago Press.
Lonnqvist, A., Kapyla, J., Salonius, H., & Yigitcanlar, T. (2014). Knowledge that matters. *European Planning Studies*, 22(10), 2011–2029. 10.1080/09654313.2013.799781
Leydesdorff, L., & Ivanova, I. (2016). Open innovation and triple helix models of innovation. *Journal of Open Innovation*, 2(1), 11. 10.1186/s40852-016-0035-x
Makkonen, T., & Weidenfeld, A. (2016). Knowledge-based urban development of cross-border twin cities. *International Journal of Knowledge-Based Development*, 7(4), 389–406. 10.1504/IJKBD.2016.081456
Markusen, A. (1996). Sticky places in slippery space. *Economic Geography*, 72(3), 293–313.
Massey, D. (1991). A global sense of place. *Marxism Today*, 35(6), 24–29.

Merrifield, A. (1993). Place and space. *Transactions of the Institute of British Geographers*, 18(4), 516–531.
Montgomery, J. (1998). Making a city. *Journal of Urban Design*, 3(1), 93–116.
Ophir, A., & Shapin, S. (1991). The place of knowledge a methodological survey. *Science in Context*, 4(1), 3–22.
Pancholi, S., Yigitcanlar, T., & Guaralda, M. (2014). Urban knowledge and innovation spaces. *Asia Pacific Journal of Innovation and Entrepreneurship*, 8(1), 15–38. 10.1108/APJIE-09-2013-0043
Pancholi, S., Yigitcanlar, T., & Guaralda, M. (2015a). Place making facilitators of knowledge and innovation spaces. *International Journal of Knowledge-Based Development*, 6(3), 215–240. 10.1504/IJKBD.2015.070147
Pancholi, S., Yigitcanlar, T., & Guaralda, M. (2015b). Public space design of knowledge and innovation spaces. *Journal of Open Innovation*, 1(1), 13. 10.1186/s40852-015-0013-0
Pancholi, S., Yigitcanlar, T., & Guaralda, M. (2017). Governance that matters. *Journal of Place Management and Development*, 10(1), 73–87. 10.1108/JPMD-08-2016-0054
Pancholi, S., Yigitcanlar, T., & Guaralda, M. (2019). Place making for innovation and knowledge-intensive activities: The Australian experience. *Technological Forecasting and Social Change*, 146, 616–625. 10.1016/j.techfore.2018.07.025
Peck, J. (2010). *Constructions of neoliberal reason*. Oxford: Oxford University Press.
Phillips, S. M., & Yeung, H. W. (2016). A place for R&D? *Urban Studies*, 40(4), 707–732.
Porter, M. E. (2000). Location, competition and economic development. *Economic Development Quarterly*, 14(1), 15–34.
Reynolds, E. B., & Uygun, Y. (2017). Strengthening advanced manufacturing innovation ecosystems. *Technological Forecasting and Social Change*. 10.1016/j.techfore.2017.06.003
Saxenian, A. (1994). *Regional advantage*. Cambridge: Harvard University Press.
Scott, A. (2006). Creative cities. *Journal of Urban Affairs*, 28(1), 1–17.
Stehlin, J. (2016). The post-industrial shop floor. *Antipode*, 48(2), 474–493.
Storper, M., & Venables, A. (2004). Buzz: face-to-face contact and the urban economy. *Journal of Economic Geography*, 4(4), 351–370.
Tallman, S., Jenkins, M., Henry, N., & Pinch, S. (2004). Knowledge clusters and competitive advantage. *Academy of Management Review*, 29(2), 258–271.
Van Winden, W., de Carvalho, L., Van Tuijl, E., Van Haaren, J., & Van den Berg, L. (2013). *Creating knowledge locations in cities*. London: Routledge.
Witt, U. (2016). What kind of innovations do we need to secure our future? *Journal of Open Innovation*, 2(1), 17. 10.1186/s40852-016-0037-8
Yigitcanlar, T., Baum, S., & Horton, S. (2007). Attracting and retaining knowledge workers in knowledge cities. *Journal of Knowledge Management*, 11(5), 6–17.
Yigitcanlar, T., O'Connor, K., & Westerman, C. (2008a). The making of knowledge cities. *Cities*, 25(2), 63–72.

Yigitcanlar, T., Velibeyoglu, K., & Martinez-Fernandez, C. (2008b). Rising knowledge cities. *Journal of Knowledge Management*, 12(5), 8–20.

Yigitcanlar, T. (2009). Planning for knowledge-based development. *Journal of Knowledge Management*, 13(5), 228–242.

Yigitcanlar, T., & Sarimin, M. (2011). The role of universities in building prosperous knowledge cities. *Built Environment*, 37(3), 260–280.

Yigitcanlar, T., & Dur, F. (2013). Making space and place for knowledge communities. *Australasian Journal of Regional Studies*, 19(1), 36–63.

Yigitcanlar, T., & Lonnqvist, A. (2013). Benchmarking knowledge-based urban development performance. *Cities*, 31(1), 357–369.

Yigitcanlar, T., & Lee, S. (2014). Korean ubiquitous-eco-city. *Technological Forecasting and Social Change*, 89(1), 100–114.

Yigitcanlar, T., Guaralda, M., Taboada, M., & Pancholi, S. (2016). Place making for knowledge generation and innovation. *Journal of Urban Technology*, 23(1), 115–146.

Yigitcanlar, T., Edvardsson, I., Johannesson, H., Kamruzzaman, M., Ioppolo, G., & Pancholi, S. (2017a). Knowledge-based development dynamics in less favoured regions. *European Planning Studies.* 10.1080/09654313.2017.1358699

Yigitcanlar, T., Sabatini-Marques, J., Da Costa, E., Kamruzzaman, M., & Ioppolo, G. (2017b). Stimulating technological innovation through incentives. *Technological Forecasting and Social Change.* 10.1016/j.techfore.2017.05.039

Yin, R. K. (2003). *Case study research, design and methods.* Newbury Park: Sage.

Yin, R. K. (2011). *Qualitative research from start to finish.* London: Guilford Press.

Yun, J. J. (2015). How do we conquer the growth limits of capitalism? *Journal of Open Innovation*, 1(1), 17.

Yun, J. J., Won, D., & Park, K. (2016a). Dynamics from open innovation to evolutionary change. *Journal of Open Innovation*, 2(1), 7.

Yun, J., Lee, D., Ahn, H., Park, K., Lee, S., & Yigitcanlar, T. (2016b). Not deep learning but autonomous learning of open innovation for sustainable artificial intelligence. *Sustainability*, 8(8), 797.

Yun, J. J., Won, D., Jeong, E., Park, K., Lee, D., & Yigitcanlar, T. (2017). Dismantling of the inverted u-curve of open innovation. *Sustainability*, 9(8), 1423.

Zhao, F. (2005). Exploring the synergy between entrepreneurship and innovation. *International Journal of Entrepreneurial Behaviour and Research*, 11(1), 25–41.

CHAPTER 4

Societal Integration of Innovation Districts

4.1 INTRODUCTION

Over the last two decades, scholars have arrived to a unanimous concordance over knowledge-based urban development as the most sustainable path towards the future (Knight, 1995; Yigitcanlar, 2010; Van Winden *et al.*, 2013; Carrillo *et al.*, 2014; Lonnqvist *et al.*, 2014). Henceforth, globally cities are investing in their 'knowledge and innovation spaces' (innovation districts)—spatial congregations of knowledge-intensive activities manifested as specialised mixed-use locations aimed at production and dissemination of new ideas and knowledge (Glaesar, 1999; Scott, 2000; Breschi & Lissoni, 2001; Hutton, 2004; Evans, 2009; Evers *et al.*, 2010; Pancholi *et al.*, 2015). Thriving as the modern growth nodes of the metropolitan cities, their contribution is not only limited as economy stimulators, but also extends to stimulating technological, social, cultural, as well as environmental development (Katz & Wagner, 2014; Yigitcanlar et al., 2017). More recently developed ones like One-north (Singapore), Arabianranta (Finland), Digital Milla (Spain), and Strijp-S (The Netherlands) as well as more established ones like Silicon Valley (US), Sophia Antipolis (France), and Route 128 (Boston) are among the well-known global innovation district examples.

In order to ensure the perpetual generation of knowledge in innovation districts for gaining competitive edge over other cities, innovative place-making strategies targeted at luring and retaining talented workforce—the key producers of knowledge—are ardently sought after by policymakers (Yigitcanlar *et al.*, 2007; Pancholi *et al.*, 2015; Edvardsson *et al.*, 2016). A number of studies have propounded the coalesced role of integrating arts and technology with local culture as well as network-rich and tolerant

DOI: 10.1201/9781032657431-6

social environment to satisfy specialised lifestyle requirements of the creative class of knowledge workers (Florida, 2005; Yigitcanlar, 2008a; Katz & Wagner, 2014). Furthermore, scholars such as Zelinsky (2004) and Moultrie *et al.* (2007) highlight the important role of the physical environment in boosting innovation capabilities. However—in practice—planners and policymakers are facing few critical challenges. Economically, in the era of 'open innovation', the key challenge is to retain firms and facilitate knowledge sharing and collaboration within various tenants as well as mutually different sectors (Chesbrough, 2003; Garnsey & Heffernan, 2005; Yigitcanlar *et al.*, 2008b; Van Winden *et al.*, 2013). Organisationally, bringing consensus between various actors with conflicting mutual interests becomes challenging (Van Winden *et al.*, 2013; Pancholi *et al.*, 2017).

This research specifically focuses on the heightened key social challenges of innovation districts (Carrillo *et al.*, 2014). Firstly, studies have revealed—owing to the competition posed by the vibrant inner-city areas—that it remains a challenge to retain talent force in innovation districts that are not located in socially vibrant areas (Graham & Guy, 2002). Secondly, despite innovation districts increasingly being considered as the new face of economy and society, the process of their development is criticised for bringing certain non-anticipated societal impacts, i.e., promoting seclusion, displacement, gentrification, and social inequality by marginalising some groups (Peck, 2005, 2010; Sarimin & Yigitcanlar, 2012; Solnit, 2014; Stehlin, 2016). Lastly, innovation districts not being integrated with their surrounding areas poses a gated community disadvantage for both communities inside and outside innovation districts (Metaxiotis *et al.*, 2010; Yigitcanlar *et al.*, 2012).

Although the significance of place making in the success of innovation districts is acknowledged in the knowledge-based urban development literature, a gap lies in related scholarly works investigating its role through a societal lens. The objective of this chapter is to investigate the role and challenges faced by place making in societal integration of innovation districts within their boundaries as well as within the larger urban social fabric. The key question this study aims to address is: 'What are the contributions of place making approaches in contemporary innovation districts in terms of societal integration?' Macquarie Park Innovation District (MPID) from Sydney (Australia) is selected as the case for the investigation. The study derives its base from a qualitative review of 14 in-depth key stakeholder interviews—including government officials, planners and designers, managing agents of formal groups, community organisations, firms and institutions, and knowledge workers—supplemented with field observations, and the secondary data such as academic literature, government documents, data from the MPID website, and maps.

4.2 LITERATURE BACKGROUND

4.2.1 Transfiguration of Knowledge and Innovation Spaces

Ample amount of recent literature has discussed the transformation of innovation districts across the world. Earlier, many successful innovation districts developed as 'science parks' or 'techno-industrial complexes' generally in the vicinity of university. Due to the nature of secrecy involved in research-based functions and patenting policies, the functions carried out involved discouraged sharing and circulation of information (Katz & Wagner, 2014). The direct spatial impact of this kind of functional requirement was the seclusion of spaces or buildings from each other as well as from rest of the city—resulting in isolated locations, which were occasionally gated and accessible only by car. Silicon Valley is one of the prominent examples of this. As a recent global trend—in an attempt to revitalise the inner-city economy—many of the downtown districts have been transformed into live-work lofts (Hutton, 2004; Pratt, 2000; Evans, 2009). This involves repurposing the dilapidated buildings of older industries and workshop for their use as new office spaces. These spaces specifically attract the knowledge workers involved in creative sectors (Baum *et al.*, 2009). They also befit the requirements of start-ups and small-sized firms that look out for affordable and creative spaces in vibrant localities. With the onset of the era of 'open innovation', proximity is gaining prime preference and collaboration is emerging as the new modus operandi of innovation districts (Chesbrough, 2003). The booming of small- and medium-sized enterprises (SMEs), start-ups, and collaborative spaces is further fuelling this. As the spatio-economic manifestation of 'open innovation', innovation districts are also attempting to transfigure—referred to as 'urban turn' by Van Winden *et al.* (2013)—from their earlier introvert, secluded, and mono-functional models to the current extrovert, more connected and mixed-use models boasting a mix of sectors and functions. Physically, they are advancing towards open layouts, connected precincts, creative environment, and collaborative culture.

Organisationally, innovation districts are emerging as public-private-academia-community partnerships—i.e., quadruple helix model partnership. Studies have been increasingly advocating the integration of social and cultural aspects in addition to economic motives of innovation districts. Studies like Yigitcanlar (2008a), Kunzmann (2009), and Carrillo *et al.* (2014) have propounded that it is necessary to have a balanced and holistic economic, political, physical, and social development for the successful knowledge-based urban development. Other studies such as Saxenian (1994) and Van Winden *et al.* (2013) have also emphasised the role of strengthening social networks in the form of formal and informal connections for the new knowledge generation and success. Katz and Wagner (2014) postulate that innovation ecosystem is nourished by a synergistic relationship between people, firms, and place—referred as economic, networking, and physical assets of innovation districts or innovation districts. Numerous research have shown that the integration of local culture shapes creativity as well as gives uniqueness and competitive edge to locations such as innovation districts that sets them apart from other locations (Porter, 2000; Livingstone; 2003; Chang & Huang, 2008; Evans, 2009; Meusburger *et al.*, 2009). Proximity in terms of society, organisation, cognition and institution or 'relational proximity' enhances knowledge spillover and results in a social ecosystem of learning (Storper & Venables, 2004; Berkes, 2009; Bathelt *et al.*, 2013). In addition to their economic merits, the specialised role of innovation districts in providing social equity and inclusion for a democratic society achieved through strong social and human capitals has been emphatically accentuated (Fernandez-Maldonado & Romein, 2010; Yigitcanlar, 2011; Yigitcanlar & Dur, 2013; Pancholi *et al.*, 2015). Moreover, recent works have also advocated public participation as the factor of social change and one that brings innovation (Gonzalez & Carrillo, 2012; Pancholi *et al.*, 2017).

4.2.2 Emerging Challenges and Conflicts

Despite the above-mentioned physical and spatio-economic transformations, these locations are facing a few challenges. Politico-economically, consequent to the evolution of innovation districts into hybrid typologies and adoption of quadruple helix models, new actors have come into foreplay. Dynamics of organisational processes and ensuring coordination within these actors with conflicting mutual interests stands as a challenge for success of innovation districts (Van Winden *et al.*, 2013; Lonnqvist *et al.*, 2014). More importantly, recent research studies have highlighted the growing need for reconfiguration of governance to 'relational governance' based on state-society relations and shift of its role from its authoritarian to a less controlled one (Henton & Held, 2013; Pancholi *et al.*, 2017). In economic terms, referring to the innovation district-level integration, to what extent collaborations actually happen remains questionable. Although few research studies have revealed that firms in innovation districts generally display a strong connection with a local anchor such as a university; it has also been proved—in different contexts—that this remains limited to strong international but weaker local networks (Bakouros *et al.*, 2002; Garnsey & Heffernan, 2005; Yigitcanlar *et al.*, 2008b; Van Winden *et al.*, 2013). In addition, intertwining of activities in diverse sectors such as ICT, manufacturing, creative industries, as well as sustenance of diversified functions in newly emerging mixed-use developments is challenging (Evans, 2009; Van Winden *et al.*, 2013). There have also been certain social impacts of this transformation that the study in this chapter, peculiarly, aims to look into.

First of all, in spatial and social terms, it is emerging as a challenge—even for established innovation districts—to develop an environment that retains the knowledge workers. A growth in the exodus of businesses from isolated innovation districts towards inner-city areas is increasingly demonstrated by research studies (Graham & Guy, 2002; Van Winden *et al.*, 2013; Katz & Wagner, 2014). Graham and Guy (2002), in their study, draw attention to the case of Silicon Valley and the reconfiguration of San Francisco downtown as 'technopole'. The key factors for this shift as enumerated are the quality of life factors, i.e., vibrancy of downtown that attracts the knowledge workers, availability of smaller office spaces, and drop-in crime rate. Secondly, the situation has also been criticised by scholars for giving rise to certain socio-cultural conflicts. At a wider scale, studies have denounced how the integration of culture with technology stays partial and only limited to the integration of physical assets, i.e., heritage and older buildings—anticipated to adding on to the economic value of the space as well as the creativity quotient of the area (Stehlin, 2016). In the area of exodus, a subsequent gentrification happens, pushing the existing users to the cheaper areas as a consequence to the land prices soaring high (Peck, 2010). In parallel to this, the development of 'mono-culture' in the city that marginalises groups other than knowledge workers has also been denounced—that results in a negative impact on the city's polyvalent character by making it less democratic and diverse (Solnit, 2014). Henceforth, place-making strategies are acknowledged to play a crucial role in innovation districts (Florida, 2005; Katz & Wagner, 2014; Pancholi *et al.*, 2014; Yigitcanlar *et al.*, 2016). Thirdly, the lack of integration of innovation districts with their surroundings also generates a gated obstacle, limiting the quality of life and place offerings of either the KI Sot the surrounding to be fully appreciated (Esmaeilpoorarabi *et al.*, 2016a, 2016b).

4.2.3 Redefined Role of Place Making

In order to understand the specialised role of place making, it is necessary to define the 'place' considering the above challenges as well as the globalised context of innovation districts. Many eminent scholars in the cross-disciplinary literature have propagated the holistic and multidimensional definition of place. A comprehensive understanding of place, therefore, does not limit itself to tangible dimensions or hard layer but also includes intangible or soft layers such as experienced space in the form of socioeconomic processes and networks, as well as meanings attached by its users and their perceptions (Lefebvre, 1991; Montgomery, 1998; Cresswell, 2004; Funke, 2007; Ho & Douglass, 2008; Healey, 2010; Arefi, 2014; Pancholi *et al.*, 2014; 2015; 2017). In his recent work, Healey (2010) asserts that sense of place is the assimilation of physical experiences and imaginative constructions, which results in the attachment of meanings and values. Furthermore, in globalised spaces like innovation districts, multi-layered space boasts a unique identity that is produced as a result of the intersection of multiple identities, cultures, and histories at a point (Massey, 1991). It takes into consideration the spatio-temporality attached to knowledge locations as well as the dynamic character of globalised space shaped by myriad of connections and networks (Castells, 2000; Van Winden *et al.*, 2013). The role of place making in innovation districts, thus, has extended from creating a physically integrated, dynamic, and creative urban environment to also developing a functionally networked, globally tolerant, as well as culturally vibrant societal environment (Yigitcanlar & Bulu, 2015).

4.3 EMPIRICAL INVESTIGATION

4.3.1 Background

From its original history as the land of aboriginals to a rural hamlet that changed to market gardens, the area of MPID has seen a series of land use changes every 10 years. In the 1960s, the selection of the site by the state government for the establishment of Sydney's third university—Macquarie University—was one of the prime defining moment in its history. Later on, conceived to be developed on the lines of the Stanford model—anticipating knowledge exchange between university and businesses—the area was identified as industry growth area. To accomplish that, earlier there was a mandatory requirement for businesses to have a research and development (R&D) component—a regulation that was changed later on (Interviewee#1). The companies were keen to choose the site as their headquarters by consolidating their offices spread across different sites into one primarily due to the availability of big blocks of land and also due to its connectivity and accessibility from CBD. Housing started growing at the periphery of university and then the Macquarie Centre was opened in 1981. Construction of some significant infrastructure, such as the Chatswood to Epping rail line, further added to its popularity.

Today, MPID has established itself to a nationally acclaimed, Australia's largest research and business hub (see Figure 4.1). With Macquarie University—one of Australia's leading research universities—as its key anchor, it is home to many global players as well as head office locations for many of Australia's top 100 companies. A few of its key tenants across pharmaceutical, technology, electronics, and telecommunications industries are Johnson & Johnson, Microsoft, Sony, Optus, Cochlear, and Foxtel. Ranked as the area with Australia's 10th highest economic output (in any sector) and with its exceptional growth rate of 6.8%—highest in Sydney—it is soon predicted to outnumber other locations in Australia (PWC, 2014).

Spatially at a distance of only 12 kilometres from city centre, it boasts a significant location on the Global Economic Corridor of Sydney. This corridor extends from Sydney Airport and Port Botany in the south through the major employment centres of the Sydney Central Business District (CBD), North Sydney, Chatswood, MPID, and towards Parramatta and Norwest Business Park. The adjacent location of Lane Cove National Park provides an open, green environment and ample options for cycling or walking. The area has 32% more managers and 54% more professionals as compared to the NSW average (ABS, 2011). Socially, the presence of a young and culturally diverse population due to its proximity to the university has been advantageous in the growth of knowledge-based industries. Demographically, the area represents a population with a continuous growth rate in the last decade, boasting a younger demographic profile with a large proportion, i.e., 45.6% of the population aged 20–34—out of which 25.6% of the population consists of a younger workforce; 40% of people hold tertiary

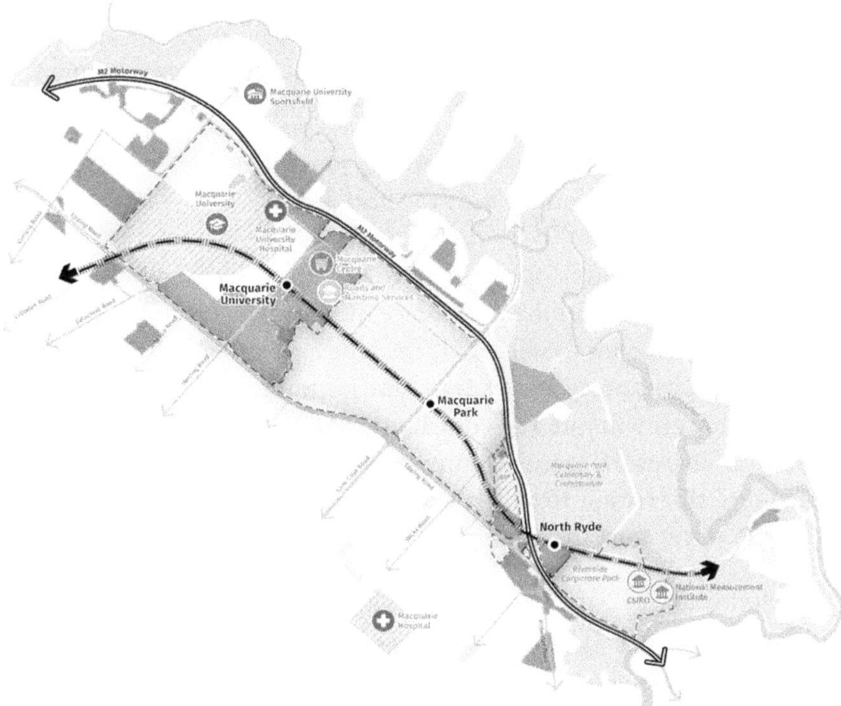

FIGURE 4.1 Layout of Macquarie Park Innovation District (Macquarie University, 2017, p. 10).

qualifications compared to Greater Sydney's average of 24.1%, depicting a well-educated population with higher qualifications. Around 18% of people are overseas born, compared to Greater Sydney's average of 20.1%, reflecting the multi-culturally rich and tolerant society (ABS, 2011).

Being recognised as a 'specialist centre' under NSW Government's Metropolitan Strategy for Greater Sydney, the strategy plans that this area will continue to grow as an internationally significant economic hub. In MPID, the Herring Road and the neighbouring north Ryde have been chosen as a priority precinct—previously called urban activation precincts. The key objectives under the Herring Road Urban Activation Precinct proposal (HRUAP) are to provide (NSW Government, 2014):

- Land use re-zoning to mixed land uses with higher height allowance to activate the precinct;
- Higher-density urban community;

- Access to soft and hard infrastructure;
- Safe, convenient, and accessible pedestrian-friendly environment by providing better connectivity and fine-grained roads;
- Strengthening activity, landscaping, amenities, community facilities, green spaces, and places to meet.

All these factors working together have set up the context to establish MPID as a frontrunner in the global race of knowledge economy. To sum up, the primary reasons to select MPID as the case study are as follows: (a) its significant position in innovation landscape in Australia owing to its high economic growth rate and contribution to GDP; (b) its characteristics of being a live-work-learn community with a variety of uses and stakeholders all existing together; and (c) the future plans and proposals that aim to put it forward as a model case of societal integration.

4.3.2 Methodology and Research Design

The study adopted a semi-structured interview-based qualitative analysis approach for carrying out empirical investigations in the selected case. The interview findings were compared and also integrated with the data collected from primary and secondary sources—i.e., policy and plan documentations obtained from government organisations, planning and design firms, developers, research institutes, and on-site tenant firms. Other primary data sources such as field observations, photographs, physical plans, and maps also contributed to the analysis as references. In order to arrive at an integrated final understanding, the perceptions of a range of key stakeholders of the project were taken into consideration to conduct a total of 14 interviews (Table 4.1). Interviewees were grouped under five major groups using a purposeful sampling technique, i.e., selection for each group was done by identifying knowledgeable individuals at key positions associated to the case and considering their knowledge in correspondence with different dimensions of the conceptual framework. These groups include the followings that are considered as the key stakeholders: (a) government officials, (b) planners and designers, (c) networking groups, (d) firms and institutions, and (e) knowledge workers. Interviews were undertaken in the second half of 2015; each lasted about 45–60 minutes, was digitally recorded, and transcribed into text manually. An inductive approach of content analysis—informed by the phenomenographic methodology—was adopted to analyse findings.

TABLE 4.1 List of Interviewees

Group	Category	No	Position	Relevance with the Site
Group 1	Government officials	Interviewee#1	Local council executive	Key role in planning and execution
		Interviewee#2	Local council manager	Key role in local economic development plans
		Interviewee#3	Local council senior strategist	Key role in local innovation strategy development
Group 2	Planners and designers	Interviewee#4	Urban designer and planner	Key role in master planning of the innovation district
		Interviewee#5	Urban designer and architect	Key role in master planning of the university campus
		Interviewee#6	Urban designer	Key role in design of infrastructure projects
Group 3	Networking groups	Interviewee#7	Manager	Key role in a formal group
		Interviewee#8	Community team leader	Key member of the leading team
Group 4	Firms and institutions	Interviewee#9	Director	Key role in an on-site institution
		Interviewee#10	Director	Key role as collaborator in an on-site institution
		Interviewee#11	Director	Leading an on-site business
Group 5	Knowledge workers	Interviewee#12	Executive position	All workers in this group are interviewed for perceptions as daily active users
		Interviewee#13	Lead associate	
		Interviewee#14	Scientist	

For the purpose of analysis, the research adopted a multidimensional conceptual framework for place making in innovation districts (Pancholi *et al.*, 2017). It is based on the theoretical paradigm delving into place making in the globalised context of innovation district as a coherence between 'conceived', 'lived', and 'perceived' forms of space (Lefebvre, 1991; Montgomery, 1998; Castells, 2000; Arefi, 2014; Pancholi *et al.*, 2017). Lefebvre (1991), in his seminal work 'The production of space', explains place quite explicitly concluding that while place relates to physical attributes and empirically measured maps or 'conceived layer', it also refers to the experiences of users or 'lived layer' as well as the representations and meanings they attach or 'perceived layer'.

The adopted place-making framework—while it rests on the Lefebvrian triad—defines place in four layers, considering the specialised context of innovation district. They are manifested as four dimensions, i.e., feature, form, function, and image surrounded by context—as the fifth dimension (Figure 4.2). The surrounding context for any innovation district includes

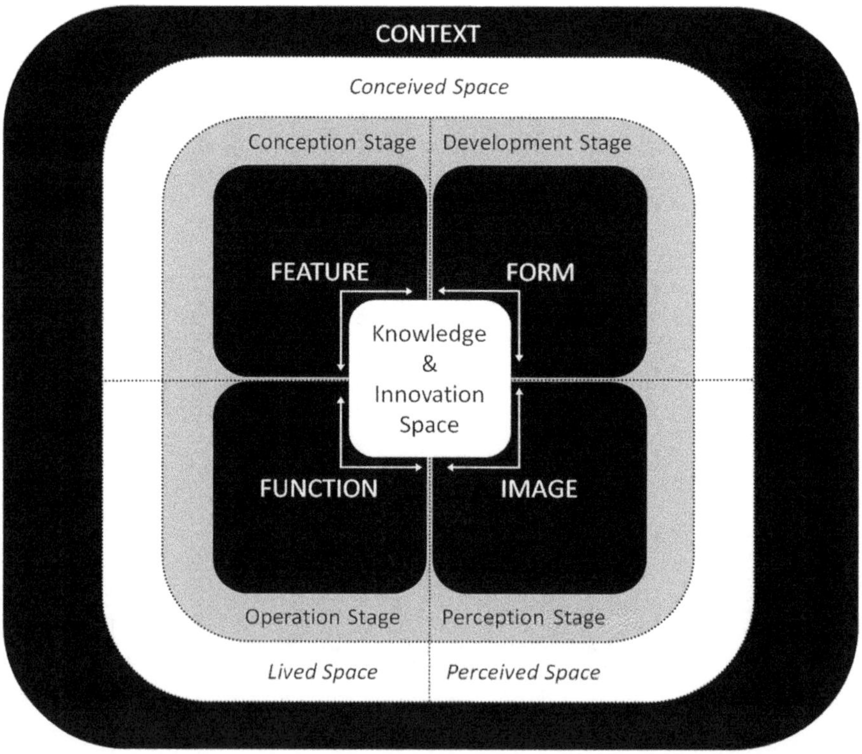

FIGURE 4.2 Conceptual framework of place making (Pancholi *et al.*, 2017, p. 77).

broader set of socio-cultural, politico-economic, and spatio-environmental conditions as already discussed under the background. Corresponding to the 'conceived' layer, there are two dimensions in framework. First is a 'feature' that refers to the conceived soft factors planned for strengthening innovation district's marketability for attracting firms and people. Second is 'form' or conceived hard factors that involve the spatial and physical aspects. Due to their key role in conception stage, data from Groups 1 and 2, i.e., government officials, planners and designers, was majorly fed into feature and form. The 'lived' layer gets manifested as a third place-making dimension, i.e., 'function' as a place. 'Function' incorporates all the socioeconomic processes and networks defining the innovation district. As major users, data from Groups 3, 4, and 5, i.e., networking groups, private firms, institutions, and knowledge workers, is considered for analysing function. The last dimension depicting the 'perceived' layer is 'image', which refers to the perceptions of users and stakeholders. Meanings associated by all the groups, their perceptions, and existing coordination between them are considered for analysing images.

Under each of the dimensions, a specific focus is placed on societal aspects. Here, society refers to both—the knowledge worker community and local community related to the innovation district. It is to be noted here that we consider place as a production of space over time, defined by its past situation, current scenario, and future opportunities and threats. Henceforth, to have a complete picture, the research attempts to investigate the area spatio-temporally across a timeline analysing the past, present, and future as well as conceived, perceived, and lived dimensions. Each of the sections under the individual dimension, hence, explores the history, current strengths, and limitations and the future proposals.

4.3.3 Feature

This section explores the major aspects affecting the brand featuring of the area on a contemporary innovation-scape. Based on the interviews, various factors explain the success of MPID in attracting companies to locate there. Key ones are economy-related, i.e., availability of cheap and large blocks of land with ample space for parking. As Interviewee#2 exclaims, *the biggest thing for them (companies) was it's the cheapest place*. He further adds, *MPID had a real competitive advantage on the CBD and all the places because it had lots and lots of car parking*. Over the years, other factors that added on to its popularity are its proximity to the CBD; strong rail and road connection; existence of Macquarie

University; Macquarie shopping centre complemented by image and latent branding that happened due to the clustering of global companies. However, interviews also reveal that with the changing preferences of knowledge workers, few of the businesses find it challenging to attract people to work in the MPID office—particularly when it involves a shift from working in CBD. Herewith, the main comparative reasons are the driving distance and lack of vibrant environment. Another key issue affecting the employability is the congestion. Referring to a study conducted in 2015, Interviewee#7 highlights that traffic congestion is identified as having a major impact on the productivity of employees by about 95% of the businesses surveyed in the area (Connect Macquarie, 2015). Yet, surprisingly, despite the time taken by public transport to be half the time taken by the car for the same journey, only 25.7% of the people use public transport (ABS, 2011).

Henceforth, to act as a key feature that addresses these issues, a formal group, 'Connect Macquarie', is established—as a joint initiative funded by NSW State Government, City of Ryde, and the businesses. Its major aim is to solve the transportation issues faced by knowledge workers, businesses, students, and the residents of MPID by tailoring out the best possible customised solutions on the basis of each company's and people's goal, current commuting pattern, and bottlenecks. Recent initiatives for employees and society include co-hop carpooling, setting up of bike committee, purpose-built tools and apps, trip planners, centralised transport information, special discounts, and free safety equipment. In addition—while earlier attraction factors centred on the economy—planners are currently firmly focusing additional efforts to make the societal aspects its future strength and appeal. With an aim to develop a live-work-learn-play community, the HRUAP proposal will enhance housing in an economically, socially, and environmentally sustainable manner by delivering up to 5,800 new homes by 2031. Societally, this includes the redevelopment of Ivanhoe Estate—an on-site social housing estate—enhancing the current 259 existing social dwellings into a mix-housing estate neighbourhood with at least 556 social housing dwellings (NSW Government, 2014). This will develop the site into *a true innovation district* with desired social/public amenities and integration (Interviewee#3).

4.3.4 Form

Originally a typical business park characterised by low-scale developments, the form of MPID is constantly evolving. A look into the map of

MPID reveals a mix of building forms. The low-scale development and its location adjacent to a national park give an open environment and opportunity for people to cycle or walk. However, there are a few pitfalls as observed and highlighted in interviews. Lack of collaborative spaces and connectivity, low density, strictly zoned land uses, and—even more importantly—a vibrant public realm are the primary ones. Large, isolated corporate campuses behind a boom gate—with their own cafes and restaurants—exist as a current norm due to lack of ineffective implementation of initial planning. Interviewee #6—an urban designer—calls it *a private internalised world not a public rich environment.* Additionally, the undulating topography and lack of effective connections poses a challenge to the walkability of the site. Interviewee#6 affirms this, *it's not a very nice environment to explore because it has minimal shade, narrow surpass, traffic dominant streets, pedestrian crossing, really challenging on topography.*

To make the precinct more people-oriented and enhance its vibrancy, recently proposed plans also aim to address these identified issues (Florida, 2005; Yigitcanlar, 2008a). Herewith, new proposals are on board for developing an innovation district in the university, entertainment precinct, and high-density housing with a mix of affordable housing (NSW Government, 2014). The university is also strongly putting new plans in place, simultaneously aimed *to create something more energising* and *to create transparency and accessibility across the university* as put into words by Interviewee#5. To make the university more integrated, accessible, and people-oriented, a few of the key physical initiatives laid down by university are as follows: (a) land use re-zoning to mixed use; (b) more commercial enterprises close to the boundaries; (c) visual display of creativity around the fringes; and (d) providing more pedestrian-friendly campus by clearly identifiable entry statements and engaging spaces. The aim is to ensure knowledge exchange into a creative environment as well as developing a dynamic and collaborative arena. In the past, a conflict of opinion existed regarding height control between government and planners. Though architects considered it a drawback spatially in terms of the density of the area, the government's point of view was appreciably driven in the direction of keeping the societal assurance. Interviewee#4 enunciates, *the council said that you must have height control because the surrounding people, they want a certainty on what's happening there.*

4.3.5 Function

MPID, with its flourishing economic output, is a successful name on the innovation-scape of Australia, as well as globally. Under 'function' as the theme, the research aimed to investigate the level and strength of formal and informal networks at a local level, defining its social layer (Bathelt *et al.*, 2013; Katz & Wagner, 2014). Despite its booming success, the interviews reveal that the park did not see much of the collaboration happening until the last four years. Interviewee#2 calls it *a maturing process* and comprehends, *what we have in MPID is more a case of collocation rather than genuine clustering or collaboration.* Interviewee#12 further confirms, *there's plenty of knowledge in these companies but that's all about private enterprises. So you really need that air to collaborate.* Investigating one of the basic reasons Interviewee #5—an architect—elucidates that Macquarie University, being quite young, had an initial delay in establishing a collaborative innovation ecosystem. In addition, the lack of a vibrant public realm is another major lacking aspect pointed out by most of the interviewees. Interviewee#5 calls attention to lack of amenity and entertainment by comparing the situation to Silicon Valley, *when you work in the city, you just walk out the door, go, we'll have a lunch there and there're plenty of options. At here, all the food is up at Macquarie Centre, it's a long way away in the middle of no way, past car parks and so forth. And that's the problem that they find in California, in Silicon Valley, is that all the development down there are sort of backend for Stanford University, but there is no amenity. So, all the young people, they'll live in San Francisco, they'll go where the bars are, where the girls are. They still want to have lives. So, they are having a huge problem with them.*

Recently, with the university opening its doors and a few strong initiatives from the local council, collaboration levels are growing higher, reflecting the potential to be strengthened further. As Interviewee#9 exclaims, the university in recent years has changed its approach from *what can you do for us to what can we do for you?* He adds, *The key would be we can help you grow your business.* The corporate team of the university organised a network, giving a common platform to companies. Interviewee#10 enunciates, *it's hard for them (companies) whereas we can introduce them naturally in a very relaxed way. And they can follow up by doing business together. It's been creating more outputs through these networking assets.* Like formal networks, it is equally crucial in the strengthening of informal networks within the community—including local community and knowledge workers—for developing sense of place and exchange of knowledge

(Gonzalez & Carrillo, 2012; Van Winden *et al.*, 2013; Pancholi *et al.*, 2017). Community participation and engagement are the key drivers here. The university integrates them by access to amenities such as sport centres, swimming pools, playgrounds, and so on as well as organising a number of joint programmes like sports events, public workshops, festivals, and seminars. In the absence of a direct common interactive platform between businesses and the community, programmes like Programme for After Class Enrichment (PACE) run by university give an opportunity to students to work not only with industry partners but also the community groups. However, apart from university-led networking initiatives, due to the lack of a common management, the area lacks organisation of common events or concerts that involve the local community.

4.3.6 Image

Image as perceived and the factors that develop a sense of place in users, i.e., their attachment and repulsion factors related to area, are investigated under this section (Yigitcanlar, 2010; Healey, 2010; Pancholi *et al.*, 2014). The analysis of knowledge workers' perceptions highlighted the green environment and availability of a shopping centre as major contributors to the image of the innovation district. Referring to the shopping centre, Interviewee#13 believes, *if you want to come away from your desk for an hour during lunch hours, this is a nice place to hang out.* However, issues like commuting, housing affordability, high rentals, lack of common events, as well as day-to-day activities emerge as some of the key concerns. Discussing the current situation, one of the knowledge workers, Interviewee#14, highlights, *commuting is the hardest part. It takes at least an hour whether you take private or public transportation.* For companies, it is a challenge, too. As Interviewee#11 underlines, *it takes an hour or good to get there for them (knowledge workers). So, we pay them in bonus. Because the city is very central, people prefer living in the city.* The transportation-focused initiatives like Connect Macquarie, therefore, hold a good amount of potential. One of the key persons from the group, Interviewee#7, considers the Connect Macquarie Park initiative as having a *pretty good success rate.* She further adds, *the businesses that have joined are achieving a much lower drive alone share than the businesses that haven't joined.*

The local community has a great sense of pride in belonging to the area, owing to its access to amenities and other assets. Interviews also reveal a high level of satisfaction in the community group for the recently announced plans related to the future development of MPID as a priority precinct. Interviewee#8 apprehends, *honestly some people were relieved, they finally had an answer, they weren't living in uncertainty anymore, may not the answer they wanted, but at least they can be certain of the plan for their future, their lives.* He further emphasises how a significant role is played by the efficiency of communication by stating, *when they announced it, they door knocked the entire community and they hand delivered the letter, it kind of answered all the questions straight away, it was translated in to different languages, it had a clear information, and staff on the site was very understanding, very compassionate very patient with people, I don't think we could have asked for much more in terms of how the government has handled itself.* Amongst the key concerns raised by the community group that relate to a striking balance in a diverse mix of different, new groups of community, affordability of new units, units owned by investors rather than owners leading to a hike in price and loss of sense of community, pressure on existing amenities. Transparency, effective communication, and public participation are a few of the key aspects that help develop trust amongst the community.

4.3.7 Key Challenges on the Path Forward

A summary of anticipated challenges in the process of implementation of place-making initiatives for societal integration under new proposals—as highlighted by each group—is presented in this section.

Groups 1 and 2 believe that the key challenge is to maintain the current economic growth rate. Introducing residential may pose a threat of losing out on the commercial spaces to the residential ones. As Interviewee#4 suggests, *so the next step here is how to allow residential to happen within MPID but still have the primacy of the employment.* With much higher returns and less complex processes involved, developers prefer to invest in residential. Interviewee#3 and Interviewee#5 both exemplified the case of the immediate neighbourhood areas, Chatswood and North Sydney—that once held a high potential of growing into a hub—where offices started moving out as residential developments took over the place. Another challenge is to develop a public realm that aids in retaining the knowledge workers. Interviewee#6 asserts, *it needs some sort of collaborative spaces that's accepted, but we also need public realm that's going to encourage I suppose the high-end workers to leave the city and not see it as a major disadvantage. You got to deal with public realm in order to get people out of the buildings and engaging with each other.*

However, for any kind of such developmental changes to happen, one of the key conflicts during the execution of a project as faced by Group 2 is to have a consensus between the local council and the state government. Interviewee #4 highlights, *we are in collision between the states—where's the state sitting and where the local is sitting*. He reckons this decelerates the pace of the project by leaving it to advance *three steps forward and two steps backward*.

According to Group 3, the key challenge that Interviewee#8 anticipates will be maintaining affordability, *because this area is in high demand, the prices are going up, up, up. And it just pushes people out further*. It is also inevitable to provide a range of housing options including social housing and a fine mix between them. Interviewee#8 adds, in this regards, *what we really want to see is a true representation of diversity and a true mix of estate. In the build form of what is proposed here there will be genuine diversity, so it won't be like here is the social housing tower and here is the private housing tower*. While it is crucial to have a fine mix of diversity in the society, the challenge will be to ensure strong informal connections between them for developing a real sense of community. Interviewee#8 exclaims, *the new neighbours will be a diverse mix of culture, income brackets and people. Bringing them all together and expect that to work is not just going to happen by osmosis, you need an organization or a group to take responsibility for actually making community happen. Building community doesn't just happen magically*. However, formal networking initiatives such as the Connect Macquarie face the challenge of developing a trust amongst users of the area to be a part of this initiative. As informed by Interviewee#7, currently there are 15 companies that are member and another approximately 100 are being approached to join. The need for strengthening of joint activities between the university and companies that benefit both, such as training workshops, is highlighted by one of the knowledge workers.

4.4 DISCUSSION AND CONCLUSION

The research findings suggest that the case of MPID is a clear demonstration of the need of societal integration for the success of any innovation district. The interviews with key experts indicate that a vibrant people environment created by the integration of community is also the key to the retainment of knowledge workers. As put in the nutshell by Interviewee#4, *what's friendly to families living in places is also friendly to employees*. This also helps to develop a better sense of place in the community. The initiatives by the government are a strong and clear depiction of how it is inevitable to integrate the community as a part of place-making initiatives in the innovation district. One of the key pioneering initiatives as a societally integrated innovation district is the proposal for redevelopment of the Ivanhoe Estate to integrate more social housing with other residential types and developments of the priority precinct. Such initiatives can help to combat social issues such as gentrification and marginalisation. Research demonstrates that the commitment of government to prioritise the community's interests, overruling the narrow interests of other stakeholders—as exemplified in the conflict with designers over height control regulations—is one of the factors that has kept the project adhered with its societal integration motive. Moreover, community consultation emerges as playing an integral role. Three main requisites for communication are ensuring it is transparent, personalised, and explicit. Interviewee#8 summarises, *we did the research, with such teams like Macquarie University and we know having an answer just had a caustic effect on people's well-being.* Networking assets emerge as the key to sustainable success and collaboration, as synopsised by Interviewee#10: *Final comment on this is, the whole concept of open innovation is much stronger as compared to keeping everything secret. Companies are recognising they need new ideas and to work in collaboration rather than just relying on themselves. That's a trend now picking up everywhere.* Contextually, the presence of a multicultural and diverse demography provides a rich societal base and contributes in creating a globally integrated innovation district. In other dimensions of place making, the major implications for societal integration as derived from the lessons learnt are as follows (Figure 4.3):

- Feature:
    - *Social marketing*: Significant focus on social aspects and community should be placed in the way the innovation district is being featured and marketed—locally and globally.
    - *Awareness:* It is necessary to make the private stakeholders and companies aware of the benefits that societal integration can bring to their talent retainment and innovation district. A clear idea of the innovation district concept needs to be communicated to the society, too.

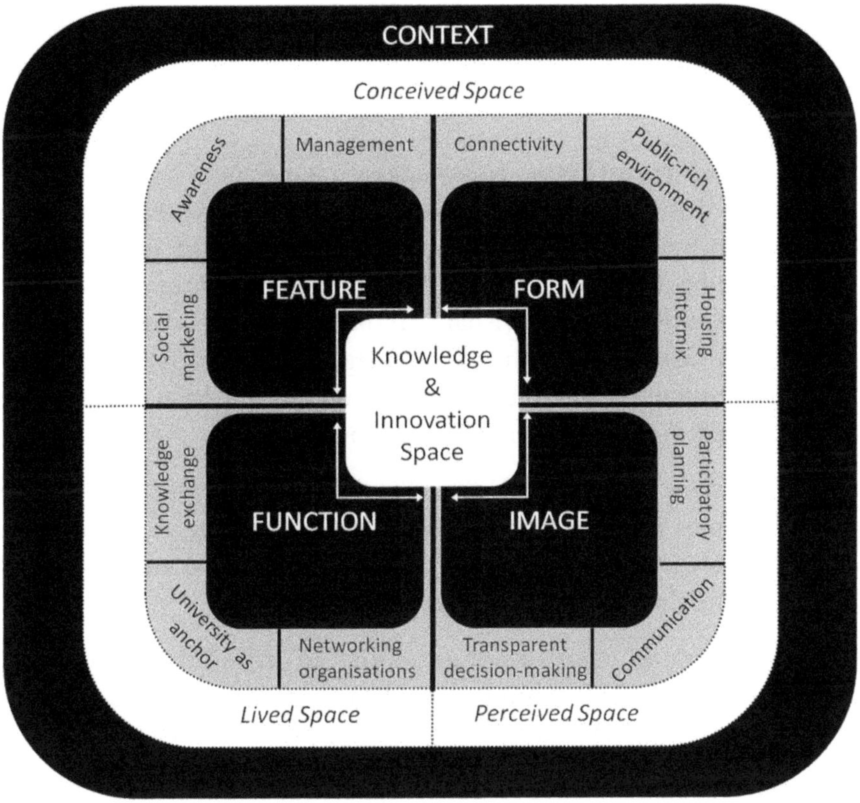

FIGURE 4.3 Conceptual framework of societal integration through place making.

- *Management:* For bringing the various stakeholders together, establishing a common management with equitable representation from all groups is desirable.

- Form:

  - *Connectivity:* The flow and connectivity within the site and between the buildings is critical to avoid isolation of campuses and a collaborative environment.

  - *Public-rich environment:* Every innovation district needs to have an active public realm catering to various needs and requirements. A step ahead would be to creatively engage them by interactive physical displays, exhibitions, and so on.

  - *Housing intermix*: A genuine and fine intermix of diversity in housing options will not only cater to the needs of various segments of society, but also will effectively integrate them as a community.

- Function:
    - *Knowledge exchange*: Knowledge exchange between the innovation district and society should be mutual. Functionally, it refers to bringing out the research happening behind the walls and making people aware of and participative in it. Organising useful talks, seminars, and visual demonstrations are a few examples. This also means that knowledge workers and the global community should be made aware of the local culture and society to facilitate their integration.
    - *University as anchor*: Universities—being accessible by all—have a major role to play as the moderator between society and the private sector—by acting as a common platform and organising events that bring them together.
    - *Networking organisations as catalyst*: In order to build a cohesive and resilient community, establishment of formal and informal groups that help in lubricating the process is crucial.
- Image:
    - *Transparent decision making*: During various stages of planning, it is necessary to keep community well informed, which, on the one hand, develops a sense of trust and assurance in them regarding the plans and, on the other, lubricates the process of development.
    - *Effective communication*: Initiatives such as answering the queries in clear messages translated into multiple languages and a passionate communicating team are a few examples of effective communication.
    - *Perpetual participation*: Ensuring participatory planning by integrating inputs in future plans and proposal from the businesses, knowledge workers, and local community is crucial. More importantly, this participation needs to extend from planning stages to post-development stages for sustaining the sense of ownership. Assigning significant networking roles to their groups after development is one of such examples.

However, the research also reveals certain threats that need to be considered by the policymakers in their initiatives towards societal integration of the innovation district. These future challenges and proposed development path to address them are as follows:

- *Economically, managing a balance between economic growth and societal integration:* Research has highlighted few economic threats such as demand-driven market bend towards residential, developers' lack of interest in commercial development, and so on. Special economic incentives for commercial investment, provision of efficient soft and hard innovation, and networking infrastructure with specific support to start-ups can act as assurance for the firms and commercial spaces to retain while allowing integration of more residential and other activities;
- *Physically, commercial risks involved in development of diverse and affordable spaces:* Market demand towards high-end residential in such a location may discourage developers to invest in diverse housing and spaces for other use. Renegotiating regulations and developmental incentives—such as flexible floor space index or relaxation in height controls—can help as a potential strategic tool to encourage the growth of diverse housing options at the site. Similarly, maintaining affordability is another challenge. Policies should be in place to integrate a fair number of affordable units and securing properties from being converted into investor-owned ones;
- *Socially, building sense of community in diverse people:* Spatially locating the diverse groups with different interests and socio-economic backgrounds together is not enough. It is inevitable to establish special groups dedicated for community building after the development. The organisation of frequent formal and informal events displaying arts, technology, and local culture is helpful in lubricating the relationship within a diverse community and with society. More importantly, marketing plays a big role in making them a success or failure;
- *Politically, coordination between stakeholders:* Research demonstrates that discordance between various tiers of government over a few issues impede the pace of development. It also portrayed an

indecisive image of vision in front of other stakeholders. For strengthening coordination, a strong and committed leadership guided by a clear knowledge-based urban development vision will play a key role in bringing the various groups together and driving the project forward. A comprehensive understanding of local dynamics, followed by a convincing communication and effective execution, are vital for leading such projects with multiple stakeholders.

In conclusion, our research demonstrates that societal integration emerges as one of the key aspects of place making in an innovation district. As Interviewee#8 accentuates, *something that we should think about is how we create a city that is prosperous economically but also is inclusive in its prosperity. So it's an inclusive city, it's a city that is accessible and affordable.* While physical integration of innovation districts with their surroundings is necessary for providing the quality of life and vibrancy of environment as desired by the knowledge workers as well as shaping creativity and uniqueness, a holistic approach towards societal integration is necessary for reaping the benefits of knowledge produced in these specialised environment for societal development. For policymakers, it is necessary to expand their definition of society from only considering knowledge workers to also accommodating the local community in the place-making initiatives in innovation districts. One of the key practical ways for societal integration is by making the processes related to the innovation district—i.e., political, functional, design-related, and societal—open and transparent. Effective societal integration leads to a fruitful exchange of knowledge, as it is not limited to only one group but is shared with the society. This will also ensure reduction in any kind of conflicts arising from marginalisation of certain groups by setting a harmony between the innovation district and its surroundings. By doing so, the key objectives of knowledge-based urban development will be accomplished in a true sense (Pancholi *et al.*, 2018).

ACKNOWLEDGEMENTS

This chapter, with permission from the copyright holder, is a reproduced version of the following journal article: Pancholi, S., Yigitcanlar, T., & Guaralda, M., (2018). Societal integration that matters: Place making experience of Macquarie Park Innovation District, Sydney. City, Culture and Society, 13(1), 13–21.

REFERENCES

ABS. (2011). Australian Bureau of Statistics census of population and housing. Retrieved from 10.4225/87.2.8.1 (Accessed 12 Oct 2016).

Arefi, M. (2014). *Deconstructing placemaking*. New York: Routledge.

Bakouros, Y. L., Mardas, D. C., & Varsakelis, N. C. (2002). Science park, a high-tech fantasy?. *Technovation*, 22(2), 123–128. 10.1016/s0166-4972(00)00098-6

Bathelt, H., Feldman, M., & Kogler, D. F. (2013). *Beyond territory*. London: Routledge.

Baum, S., O'Connor, K., & Yigitcanlar, T. (2009). The implications of creative industries for regional outcomes. *International Journal of Foresight and Innovation Policy*, 5(1/2), 44–64. 10.1504/ijfip.2009.026564

Berkes, F. (2009). Evolution of co-management. *Journal of Environmental Management*, 90(5), 1692–1702. 10.1016/j.jenvman.2008.12.001

Breschi, S., & Lissoni, F. (2001). Knowledge spillovers and local innovation systems. *Industrial and Corporate Change*, 10(4), 975–1005. 10.1093/icc/10.4.975

Carrillo, F., Yigitcanlar, T., Garcia, B., & Lonnqvist, A. (2014). *Knowledge and the city*. New York: Routledge.

Castells, M. (2000). *The rise of network society*. Malden: Blackwell.

Chang, T., & Huang, S. (2008). Geographies of everywhere and nowhere. *International Development Planning Review*, 30(3), 227–247. 10.3828/idpr.30.3.1

Chesbrough, H. (2003). The era of open innovation. *MIT Sloan Management Review*, 44(3), 35–41.

Connect Macquarie. (2015). Annual report. Retrieved from 10.13140/RG.2.1.4557.3529 (Accessed 30 Aug 2017).

Cresswell, T. (2004). *Place: a short introduction*. London: Wiley-Blackwell.

Edvardsson, I., Yigitcanlar, T., & Pancholi, S. (2016). Knowledge cities research and practice under the microscope. *Knowledge Management Research & Practice*, 14(4), 537–564. 10.1080/14778238.2016.1255081

Esmaeilpoorarabi, N., Yigitcanlar, T., & Guaralda, M. (2016a). Place quality and urban competitiveness symbiosis? *International Journal of Knowledge-Based Development*, 7(1), 4–21. 10.1504/ijkbd.2016.075807

Esmaeilpoorarabi, N., Yigitcanlar, T., & Guaralda, M. (2016b). Towards an urban quality framework. *International Journal of Knowledge-Based Development*, 7(3), 290–312. 10.1504/ijkbd.2016.078798

Evans, G. (2009). Creative cities, creative spaces and urban policy. *Urban Studies*, 46(5-6), 1003–1040. 10.1177/0042098009103853

Evers, H., Gerke, S., & Menkhoff, T. (2010). Knowledge clusters and knowledge hubs. *Journal of Knowledge Management*, 14(5), 678–689. 10.1108/13673271011079237

Fernandez-Maldonado, A., & Romein, A. (2010). The role of organisational capacity and knowledge-based development. *International Journal of Knowledge-Based Development*, 1(1), 79–96. DOI:10.1504/IJKBD.2010.031306

Florida, R. (2005). *Cities and the creative class*. New York: Routledge. DOI:10.4324/9780203996619

Funke, J. (2007). The perception of space from a psychological perspective. In *Experiencing New Worlds*, edited by Wassmann, J. & Stockhaus, K., 345–357. Oxford, New York: Berghahn. DOI:10.2307/j.ctv1fj10n8.27

Garnsey, E., & Heffernan, P. (2005). High-technology clustering through spin-out and attraction. *Regional Studies*, 39(8), 1127–1144. DOI:10.1080/00343400500328017

Glaesar E. (1999). Learning in cities. *Journal of Urban Economics*, 46(2), 254–277. DOI:10.1006/juec.1998.2098

Gonzalez, O., & Carrillo, F. (2012). Cities-benchmarking algorithm. *International Journal of Knowledge-Based Development*, 3(4), 367–387. DOI:10.1504/IJKBD.2012.051129

Graham, S., & Guy, S. (2002). Digital space meets urban place. *City*, 6(3), 369–382. DOI:10.1080/1360481022000011154

Healey, P. (2010). *Making better places*. New York: Palgrave Macmillan. DOI:10.1057/9780230275566

Henton, D., & Held, K. (2013). The dynamics of Silicon Valley. *Social Science Information*, 52(4), 539–557. DOI:10.1177/0539018413487629

Ho, K., & Douglass, M. (2008). Globalisation and liveable cities. *International Development Planning Review*, 30(2), 199–213. DOI:10.3828/idpr.30.2.3

Hutton, T. (2004). The new economy of inner cities. *Cities*, 21(1), 89–108. DOI:10.1016/j.cities.2003.11.001

Katz, B., & Wagner, J. (2014). *The rise of innovation districts*. Washington: The Brookings Institution. DOI:10.7864/JCDR/2015/12626.6681

Knight, R. (1995). Knowledge-based development. *Urban Studies*, 32(2), 225–260. DOI:10.1080/00420989550012107

Kunzmann, K. R. (2009). The strategic dimensions of knowledge industries in urban development. *DISP-The Planning Review*, 45(4), 40–47. DOI:10.1080/02513625.2009.10556974

Lefebvre, H. (1991). *The production of space*. Malden: Blackwell. DOI:10.1002/9780470693222

Livingstone, D. (2003). *Putting science in its place*. Chicago: University of Chicago. DOI:10.7208/chicago/9780226487267.001.0001

Lonnqvist, A., Kapyla, J., Salonius, H., & Yigitcanlar, T. (2014). Knowledge that matters. *European Planning Studies*, 22(10), 2011–2029. DOI:10.1080/09654313.2014.914563

Macquarie University (2017). Submission to Greater Sydney Commission, Draft North District Plan, Sydney, Macquarie University. DOI:10.26183/5f08650b2fb2c

Massey, D. (1991). The political place of locality studies. *Environment and Planning A*, 23(3), 267–281. DOI:10.1068/a230267

Metaxiotis, K., Carrillo, F., & Yigitcanlar, T. (Eds.) (2010). *Knowledge-based development for cities and societies*. Hersey: IGI Global. DOI:10.4018/978-1-61520-929-3

Meusburger, P., Funke, J., & Wunder, E. (2009). *Milieus of creativity*. Dordrecht: Springer. DOI:10.1007/978-1-4020-8870-6

Montgomery, J. (1998). Making a city. *Journal of Urban Design*, 3(1), 93–116. DOI:10.1080/13574809808724455

Moultrie, J., Nilsson, M., Dissel, M., Haner, U. E., Janssen, S., & Van Der Lugt, R. (2007). Innovation spaces. *Creativity and Innovation Management*, 16(1), 53–65. DOI:10.1111/j.1467-8691.2007.00423.x

NSW Government (2014). Herring Road Macquarie Park urban activation precinct planning report. Retrieved from DOI:10.26190/5c5d5d5c546dc (accessed 7 Apr 2016).

PWC (2014). *Big city analytics*. Sydney: PriceWaterhouseCoopers. DOI:10.1002/jtr.1964

Pancholi, S., Yigitcanlar, T., & Guaralda, M. (2014). Urban knowledge and innovation spaces. *Asia Pacific Journal of Innovation and Entrepreneurship*, 8(1), 15–38. DOI:10.1108/APJIE-03-2014-0021

Pancholi, S., Yigitcanlar, T., & Guaralda, M. (2015). Place making facilitators of knowledge and innovation spaces. *International Journal of Knowledge-Based Development*, 6(3), 215–240. DOI:10.1504/IJKBD.2015.070338

Pancholi, S., Yigitcanlar, T., & Guaralda, M. (2017). Governance that matters. *Journal of Place Management and Development*, 10(1), 73–87. DOI:10.1108/JPMD-09-2016-0077

Pancholi, S., Yigitcanlar, T., & Guaralda, M. (2018). Societal integration that matters: place making experience of Macquarie Park Innovation District, Sydney. *City, Culture and Society*, 13(1), 13–21. DOI: 10.1016/j.ccs.2017.09.004

Peck, J. (2005). Struggling with the creative class. *International Journal of Urban and Regional Research*, 29(4), 740–770. DOI:10.1111/j.1468-2427.2005.00611.x

Peck, J. (2010). *Constructions of neoliberal reason*. Oxford: Oxford University Press. DOI:10.1093/acprof:oso/9780199580576.001.0001

Porter, M. E. (2000). Location, competition, and economic development. *Economic Development Quarterly*, 14(1), 15–34. DOI:10.1177/089124240001400105

Pratt, A. (2000). New media, the new economy and new spaces. *Geoforum*, 31(4), 425–436. DOI:10.1016/S0016-7185(00)00016-0

Sarimin, M., & Yigitcanlar, T. (2012). Towards a comprehensive and integrated knowledge-based urban development model. *International Journal of Knowledge-Based Development*, 3(2), 175–192. DOI:10.1504/IJKBD.2012.047428

Saxenian, A. (1994). *Regional advantage*. Cambridge: Harvard University Press. DOI:10.4159/harvard.9780674330198

Scott, A. (2000). *The cultural economy of cities*. New York: Sage. DOI:10.4135/9781446218509

Solnit, R. (2014). Resisting monoculture. *Guernica*. Retrieved from https://www.guernicamag.com/daily/rebecca-solnit-resisting-monoculture/ (accessed 9 Sep 2016).

Stehlin, J. (2016). The post-industrial shop floor. *Antipode*, 48(2), 474–493. DOI:10.1111/anti.12201

Storper, M., & Venables, A. (2004). Buzz: face-to-face contact and the urban economy. *Journal of Economic Geography*, 4(4), 351–370. DOI:10.1093/jnlecg/lbh027

Van Winden, W., Carvalho, L., Van Tuijl, E., Van Haaren, J., & Van Berg, D. (2013). *Creating knowledge locations in cities*. London: Routledge. DOI:10.4324/9780203721281

Yigitcanlar, T., Baum, S., & Horton, S. (2007). Attracting and retaining knowledge workers in knowledge cities. *Journal of Knowledge Management*, 11(6), 6–17. DOI:10.1108/13673270710830375

Yigitcanlar, T., Edvardsson, I. R., Johannesson, H., Kamruzzaman, M., Ioppolo, G., & Pancholi, S. (2017). Knowledge-based development dynamics in less favoured regions: insights from Australian and Icelandic university towns. *European Planning Studies*, 25(12), 2272–2292. https://doi.org/10.1080/09654313.2017.1358699

Yigitcanlar, T., O'Connor, K., & Westerman, C. (2008a). The making of knowledge cities. *Cities*, 25(2), 63–72. DOI:10.1016/j.cities.2008.01.003

Yigitcanlar, T., Velibeyoglu, K., & Martinez-Fernandez, C. (2008b). Rising knowledge cities. *Journal of Knowledge Management*, 12(5), 8–20. DOI:10.1108/13673270810901626

Yigitcanlar, T. (2011). Position paper: redefining knowledge-based urban development. *International Journal of Knowledge-Based Development*, 2(4), 340–356. DOI:10.1504/IJKBD.2011.043408

Yigitcanlar, T. (2010). Making space and place for the knowledge economy. *European Planning Studies*, 18(11), 1769–1786. DOI:10.1080/09654313.2010.526174

Yigitcanlar, T., Metaxiotis, K., & Carrillo, F. (Eds.) (2012). *Building prosperous knowledge cities*. Cheltenham: Edward Elgar. https://www.elgaronline.com/view/9780857930569/9780857930569.xml

Yigitcanlar, T., & Dur, F. (2013). Making space and place for knowledge communities. *Australasian Journal of Regional Studies*, 19(1), 36–63. https://www.anzrsai.org/assets/Uploads/AJRS/Vol191/Yigitcanlar-and-Dur.pdf

Yigitcanlar, T., & Bulu, M. (2015). Dubaization of Istanbul. *Environment and Planning A*, 47(1), 89–107. DOI:10.1068/a130203p

Yigitcanlar, T., Guaralda, M., Taboada, M., & Pancholi, S. (2016). Place making for knowledge generation and innovation. *Journal of Urban Technology*, 23(2), 115–146. DOI:10.1080/10630732.2016.1172384

Zelinsky, M. (2004). *The inspired workplace*. Beverly: Rockport Publishers.

3

Place Quality

This part of the book explores the notion of place quality regarding innovation districts. Place quality characterises the intersection of key elements that people experience in a place. The part clarifies that place quality is an extremely complicated phenomenon containing objective and tangible physical factors of a place, to subjective and intangible feelings of well-being, fulfilment and satisfaction. Place quality is also closely related to the ensemble concept of liveability and sustainability, which combine economic, environmental, social and personal dimensions. This part aims to respond to a rising question—how innovation districts can effectively shape such complex place quality as they are ever more responsible for not only empowering knowledge industries and workers but also contributing to the sustainability of cities and societies. Since innovation districts' level is the scale of real day-to-day life and has been connected to touchable places and actual people, it is difficult to measure it solely through quantitative measures. This part combines both qualitative and quantitative measures and methods to define and evaluate place quality in innovation districts.

DOI: 10.1201/9781032657431-7

CHAPTER 5

Place Quality in Innovation Districts

5.1 INTRODUCTION

Over the last two decades, a global phenomenon emphasising a knowledge-based and technology-driven economy has occurred. In this new economy—characterised by digital disruption, knowledge-based activities, and creativity—knowledge has become central for driving economic growth (Baum *et al.*, 2009; Lonnqvist *et al.*, 2014; Cooke, 2017). This economy is not only about the processes of producing knowledge-intensive products and increasing profit, but also about creating an intangible value. Knowledge economy emphasises on education and skills acquisition, while providing information systems and networking infrastructures to facilitate the generation and spill-over of knowledge (Van Winden *et al.*, 2007; Carrillo *et al.*, 2014). The process of new economic growth, accordingly, is tangled with knowledge industries, knowledge workers, innovative atmospheres, networking systems, and innovative products—where their qualitative attributes are also paramount (Scott, 2006; Murphy *et al.*, 2015). Within this new economic system, intangible experience is often more central than tangible material gains (Yigitcanlar *et al.*, 2008b).

Earlier speculations argued that digital technologies and their disruption would make place irrelevant and cause the death of public spaces and traditional cities (Webber, 1964; Hall, 1996). In spite of ubiquitous technology—that allows knowledge workers and industries to locate wherever they want—geographic space still has a critical role in the new economy. The insurgence of third places is evident in gentrified contexts, where boundaries between work and private environments are blurred (Oldenburg, 1999). Today, the vitality of knowledge industries and

DOI: 10.1201/9781032657431-8

workers mainly relies on digital interactions. However, they still value face-to-face interactions and networking, which led to the formation of creative atmospheres and generated experiences of unique settings and locations (Storper & Venables, 2004). Planning our cities, in the age of global knowledge economy, requires a knowledge-based urban development (KBUD) approach (Yigitcanlar & Velibeyoglu, 2008; Yigitcanlar *et al.*, 2017). It is a physical response to the new socioeconomic paradigm (Frenkel *et al.*, 2013a; Yigitcanlar & Bulu, 2015). KBUD, as a sustainable socio-spatial strategy, firstly emerged at the global best practices such as Silicon Valley, Cambridge Science Park, and Sophia Antipolis, and then adopted by leading cities in Europe, North America, Australia, and Asia—e.g., Austin, Barcelona, Boston, Delft, Manchester, Melbourne, Singapore, Toronto (Yigitcanlar *et al.*, 2008a; Yigitcanlar & Sarimin, 2015). These cities have planned their digital infrastructures, green technologies, and other infrastructural KBUD requirements, as ways to revitalise stagnant urban environments, provide economic opportunities, and strengthen their global competitiveness (Benneworth & Ratinho, 2014; Lee *et al.*, 2014). Studies highlight that the neoclassical approach has its limitations on dealing sufficiently with the needs of knowledge industries and workers (Zaheer & Nachum, 2011; Frenkel *et al.*, 2013b). This directs attention to the social impacts of innovation districts (Mendez & Moral, 2011).

Innovation districts—e.g., creative hubs; innovation districts; science, knowledge or research precincts; technology parks; and the like—where innovation activities that cluster together are not isolated from a knowledge community-based revolution (Evers, 2008). The new generation of innovation districts provides plenty of third places for living, learning, playing, and networking—rather than solely focusing on workspaces (Yigitcanlar, 2010). These districts are being increasingly recognised as not only economic engines, but also as the home of creative class (Yigitcanlar & Velibeyoglu, 2008; Van Winden *et al.*, 2013). They are platforms of flourishing work conditions with high-quality services that offer a distinctive urban image of the locality—e.g., cafes and bars in a historic or bohemian urban areas (Florida, 2005; McCann, 2008). They demonstrate that place quality strongly influences location choices of knowledge industries and workers. These aspects need to be precisely taken into consideration in the planning, design, development, and management stages of these districts (Pancholi *et al.*, 2017).

Although there has been considerable amount of literature underlining the importance and influence of place quality on KBUD, our understanding about the impacts of place quality at the district (cluster) scale is still limited. Indeed, many elements of place quality are relevant to more than one scale—e.g., regional, city, district, neighbourhood scales. However, district (cluster) scale is recognised as specifically important, because it is the scale where the actual quality of a space is designed and realised, and day-to-day activities occur (Trip, 2007; Durmaz, 2015).

Against this backdrop, this chapter aims to identify indicators of place quality and their impacts on shaping the appealing of innovation districts. The research intends to address the question of: How does place quality contribute to the attractiveness of innovation districts? In order to tackle this issue, the research undertakes an in-depth review of the literature, and places four global best practice innovation districts under the microscope—i.e., One-North (Singapore), Arabianranta (Helsinki), DUMBO (New York), MPID (Sydney). These innovation districts are investigated through an inductive case study approach that applies descriptive and explanatory methods.

Following this introduction, innovation district and place quality concepts are investigated thoroughly from the literature. Then, the methodological approach is elaborated. This is followed by introducing the theoretical framework for place quality in innovation districts. Four best practice cases that are the focal point of the investigation are also presented. Afterwards, results of the empirical study are revealed. Finally, the key findings are highlighted, and potential impacts of introduced indicators on the success of investigated best practices are discussed.

5.2 LITERATURE BACKGROUND

To date, the most popular definition of the clustered districts is the one proposed by Porter (1998). He defined a cluster as 'a critical mass of companies in a particular field in a particular location, whether it is a country, a state or region, or even a city' (Porter, 1998, p. 10). As such, an innovation district is a local industrial specialisation that is generally organised around universities, research institutions, and knowledge-based industries, with a high internal and external networking and knowledge sharing capabilities (Evers, 2008; Millar & Ju-Choi, 2010). Even though some studies focus more on city and regional aspects of districts (Cooke, 2001; Baum *et al.*, 2009; Brocker *et al.*, 2012), innovation districts mainly refer to district- or neighbourhood-scale spaces, such as science and technology parks, research hubs, industrial precincts, and creative clusters, where knowledge- and innovation-based activities beneficially agglomerate together (Heebels & Van Aalst, 2010; He & Gebhardt, 2014; Pancholi *et al.*, 2014; Durmaz, 2015). While placeless nature and the low transaction cost of knowledge-based products seem to invalidate the initial theory of clustering (Porter, 1990, 1998), knowledge-based industries still tend to follow a place-based clustering pattern. Innovative capacity of knowledge industries increases when firms share ideas, products, and services in tangible and face-to-face basis (Homan, 2014). These clustered districts also restructure economic capability of cities and regions and develop new businesses and alliance between local governments, universities, and knowledge industries and workers (Parker, 2010).

Innovation districts typically appear in three forms: (a) science districts, where knowledge-intensive service sectors—e.g., universities and R&D centres—are located around business districts; (b) high-tech districts, which focus on the development of high-tech manufacturing activities—e.g., ICT or biotechnology; and (c) creative districts, which are shaped based on cultural knowledge generation— e.g., movie-making, media, arts, design (Brocker *et al.*, 2012). Nevertheless, innovation districts are in a transition phase towards mixed functional types to provide better support to the expectations of knowledge industries and workers (Cooke, 2001). While mixed functional generation of innovation districts mainly contain universities, R&D centres, and business districts inside or around their boundaries, it seems the current typology contains only two broad types of districts: (a) creative and (b) high tech. These types attempt to present a welcoming place to live, work, and play as well as providing a sense of life and beauty for knowledge industries and workers. As knowledge industries and workers are the driving force of the knowledge economy, concentration of talent has become as significant as concentration of infrastructure for flourishing the growth of innovation districts (Esmaeilpoorarabi *et al.*, 2016a). Thus, to systematically foster, attract, and retain knowledge industries and workers, the following questions have been repeatedly asked: Where do knowledge industries and workers choose to locate, and why?

Today, knowledge workers increasingly prefer to be located in urban innovation districts—especially for specific place quality and day-to-day vibrant lifestyle offerings. In other words, knowledge workers prefer authentic locations, which fulfil their sophisticated lifestyles and fit their creative identity—rather than just where the high-paying jobs are (Yigitcanlar *et al.*, 2007). Literature highlights these characteristics of urban environment as quality of place, and considers three main components: (a) what is there: the built environment plus specific amenities; (b) who is there: the diversity of population; and (c) what is going on: street life, buzz (Florida, 2005; Kloosterman & Trip, 2011). From this perspective, place quality mostly relies on the intangible conditions of a place, or soft factors, including quality of life, urban ambiance, cultural and social characteristics, as well as levels of diversity, tolerance, and openness of the population (Bereitschaft & Cammack, 2015). However, some scholars emphasise the importance of classic conditions of place, or hard factors, to describe place quality and locational choices of knowledge industries and workers—e.g., investment availability, job opportunity, cost of living (Darchen & Tremblay, 2010; Lawton *et al.*, 2013).

There is also a third approach, which claims that attracting knowledge workers involves a mix of a hard-labour market and economic factors as well as soft quality-based factors (Brown & Mczyski, 2009). In recent years, this approach has gained high popularity (Bontje & Musterd, 2009; Boren & Young, 2013; Durmaz, 2015). Such a balanced viewpoint is also more likely to support the desired urban policy and planning outcomes. New urban strategies such as KBUD, knowledge cities, smart cities, city branding, and knowledge clustering that invest in quality-based factors as well as hard factors are potentially more effective approaches (Sarimin & Yigitcanlar, 2012; Carrillo *et al.*, 2014; Yigitcanlar, 2016). These studies have reviewed the multidisciplinary literature that confirms the importance of place quality in attracting knowledge industries and workers, and improving the competitive capabilities among innovation districts. However, there is still no general agreement on the definition and evaluation methods of place quality. Therefore, the empirical study reported in this chapter adopted the third approach to provide a comprehensive understanding of place quality attributes in innovation districts.

5.3 METHODOLOGY

5.3.1 Research Design

This research utilises a case study approach to investigate attributes of place quality in order to understand how place quality contributes to the attractiveness of innovation districts. The study takes contextual issues of each case into account and relies on multiple sources of evidence (Yin, 2011). The two most common approaches to this case study research are an inductive approach, based on Grounded Theory (Glaser & Strauss, 1967), and deductive or testing approach (Yin, 2011). The basic difference between these approaches is that while Grounded Theory relies on data to generate new theories—there is no initial preconceived framework of concepts and hypotheses—Yin's approach develops a theory at the beginning of the research by focusing on testing and validating the theory. There is also another proposed approach by Eisenhardt (1989), which lays in-between these two approaches that is inductive but has more pre-planned structure. This research mainly follows the third approach (Eisenhardt, 1989), utilising a planned conceptual framework for an inductive theory building. As the conceptual framework solely clarifies the general aspects of place quality in innovation districts, inductive findings from case studies will help to identify and classify exact effective indicators from the real world.

Four global best practice innovation districts from different geographical contexts have been selected—One-North from Southeast Asia, Arabianranta from Europe, Down Under the Manhattan Bridge Overpass (DUMBO) from North America, and Macquarie Park Innovation District (MPID) from Oceania. These cases were carefully chosen as they are all: (a) from a different continental context to reflect a global viewpoint as well as supporting the generalisation of results; (b) located in renowned cities with strong KBUD achievements; (c) are both common and distinctive in their forms, functions, ambiances, and images to satisfy the logic of an in-depth comparative analysis; (d) relatively the same size to avoid the size-related influences; (e) globally well-known and economically prosperous cases to avoid economic-related influences; (f) representative of the new generation of innovation districts, which are mixed-used and people-centred, and offer various job opportunities; and (g) cover both general typology of innovation districts—creative and high-tech industries.

This study, firstly, identified the academic literature that is directly focusing on the investigated innovation district case studies. The Web of Science, Scopus, and Google Scholar databases are used to search the scholarly literature—mainly journal articles. The objective of using these databases is to obtain a multidisciplinary selection of architectural, social, and urban studies. The search is performed using the name of each case study as a text word term along with 'urban' OR 'architecture' OR 'social'. The results are limited to articles published between January

2000 and June 2017. This is due to gathering the most updated information about the place quality in each case study and to monitor their development process during the last two decades—17 and half years to be precise. Retrieved literature is assessed by the research team to identify the relevant studies. Articles have been included only if the paper has directly analysed the case study at district, city, or regional levels—in addition to being available online as a full-text article. The research team also attempted to select a balanced number of sources for the four cases. For instance, more academic articles are written about One-North and Arabianranta, which are highly relevant and reliable. To narrow the gap for DUMBO and MPID, related books and their chapters have also been included in the review. In total, 48 academic references on the 4 case studies are selected and reviewed (Table 5.1)—on top of these, 74 scholarly literature pieces are also reviewed on place quality in innovation districts (Table 5.2).

Secondly, in addition to the academic literature, innovation districts' master plan reports and websites contents are also retrieved. The composition of sources for each innovation districts is shown in Table 5.1. These sources are reviewed through an exploratory method

TABLE 5.1 Literature on Investigated Innovation Districts

	Academic Literature	Master Plan Report	Official Website
One-North	15 (journal articles)	One-North Master Plan 2001–2021	www.jtc.gov.sg/industrial-land-and-space/pages/one-north.aspx
Arabianranta	15 (journal articles)	Walking around Arabianranta Helsinki Plans 2009	www.arabianranta.fi
DUMBO	10 (journal articles, book chapters)	Brooklyn Tech Triangle Strategic Plan 2013, and the 2015 update	www.brooklyntechtriangle.com
MPID	8 (journal articles, book chapters)	Macquarie Park Corridor, Development Control Plan 2014, Macquarie Park Investment Prospectus 2015, and Macquarie Park Innovation District Roadmap 2017	www.macquariepark.com.au www.mpid.com.au

TABLE 5.2 General Factors of Place Quality Cited in the Literature

Context	Form	Function	Ambiance	Image
• Baum *et al.* (2009) • Carrillo *et al.* (2014) • Carvalho & Van Winden (2017) • Cooke (2017) • Esmaeilpoorarabi *et al.* (2016a, 2016b) • He & Gebhardt (2014) • Homan (2014) • Lonnqvist *et al.* (2014) • Mian *et al.* (2012) • Pancholi *et al.* (2015, 2017, 2019) • Parker (2010) • Pelkonen (2005) • Trip (2007) • Yigitcanlar & Dur (2013) • Yigitcanlar *et al.* (2007, 2008b, 2017)	• Brown & Mczyski (2009) • Carrillo *et al.* (2014) • Da Cunha & Selada (2009) • Durmaz (2015) • Flew (2012) • Frenkel *et al.* (2013a, 2013b) • He & Gebhardt (2014) • Kloosterman & Trip (2011) • Pancholi *et al.* (2015, 2017, 2019) • Smith *et al.* (1997) • Yigitcanlar & Dur (2013) • Yigitcanlar *et al.* (2007, 2008b) • Zaheer & Nachum (2011)	• Benneworth & Ratinho (2014) • Carrillo *et al.* (2014) • Cooke (2001) • Da Cunha & Selada (2009) • Darchen & Tremblay (2010) • Grant & Buckwold (2013) • Hazelkorn & Murphy (2002) • He & Gebhardt (2014) • Kloosterman & Trip (2011) • Lawton *et al.* (2013) • Mian *et al.* (2012) • Murphy *et al.* (2015) • Musterd & Gritsai (2013) • Pancholi *et al.* (2015, 2017, 2019) • Porter (1990, 1998) • Scott (2010) • Yigitcanlar *et al.* (2008b) • Yigitcanlar & Dur (2013)	• Bereitschaft, & Cammack (2015) • Bontje & Musterd (2009) • Carrillo *et al.* (2014) • Clifton & Cooke (2009) • Cooke (2001) • Evers (2008) • Florida (2005) • Hazelkorn & Murphy (2002) • He & Gebhardt (2014) • Heebels & Van Aalst (2010) • Homan (2014) • Murphy *et al.* (2015) • Musterd & Gritsai (2013) • Storper & Venables (2004) • Yigitcanlar *et al.* (2008b) • Yigitcanlar & Dur (2013)	• Carrillo *et al.* (2014) • D'Mello & Sahay (2007) • Durmaz (2015) • Florida (2005) • Hafeez *et al.* (2016) • He & Gebhardt (2014) • Homan (2014) • Pancholi *et al.* (2015, 2017, 2019) • Sepe (2010) • Trip (2007) • Yigitcanlar *et al.* (2007, 2008b) • Yigitcanlar & Dur (2013)

to define general elements of place quality in innovation districts. This led to the design of a proposed conceptual framework of place quality in innovation districts (see the next section).

Thirdly, collected data are manually coded based on the elements of the conceptual framework and are analysed by a descriptive method. The descriptive method is used for a content analysis of data to explore and describe the existing place quality in each case.

Fourthly, to facilitate analysis of such amount of data and reduce the risk of biased interpretation, N-Vivo software is used for the rest of quantitative and qualitative content analyses. The quantitative content analysis—or lexical analysis—shows the most frequent factors and the phenomenon (Emmanuelle & Isabelle, 2013) that impact the main domains of the conceptual framework (i.e., context, form, function, ambiance, image) in case studies—by word count. Results build a set of indicators to be used in the qualitative explanatory analysis of the case studies.

Fifthly, for a qualitative content analysis through the explanatory method, series of indicators practiced and previous data are recoded. As a result, data have become available to easily, systematically, and accurately be classified and interpreted. The applied explanatory analysis connects and explains similarities and differences in cases to find out the exact indicators of place quality and their impacts in the real world (Zainal, 2007). For both descriptive and explanatory analysis, results are presented in a comparative way to support the deep understanding of variations and their impacts.

5.3.2 Conceptual Framework of Place Quality in Innovation Districts

This study proposes a multidimensional model as its conceptual framework to highlight the most significant hard and soft factors of place quality (Figure 5.1). In total 5 general factors or domains—i.e., context, form, function, ambiance, image—are derived by reviewing 74 scholarly literature sources and previous research that proposed relevant frameworks (i.e., Pancholi *et al.*, 2015, 2017, 2019; Esmaeilpoorarabi *et al.*, 2016b). Table 5.2 presents the list of these literature pieces that focus on the generic factors of place quality. While form and function support tangible, economic, and place-based hard factors, ambiance and image focus more on intangible and people-based soft factors. The followings, based on the academic literature, are the main characteristics of these factors:

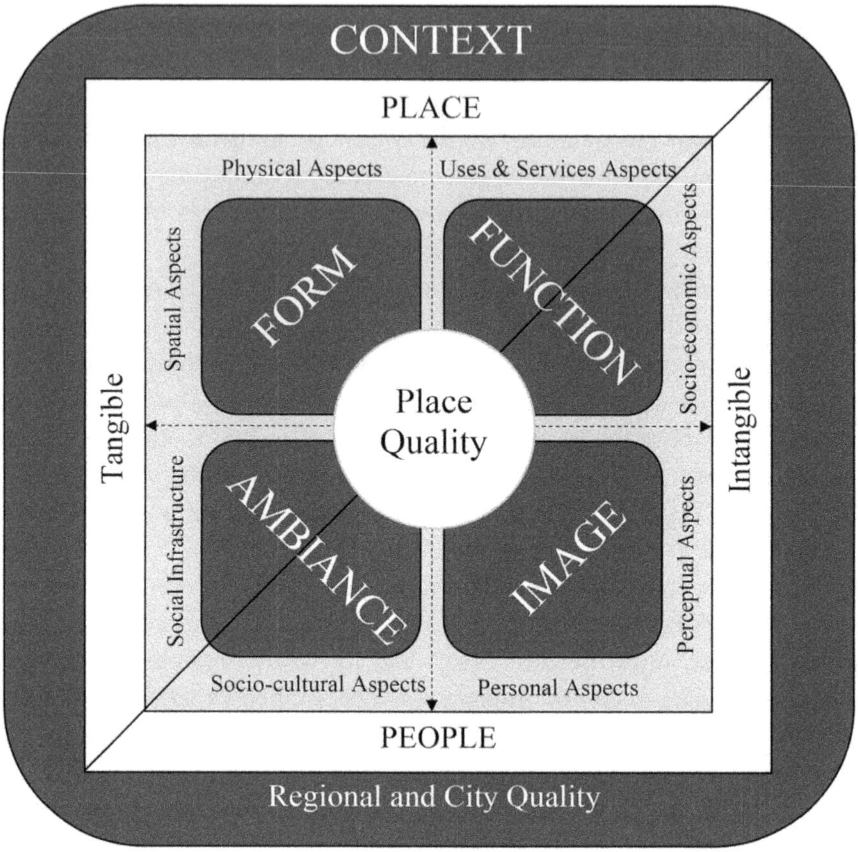

FIGURE 5.1 Conceptual framework of place quality in innovation districts.

Context: Place quality in innovation districts is not an isolated phenomenon. Place quality in the micro-scale of the district has a close link to the performance of urban quality in both the mezzo-scale of the city and the macro-scale of the region (Trip, 2007; Yigitcanlar & Dur, 2013; Durmaz, 2015; Pancholi *et al.*, 2015).

Form: Place quality is strongly related to the physical representation of innovation districts —i.e., inner city or suburban locations, urban and architectural structures and design, and amenities. Form, by both physical and social aspects, supports a thriving, dynamic, and healthy knowledge industries and workers, and guarantees an attractive physical environment for their wellbeing (Smith *et al.*, 1997; Brown & Mczyski, 2009; Kloosterman & Trip, 2011; Frenkel *et al.*, 2013a).

Function: Place quality highly relies on the type of activities, economic opportunities, labour pool, and development process, which is offered by innovation districts. Function represents the procedure of planning, constructing, and leading of districts; defines districts' typology; and characterises the relation of creative people to a thick job market (Darchen & Tremblay, 2010; Scott, 2010; Grant & Buckwold, 2013).

Ambiance: Place quality contains the socio-cultural structure of knowledge industries and workers, which creates the creative ambiance of innovation districts—i.e., diversity of opinions, people and lifestyles, multicultural attractions, and openness. The ambiance is closely linked with a variety of scenes—i.e., music events, street arts, nightlife, presence of third places—to enhance social vitalities as well as social and business interactions (Florida, 2005; Clifton & Cooke, 2009; Heebels & Van Aalst, 2010).

Image: Place quality involves memorable and imaginable perceptions or experiences of knowledge industry and workers in innovation districts. Image is an authentic identification, which distinguishes a district from others. Physical features or appearance (i.e., form), observable activities and purposes (i.e., function), and meanings or symbols (i.e., ambiance), altogether, create the specific image of an innovation districts (Florida, 2005; D'Mello & Sahay, 2007; Sepe, 2010).

5.3.3 Descriptive Analysis of the Case Innovation Districts

As the theoretical structure of the study, the conceptual framework has been employed for coding and categorising of the collected data for each case study. Coding elements contain context, form, function, ambiance, and image. The following descriptive analysis of case studies has been conducted based on the aforementioned coding system.

5.3.3.1 One-North, Singapore

Context: Due to the political history and strategic geographic location, Singapore has always been a regional trade hub with strong international economic connections. Singapore has employed this advantage to become a prominent financial player in the global knowledge economy. To increase capacities of the knowledge economy, Singapore has developed a superior intellectual property system, a flourish start-up ecosystem, a strong marketing channel to the Asian region, and a rich socio-cultural atmosphere (Yigitcanlar & Dur, 2013).

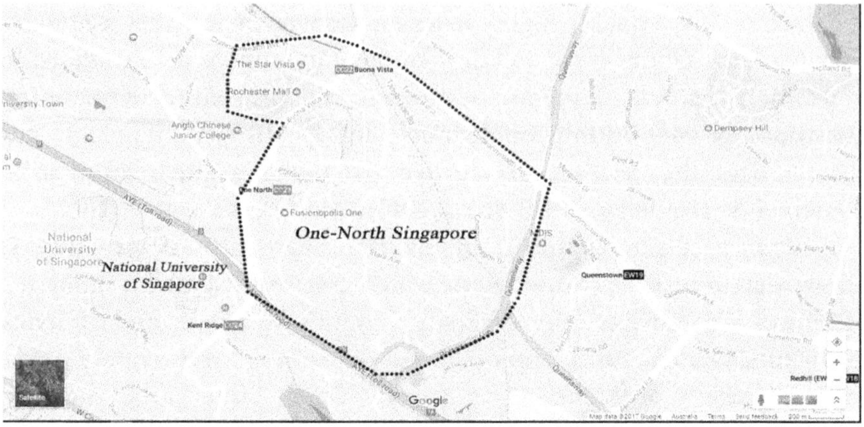

FIGURE 5.2 Google Maps (2 Oct 2017), One-North, Singapore (retrieved from https://www.google.com.au/maps/@1.3004151,103.7908645,15z).

Form: One-North, a 200-ha innovation district development, is strategically located in Singapore's technology corridor, between the Singapore CBD and the Nanyang Technological University (Figure 5.2). The location was carefully selected, considering infrastructures were already available—public transportation, roadways, R&D institutions, National University Hospital, Singapore Polytechnic, Science Parks. Zaha Hadid's master plan for One-North has targeted organic and cutting-edge architectures, natural features of the area, and mixed-use and flexible zonings. It aggregates offices, residents, shops, schools, recreational facilities, green spaces, and heritage sites in the same area. A winding green park also runs through One-North and works as a strategic element that links the seven districts of the area together by strong pedestrian networks and distinctive landscapes (Da Cunha & Selada, 2009; Pancholi *et al.*, 2015).

Function: In 1966, the Singaporean government decided to develop a Science-Culture-Business Park as an icon of Singapore's transition to a knowledge-based economy. In 2000, the management and co-ordination of One-North were delegated to a public agency as the master developer of anchor projects and infrastructures. The private sector has also developed up to 80% of One-North. To integrate public-private development, a flexible and non-continuous master plan was designed in three phases. Under this plan, One-North has provided world-class research facilities and business park spaces for biomedical sciences, ICT, media, physical sciences, and engineering sectors. The mixed land-use plan has also shaped an ideal work-live-play-learn environment (Carrillo *et al.*, 2014). However, One-North has strong technological capabilities and an expanded pool of expertise (www.jtc.gov.sg/industrial-land-and-space/pages/one-north.aspx).

Ambiance: The cultural ambiance is shaped by the presence of public art, cultural facilities and by preservation and regeneration of the existing cultural heritage, such as Rochester Park, Nepal Park, and the Wessex Estates. Furthermore, Singapore has a thick, culturally diverse, and mixed community. To support this inclusion, One-North provides different residential options in the area, such as the bohemian neighbourhoods of Holland Village and Ports-Down Road. It helps the ambiance of openness, personal freedom, and expressive life that have historically been denied to the Singaporean population. Also, Fusionopolis digital hubs, Biopolis R&D hub, and Metropolis technological hub, together, guarantee the presence of artists, inventors, and accordingly a creative atmosphere. Social and business networking is also guaranteed by a dense and dynamic urban design besides seamless networking infrastructures, which facilitate knowledge spill-overs and social interaction (Da Cunha & Selada, 2009).

Image: One-North is branded as a global talents hub. By neighbouring the historical sites, this global brand also takes benefits from the local heritages and cultural assets. For instance, the symbolic Rochester Park, with several double-storey black-and-white colonial bungalows, turned into a diverse dining and lifestyle retail hub consisting of restaurants, galleries, and spas. The authentic characteristic and image of One-North are shaped by the combination of this local heritage with modern arts and technologies in design. Private sectors, as the exact users of One-North, are also involved in this design and development, which gives them more sense of attachment and identity. Furthermore, Singapore has gained a good reputation for its low crime rate. The permeable and pedestrian-oriented design of One-North, besides the central green spine as the corridor of activities on the street level, supports this safe and secure image (Pancholi *et al.*, 2015).

5.3.3.2 Arabianranta, Helsinki

Context: Helsinki, with a long history of art and design, a thick pool of design-related professions, a high level of the talented and educated workforce, and strong international ties, has shaped a well-established innovation economy (Makkonen & Inkinen, 2014). To support this economy, the educational infrastructure in Helsinki closely co-evolves with the needs of design-driven industries such as furniture, pottery, ceramics, decoration, and fashion. Helsinki, today, is an international reference in design and applied arts—it was the European Capital of Culture in 2000—and also a European model in terms of variety, quality, and efficiency of services (Pelkonen, 2005; Van Winden *et al.*, 2007; Yigitcanlar & Lönnqvist, 2013).

Form: Arabianranta is located in an old industrial and waterfront site, originally occupied by the Arabia ceramics factory in 1874 (Figure 5.3). During the 1980s, the Helsinki Art and Design University (TiaK) was relocated to the unoccupied industrial buildings and the undeveloped shoreline was used for housing the production. The elements that inspire the urban and architectural design are based on linking the past and present, connecting the natural environment with organic urban grids, involving science and technology with art, and offering distinctive architectural style. Around these elements, U-shaped housing blocks are oriented to face the waterfront, with arms open towards the coastal park and embrace gardens, playgrounds, and public arts. Old industrial buildings have been artfully renovated and equipped with new technologies and infrastructures. Arabiantranta is also equipped with an excellent public transport, connected to the CBD and the rest of the city (Pancholi *et al.*, 2015).

Function: While the City of Helsinki and TiaK have been the main players, other private and investment companies have also participated

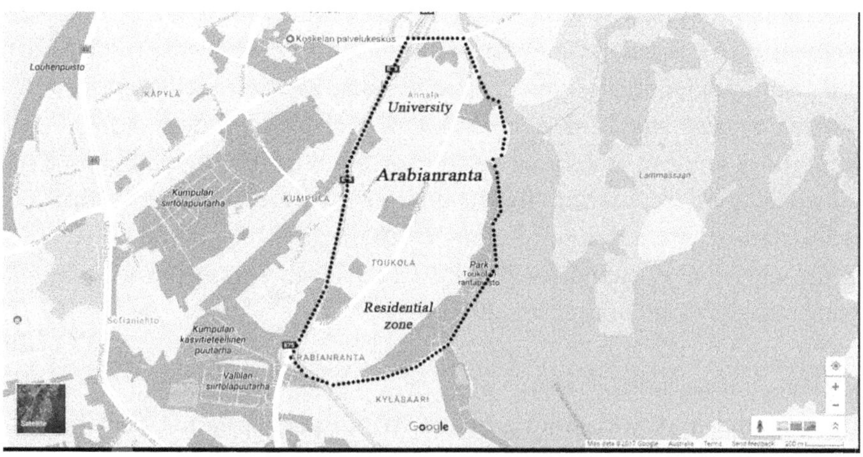

FIGURE 5.3 Google Maps (2 Oct 2017), Arabianranta, Helsinki (retrieved from https://www.google.com.au/maps/@60.2050264,24.9802493,14z).

in this large development project. Arabianranta is a mixed-use district, combining diverse residential areas, education and learning facilities, media, ICT and design businesses, commerce and services, and recreational areas. Concerning the talent environment, Arabianranta hosts 10,000 residents, 8,000 jobs, and 6,000 students to form a thick community of students, workers, artists, researchers, and residents (Pancholi *et al.*, 2015). As an advanced technology, Arabianranta is equipped with the first European fibre-optic network—Helsinki Virtual Village—to support the digital connectivity of these residents, companies, and educational institutes (www.arabianranta.fi/en/info).

Ambiance: Cultural facilities and events such as the library and information centre, the museums of design and industry, theatre organised by the Faculty of Culture of Stadia Polytechnic, and Arabia Street Festival play important roles in the area. Arabianranta also hosts Europe's first living labs—driven by Nokia, TaiK, and residents—which bridges the gap between market, research, users, and industries. The dynamic and creative ambiance of Arabianranta is assured by these living labs and also by the high presence of artists, art and design institutes, and creative companies on the site—Helsinki Pop and Jazz Conservatory, Arabus Design, Media and Art Business Centre, Designium Centre of Innovation in Design, Nokia. Furthermore, the presence of 200 pieces of artwork in the form of sculptures, installations, ceramics, and photographs within the site has created an artistic character (Carvalho & Van Winden, 2017).

Image: The initial concept for Arabianranta was a large urban park for living and leisure. This has gradually changed into the development of a multifunctional art and design city. Hence, art and design have remained a central statement in the development—even in living and leisure areas—to maintain its identity as a creative district. In order to make design integral and to shape the sense of authenticity, developers are encouraged to use distinctive architectural designs and unique art projects in public spaces. The identity and image of Arabianranta are also well grafted onto the historical memory of the site's industrial past. Regarding these creative potentials, Arabianranta aims to brand itself as 'the most leading design centre in the Baltic Sea area' (Pancholi *et al.*, 2015).

5.3.3.3 Down Under the Manhattan Bridge Overpass (DUMBO), New York

Context: New York City (NYC), as one of the most populous cities in the United States, is a major national centre for banking and finance, world trade, tourism, media, theatre, fashion, and the arts. Besides the city's world-renowned districts, bridges, skyscrapers, and parks, NYC is branded as the cultural, financial, and media capital of the world. Following the Silicon Valley, NYC is the second-leading high-tech hub in the nation, which has shaped a key economic boost, and left the city well positioned for future high-tech growth (Bowles, 2012). Particularly, the area of Brooklyn is becoming a flourishing ecosystem for this growth through an increasing number of tech, media, and creative firms; a strong socio-cultural diversity; an independent art scene; and a distinctive architectural heritage.

Form: DUMBO is located between two renowned Brooklyn and Manhattan bridges, along with the shoreline of Brooklyn, which is easily accessible by land or water (Figure 5.4). The area was originally a ferry landing, formed by large-scale reinforced concrete industrial and warehouse buildings. Dumbo was bought by 'Two Trees Management' in the late 1990s and transformed into a dense, classy residential and commercial area. Old buildings renovated into lofty, flexible spaces which first became a heaven for the existing arts community and currently a centre for media companies and technology start-ups (Hackworth, 2002). DUMBO is also close to well-reputed universities and two major high-tech districts—Brooklyn Navy Yard and Downtown Brooklyn, which together are known as 'Brooklyn Tech Triangle' (BTT). Due to the land limitations, DUMBO shares infrastructures, work, learn, and play spaces, and also residential areas with BTT.

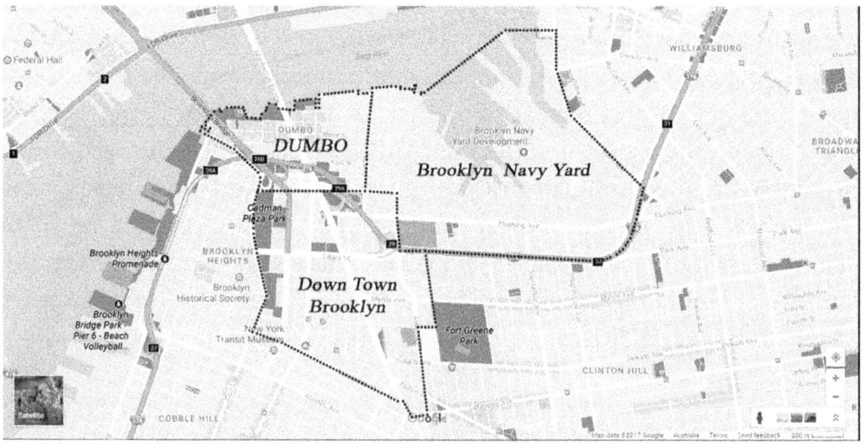

FIGURE 5.4 Google Maps (2 Oct 2017), DUMBO, New York (retrieved from https://www.google.com.au/maps/@40.6963098,-73.9799743,15z).

Function: DUMBO is home to hundreds of high-tech and creative companies and digital start-ups within only a 10-block radius (Camay *et al.*, 2012). The challenge of nurturing this explosion of innovation requires ensuring that the right type of space, amenities, and work environment are available for them. The Dumbo Improvement District (DID) is a non-profit organisation dedicated to support this long-running development and promotes DUMBO as a world-class district. The state and city governments are also investing in new infrastructures, major amenities, parks, and transportation to provide tools for a vibrant growth. Meanwhile, under BTT's master plan and DID supervision, the private sector is developing new residential and commercial constructions to overcome land limitations and create space for the district to grow. Besides, DUMBO is nurtured by BTT's wide range of knowledge and creative workers guaranteeing a thick and dynamic labour market.

Ambiance: Considering a cutting-edge cultural sense right inside a historic site, DUMBO hosts some of the world's most innovative companies, artists, and journalists. People who live in DUMBO tend to be highly creative and open-minded. They are inspired by the artistic ambiance of the DUMBO area, which is activated by fancy socio-cultural facilities and events (e.g., Barclays Centre in the Cultural District, Klompching Gallery, DUMBO Arts Festival, Flea Food). BTT plans to enhance networking between Brooklyn's educational institutions and the high-tech sector, which brings the right kind of training, enables new start-ups, and creates jobs (Harvey *et al.*, 2013).

Image: DUMBO is physically characterised by the bridges that cut across the area, causing some constraints of walking and cycling connectivity. It also creates a series of unique and dynamic public spaces—such as Cadman Plaza and world-class Brooklyn Bridge Park. These parks and plazas, combining with the view of Brooklyn Bridge and historic industrial buildings have formed the unique physical identity of DUMBO. Moreover, day-to-day life in DUMBO is often accompanied by film shootings and other entertainment activities—such as festivals, dynamic nightlife, and plenty of restaurants, shops, and bars (Harvey *et al.*, 2013). However, this appealing image of DUMBO is threatened by a high local crime rate. To turn DUMBO into a safe place, currently, more attention has been paid to lighting the area at night and to bring activities at street level (www.brooklyntechtriangle.com).

5.3.3.4 Macquarie Park Innovation District (MPID), Sydney

Context: Sydney, as the largest, most globalised and populated Australian city, is an important contributor in the global economy. Although the city is dominated by tourism, finance and insurance sectors, and business and property services, there are also a number of districts specialising in creative industries, such as health and biotechnology sectors (see Chapter 4). These districts are fostered by the world-class education and research institutions of Sydney, which provides a thick community of qualified workforce in the city. However, the high quality of living in Sydney also attracts international knowledge workers, particularly from the Asia-Pacific region (Yigitcanlar & Dur, 2013).

Form: Macquarie Park is a suburb in northern Sydney, established in the mid-1960s to rezone the Green Belt bushland for industrial activities (Figure 5.5). The initial concept was similar to the high-tech industrial areas around Stanford University that facilitated the interaction between industries and the university. Macquarie University and a surrounding business area were formed by low-scale and brutalist-style buildings, with a large footprint in the bushland. Since then, MPID has been following a long-term development plan to gain its current position as Sydney's second-largest business district. To support this growth, the urban form has become denser and equipped with leading-edge infrastructures, diverse building types, high-tech architectures, eco-friendly designs, and a range of sports facilities, parks, and open spaces

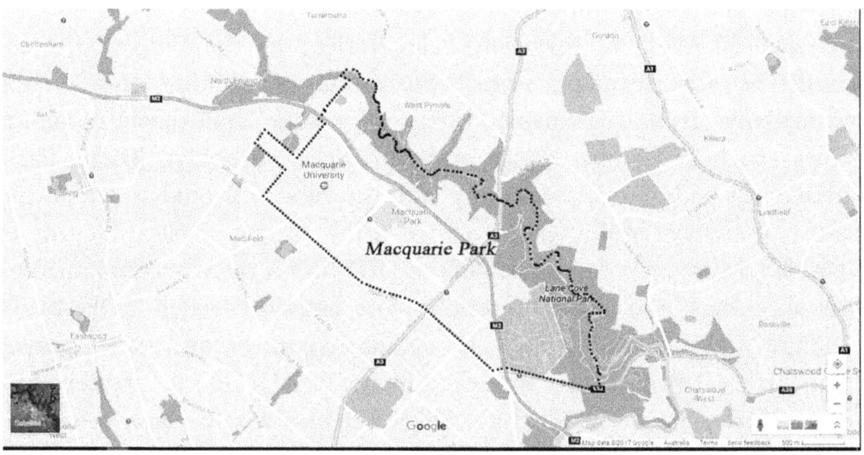

FIGURE 5.5 Google Maps (2 Oct 2017), MPID, Sydney (retrieved from https://www.google.com.au/maps/@-33.7787559,151.1087734,14z).

for leisure time. MPID is also transforming from a car-dependent district into a vibrant, pedestrian and public-transport oriented one (Pancholi *et al.*, 2017). However, in MPID, the leafy green feel and sense of doing business in a bushland are still central.

Function: During 1970, some internationally renowned high-tech companies were located in the MPID as a leading high-tech industrial area of Australia. This reputation has attracted other high-tech businesses—including electronic, scientific, computing, medical, communication, pharmaceutical, and business supply solutions—and shaped a large business district (McNeill *et al.*, 2005). While the state government initially invested in MPID, the major current developers are Macquarie University, Ryde local organisations, and the private sector (see Chapter 4). For instance, the Campus Master Plan and the Ryde Local Plan for the Macquarie Park Corridor have been designed to support the current mixed land use, affordability of property, and also the collaborative environment between research and technology. The objective is to provide 5,800 homes and 40,000 additional jobs by 2031 to help Sydney's growth needs.

Ambiance: Besides various cultural facilities such as libraries, museums, collections, and art galleries of Macquarie University, there is always something happening in bustling MPID. It is easy to combine work and play by joining in vibrant social events. Moreover, the creation of passive spaces has improved opportunities for recreation and networking with other workers to enhance the sense of community. To improve business-networking system, University's Learning Centre and also Network Management Centre cover people's technical communication needs and allow teams from across the districts to work closely (see Chapter 4). The district's open and creative ecosystem is supported not only by these flourishing business systems and high level of inventions, but also by the cultural and ethnic diversity, and the presence of art in public places (www.macquariepark.com.au; www.mpid.com.au).

Image: MPID offers plenty of lifestyle and entertainment options, from camps, barbeques, and bushwalks to a wide range of shopping, dining, and state-of-the-art sports and other leisure facilities. Activating some of these facilities and other essential services at the street level, especially around transport nodes and the new central corridor also helps to create a safe and vibrant place day and night. Despite the safe atmosphere, corporate prestige of MPID and access to such facilities, the attractive image of MPID is unique and an aesthetically pleasing environment. Public arts also help to reflect local character, cultural identity, and the natural environment to create a distinctive sense of place (see Chapter 4).

5.4 RESULTS

Descriptive analysis of the results of the case studies reinforces the initial theoretical structure of the research, which claims context, form, function, ambiance, and image have significant roles in shaping place quality in innovation districts. For a quantitative content analysis—or lexical analysis—the coded data have been uploaded in N-Vivo to calculate the word count. The word frequency of coded data has demonstrated at what degree each category—i.e., form, function, ambiance, image—has been influenced by certain elements. Due to this, research needs a limited set of indicators to simplify the next stage of the study, only the most four frequent words—or containing their synonyms—have been chosen—e.g., location, urban form, design, and amenities in the category of 'form'. Table 5.3 shows the results of the quantitative content analysis and displays the selected keywords—or indicators.

The qualitative explanatory analysis is done based on the outcome of the quantitative content analysis and the identified indicators. It clarifies how indicators shape and influence place quality in innovation districts. The collected data for each case study have been recoded again and categorised under the new set of indicators. This facilitates the comparative and explanatory interpretation of data. Results of the comparative and explanatory analyses show that although each case, certainly, has unique characteristics, all cases have a lot of similarities. Table 5.4 illustrates similar strategies and characteristics that all four case studies have followed to create an attractive place quality for their knowledge industries and workers. Nevertheless, besides commonalities among the case studies, each district has its own unique tactic and character tailoring the ongoing place quality. These distinctive tactics and characters are presented in Table 5.5. The comparative analysis of mentioned similarities and differences explains the following inductive findings.

Context: Results show that the growth of innovation districts is a co-evolutionary process between potentials of urban contexts, governmental knowledge-based strategies, and the district's development itself (Carvalho & Van Winden, 2017). Creative districts are more deeply rooted in socio-cultural abilities of their context (Arabianranta), often

TABLE 5.3 Results of the Quantitative Content Analysis with N-Vivo

Category	Most Frequent Concepts and Relevant Words	Frequency
Form	Location: Area (70), space (52), zoning (38), connectivity (49), environment (33)	242
	Urban form: Construction and establishment (84), plan (47), urban (18), boundaries (20)	169
	Design: Design (180), project (43), open spaces (47), landscape (5), sustainable (9)	284
	Amenity: Park (103), infrastructure (37), services (33), facilities (22), transport (20)	215
Function	Management: Development (95), public (56), local (33), government (57), leading (39)	280
	Land use: Housing (63), mixed (37), office (52), available (38), resident (25), land (35)	250
	Talent: Work (90), companies (32), society (37), community (31), talent (11), worker (10)	211
	Technology: Business (69), university (71), tech (42), technology (40), industrial (35)	257
Ambiance	Cultural milieu: Cultural (39), present (30), event (18), scene (14), heritage (6), festival (6)	113
	Networking: Network (21), social (18), involve (25), support (36), coordinating (35)	135
	Diversity: Diversity (23), vibrant (10), range (20), tourism (5), foreign (5), variety (7)	70
	Creativity: Art (45), create (45), research (28), innovation (64), knowledge (19), artist (10)	211
Image	Lifestyle: Living (56), street (18), dynamic (26), activities (21), dinning (6), leisure (18)	145
	Safety: Safe (14), control (25), lighting (7), crime (3), pedestrian (7)	56
	Sense of place: Home (35), attached (5), integrated (31), historic (12), view (13)	96
	Identity: Original (20), distinctive (17), unique (9), making (53), brand (6), global (19)	124

boundary-blurred and embedded in urban life (DUMBO). Technology districts are more likely to appear in isolated areas, solely by nourishing from outstanding research centres (MPID). However, all types of creative districts take advantage of the city-state strategies, which seek to create a global successful economy, a high quality of life, a supportive start-ups environment, and a dense pool of workforce (Hazelkorn & Murphy, 2002).

TABLE 5.4 Common Attributes of Place Quality in Innovation Districts

CONTEXT	
Regional and city quality	Having worldwide reputation in knowledge economy and KBUD, which support the reputation of innovation districts
	Having strong plans for developing and promoting innovation districts inside their boundaries
	Having high levels of tourist, famous universities, educated workforce, and ethnic and cultural diversity
FORM	
Location	Well-connecting to important parts of the city—i.e., CBD, airport, other districts, research centres, universities
	Surrounding by unique natural or built environment—i.e., waterfront sites, bushlands, historical or old industrial sites
Urban form	Avoiding sprawling development to support the dynamic atmosphere and sustainable urban development
	Considering the natural features of the area, a mixed-use and a flexible zoning
	Focusing on pedestrian, bike and public-transport oriented planning rather than car-dependent ones
Design	Emphasising unique, diverse, high-quality, and cutting-edge urban and architectural design—i.e., in workplaces
	Designing the ground floor more accessible to all to keep the connection with streetscape
	Respecting environmental-friendly designs and putting the natural environment in the centre
Amenities	Ensuring the accessibility to basic facilities—i.e., schools, hospitals, and elderly-/child-care facilities, well-served stores
	Providing advanced facilities—i.e., cycling and jogging areas, well-equipped sports grounds, cafés, restaurant, bars, free Wi-Fi
	Providing high-quality and smart transport systems and fast and reliable ICT infrastructure
FUNCTION	
Management	Involving public and governmental agencies to become responsible for developing infrastructures and anchor projects—i.e., roads and utilities
	Involving private sector to become responsible for building their own blocks based on the master plan
Land use	Aggregating different activities in the same area to shape a place to live, work, learn, and play with quite blurred boundaries
	Controlling the value and availability of residential and commercial properties and also workspaces

(*Continued*)

TABLE 5.4 *(Continued)* Common Attributes of Place Quality in Innovation Districts

Talent	Expanding the pool of knowledge workers through a targeted education system in allied universities and research centres
	Attracting local and foreign knowledge workers through offering variety of work opportunities and high-quality lifestyles
Technology	Attracting and retaining domestic and international knowledge-based firms and industries by offering a variety of state-of-the-art infrastructures—i.e., providing advanced construction of communications technology and digital structure
	Hosting well-known research centres that facilitate development and spill-over of knowledge
	Hosting well-known knowledge-based companies that help the reputation of the districts
AMBIANCE	
Cultural milieu	Developing social and cultural spaces which are accessible to all people—i.e., meeting places, cultural centres, museums, galleries, cinemas, libraries, theatres
	Planning events to involve large groups of people—i.e., live performance venues, cultural festivals
	Characterising the cultural environment by the presence of public art, and preservation and regeneration of the existing cultural heritage and sites
Networking	Facilitating business interactions and relationships to ease the spill-over of knowledge—i.e., developing shared workspaces for peer networks
	Facilitating social interactions to enhance the sense of community—i.e., developing social communities, public spaces
	Facilitating interaction of people and firms to develop a dynamic market and to link talent and firms
Diversity	Hosting and accepting different types of people and lifestyles—i.e., people from diverse ethnics and cultures, LGBT community, bohemians
Creativity	Enhancing the artistic, cultural, and technological creative atmosphere—i.e., high presence of artists and inventions
	Enhancing the presence of art within the site in the form of sculptures, paintings, and so on to support the creative ambiance
IMAGE	
Lifestyle	Providing different types of entertainment and activity, which support the diversity of lifestyle

(Continued)

TABLE 5.4 *(Continued)* Common Attributes of Place Quality in Innovation Districts

	Providing dynamic and active day and nightlife in the street level to support the district's buzz
Safety	Activating the street level with food, entertainment options, and dynamic events to create a safe place day and night
	Providing an efficient lighting system in streets and public areas
Sense of place	Promoting the quality of being unique or unusual—i.e., valuing monuments, historic buildings, unique architectures, distinctive buzzes, or unique natural environments
Identity	Making a brand for the district to support a unique identity and marketing it globally and effectively

Form: Inner-city districts (DUMBO, One-North) gain benefit from density, diversity, and the innovative scene of the inner urban areas, the existed infrastructures and the labour pool, as well as being located in close proximity to other business and innovation districts (Flew, 2012). Suburban districts (Arabianranta, MPID) are deprived of such conditions and needed to spend more money and time on developing their fortune. Planning a university in the centre in order to attract educated workforce, build innovative atmosphere and enrich the social life could support such development. Moreover, technology districts (MPID, One-North) are more likely to settle in vacant sites; though, creative districts (Arabianranta, DUMBO) prefer sites with unique backgrounds and architectures—e.g., old industrial sites (He & Gebhardt, 2014). Evaluating the open green spaces, architectural, and urban design are also important in enhancing wellbeing and satisfaction, safety, security, and shaping the physical identity of innovation districts (Loures *et al.*, 2007).

Function: Due to the complex socio-cultural and economic character of place quality, districts require time and dense mix of functions to evolve. In the planning stage, the local government is the main coordinator to construct the long-term objectives and define the overall urban structure and its functional programme (Kloosterman & Trip, 2011). However, in the development stage, a public-private partnership is an efficient mechanism to provide the exact requirements of users and building a 'sense of attachment'. While inner-city districts take benefits from an existing mix of functions and land uses, they usually suffer from

TABLE 5.5 Distinctive Attributes of Place Quality in Innovation Districts

	Singapore	**Helsinki**	**New York**	**Sydney**
CONTEXT				
Regional and city quality	World's best start-up ecosystems	European capital of culture (for year 2000)	The second-leading tech hub in the United States	The first-leading tech hub in Australia
	Flourished tourism system	Famed design background	Prominent business districts	Renowned universities and research centres
	Growing population of foreign and local qualified workers	Growing population of local qualified workers	Most populous city in the United States with educated population	High rate of foreign-born and educated people
	One-North	**Arabianranta**	**DUMBO**	**MPID**
FORM				
Location	Inner city	Suburban	Inner city	Suburban
	Bushland	Waterfront	Waterfront	Bushland
	Close to other districts	Close to a university	Close to other districts	Close to a university
	Started in an empty site	Old industrial site	Old industrial site	Started in an empty site
Urban form	Organic grid, following natural morphology	Both distinctive and organic urban grid	Consolidated and historic urban grid	Distinctive urban grid
	Shaped along a central green spine in seven nodes	Repetitive sprawling along a shoreline	Compacted between two bridges	Continuous sprawling beside a green line
	High density	Medium density	High density	Medium density
Design	Futuristic, organic and Southeast Asian styles	Combining modern and old styles	Luxury renovation of industrial buildings	Brutalism and high-tech styles

(Continued)

TABLE 5.5 *(Continued)* Distinctive Attributes of Place Quality in Innovation Districts

	Organic green spaces—i.e., keeping bushlands	Designed green spaces, i.e., shoreline park	Designed green spaces, i.e., shoreline park	Organic green spaces—i.e., keeping bushlands
Amenities	Entertainment villages—i.e., Rochester Park and Holland Village	Natural potentials—i.e., rich bird life	Classy leisure facilities—i.e., dogs' park, tech terrace	Natural potentials—i.e., national park
FUNCTION				
Management	Public-private partnership (planned)	Public-academia-private partnership (planned)	Public-private partnership (unplanned)	Public-academia- private partnership (planned)
	Public agency led	University led	Public agency led	Public agency led
Land use	Tech-culture-business	Art-design-business	Media-ICT-business	Science-tech-business
	High-tech district	Creative district	Creative district	High-tech district
Talent	40,000 residents, 70,000 workers	10,000 residents, 8,000 jobs, 6,000 students	10,000 workers, 57,000 students in Tech-Triangle	32,000 jobs, 40,000 students
	Qualified foreign workers and researchers	Mixed community of students	Thick pool of jobs and workers in Tech-Triangle	Thick pool of foreign and local students and workers
Technology	Dense technological environment (metropolis)	Cutting-edge Helsinki Virtual Village	Cutting-edge digital and ICT facilities	Flourish research environment
AMBIANCE				
Cultural milieu	Local cultural scene	Local cultural scene	Open cultural scene	Multicultural scene
	Historic sites and arts	Historic sites and arts	BTT's cultural district	Diverse cultural centres

Networking	Agglomerative effect (firms networking)	Living labs (users and firms networking)	Tech-ecosystem (firms and talents networking)	Passive spaces (talents networking)
Diversity	Bohemian Holland Village	Low ethnic diversity	Bohemian neighbourhood	High ethnic diversity
	High foreign tourism	Low foreign tourism	High foreign tourism	High foreign tourism
Creativity	High level of media and design industries (Fusionopolis)	High level of art and design firms and campuses	High level of media industries and film products	Low level of creative industries
	Moderate percentage of artworks and inventions	High population of artists	High population of artists and bohemians	High percentage of inventions
IMAGE				
Lifestyle	Rochester dining and lifestyle hub (youth style)	Living and leisure inside an urban park (family style)	Open, artistic, and bohemian styles (artistic style)	Living and leisure inside an urban park (family style)
Safety	Low crime rate	Low crime rate	High crime rate	Low crime rate
	High permeability by pedestrianisation	Promote social control by a U-shape design	Redevelopment of lighting and risky places	Vitalise the nightlife by a lively commercial corridor
Sense of place	Unique historic and cultural assets, and touristic nature	Unique historic, artistic, and natural asset	Unique industrial overtone and bridges' views	Unique natural environment
Identity	Talent hub	Leading design centre in the Baltic Sea area	Centre of the Brooklyn Tech Triangle	Biggest innovation district in Australia

the limitation of land (DUMBO)—unlike suburban districts. Local universities also support the mix of function, cutting-edge research, and technology environment that attracts more firms, start-ups, and talents (Sohn & Kenney, 2007; Youtie & Shapira, 2008; Mian *et al.*, 2012).

Ambiance: Soft infrastructure of arts and culture is mostly a low-cost tool to shape a creative ambiance—rather than high-cost big events or fixed high-tech infrastructures. There is a high presence of artists, bohemians, tourism, cultural events, and street arts in creative districts (Arabianranta, DUMBO) and in districts with rich cultural heritage (One-North). Therefore, they are more likely to shape an attractive ambiance by spending less money and time in comparison to high-tech suburban districts—i.e., MPID. It seems despite the placeless nature of knowledge product, knowledge industries and workers still seek to agglomerate in innovation districts; perhaps due to the ambiance of place-based, face-to-face, social and business interactions (Durmaz, 2015). Furthermore, closeness to universities shapes an open ambiance by attracting a variety of people with different backgrounds of ethnicity, race, national origin, age, social class, and sexual orientation.

Image: New generations of innovation districts, along with their mixed functions, are being considered as homes for many firms and talents. Accordingly, there is a higher chance to develop a welcoming image and a stronger place attachment (Yigitcanlar & Dur, 2013). Likewise, authenticity can be helpful in forming a distinctive image and identity in innovation districts. Authenticity is a unique sense of place in the form of physical appearance, ongoing activities, and symbolic values (Heebels & Van Aalst, 2010; Sepe & Pitt, 2014). For instance, DUMBO, with picturesque views of Brooklyn Bridge and with a unique street buzz and industrial overtone, or MPID, with a unique natural environment, are shaping unforgettable images in minds. Moreover, to enhance the competitive edges of innovation districts, and shape a globally renowned brand, having an image of safety is also essential.

5.5 DISCUSSION AND CONCLUSION

From a theoretical point of view, findings of this study support the existing claims on the significant role of place quality in shaping and enhancing competitive advantages of innovation districts (Trip, 2007; Kloosterman & Trip, 2011). Findings suggest important contributions of both hard and soft factors in creating attractive innovation districts. While traditional hard factors are no longer the sole driving force of innovation districts, Florida's (2005) soft indicators are also insufficient by themselves. In order to address this issue, the proposed conceptual framework of place quality has been developed on the basis of existing literature and practice. This framework underpins the critical place quality matters in innovation districts through its five key dimensions—i.e., context, form, function, image, ambience—along with their 17 indicators (Table 5.5). Furthermore, the framework has been applied to four global best practices to reveal how place quality contributes to the attractiveness of innovation districts.

From a practical point of view, findings explain the generic characteristics that could be implemented by other innovation districts (Table 5.4), but always keeping the local contextual charateristics in mind. These characteristics include choosing well-connected locations to CBD and universities; respecting natural and artificial surroundings; avoiding sprawl; forming a people-centred urban structure; providing a unique, diverse, and high-quality urban and architectural design; following a public-private development plan; shaping a thick pool of knowledge workers and jobs; offering state-of-the-art infrastructure; valuing the ongoing cultural and symbolic potentials; enhancing social and business face-to-face interactions through third spaces; supporting diversity of people, ideas, and lifestyles; making a dynamic street level; offering safe and secure environment; and promoting place attachment by highlighting authenticities.

Findings also describe the distinctive characteristics that shape the unique identity of innovation districts (Table 5.5). These set of characteristics mainly involve with local conditions of innovation districts, which need to be investigated individually and separately—e.g., existing business potentials and infrastructures. As an example, the research has detected the significant difference between the characteristics of the inner city and suburban districts, as well as creative and high-tech's characteristics. Davis *et al.* (2009) confirmed this claim by stating that creative industries significantly differ from high-tech ones with regard to their association with geography. Creative districts are characterised by a peculiar mixture of cultural content and technological elements, relying heavily on existing potentials of their location. In comparison with high-tech industries, creative industries are more deeply embedded into the local urban environment. In return, their production creates symbolic value that can enhance place quality and facilitate city branding (He & Gebhardt, 2014). As a result, creative districts are more likely to be found in the heart of cities, which are highly beneficial for both creative industries and cities.

Another finding demonstrates that place quality is a dynamic and ongoing phenomenon, which consumes time and requires flexible approaches. Due to the tight link between place quality and unfixed expectations of the knowledge industries and workers, it would be impossible to predict an exact place quality and develop such a detailed achievable plan. An attractive place quality is a dynamic function, which could be reshaped and modified frequently during the districts' life span. To support the dynamic nature of place quality in innovation districts, this research suggests less fixed development strategies, more cooperative plans with knowledge industries and workers, and continual revising sessions. Cooke (2001) has emphasised the requirement of such flexible and interactive strategies for promoting the knowledge economy. He argues that as the innovation processes amongst firms and scientific organisations have shifted from linear to interactive; the strategies for promoting innovation have also moved from a hierarchical to a more networked one. Therefore, an interactive model is the only solution for fostering and enhancing innovation districts in the long term.

As discussed earlier, place quality is a complex phenomenon; hence, this chapter is one of the pioneering explorations of the place quality dynamics in the context of innovation districts. A limited set of place quality indicators were selected to simplify the process of interpretation and to test their potential in explaining how place quality contributes to the attractiveness of innovation districts. Nevertheless, in interpreting the specific findings of the research, the reader must be aware of the following limitations: (a) The first and the most conflicting limitation is that the results highlighted in this study are based on a rather limited number of case studies—only four. However, while these case studies are good examples of successful innovation districts around the world, it should be also considered that characteristics of innovative districts are highly place specific, and hence difficult to generalise. (b) The results provide a general perspective and some guidelines for planning, designing, developing, and managing innovation districts. However, these guidelines need to be adapted to the unique conditions of individual districts. (c) The study employed N-Vivo software for objective coding and content analysis. However, there still might be some biased in interpretations. (d) The study attempted to use the most reliable and up-to-date data as much as possible. However, the data that is not current or valid might deflect the findings. (e) The study only considered the most significant and generic indicators. However, indicators assumed as lesser importance or more specific to a local context might need to be reconsidered. (f) The study did not consider the impacts or weighting of indicators in shaping place quality. Further studies are required to address this issue—along with a Delphi study to assess the adequacy of the indicators.

This chapter explored a possible approach to understand how place quality contributes to the attractiveness of innovation districts by defining some preliminary indicators. The proposed framework is only a starting point for future studies to find more fine-tuned methods or metrics for evaluating the place quality in innovation districts precisely—and consequently developing actions to improve performance. The research reported in this chapter has the potential to be expanded and applied to the real-world practice, especially in less successful innovation districts that are hoping to improve their competitive edges. Evaluating the place quality in real-world innovation districts practice will reveal detailed insights into their strengths, weaknesses, and potentials. Furthermore, this chapter focused on determining generic indicators of place quality to inform policy, design, plan making, and management processes in the district (cluster) scale. Nevertheless, as place quality in innovation districts is not isolated from the broader context, future studies exploring place quality matters at the city and regional scales, for improving the attractiveness of innovation districts, are also required (Esmaeilpoorarabi *et al.*, 2018).

ACKNOWLEDGEMENTS

This chapter, with permission from the copyright holder, is a reproduced version of the following journal article: Esmaeilpoorarabi, N., Yigitcanlar, T., & Guaralda, M., (2018). Place quality in innovation clusters: An empirical analysis of global best practices from Singapore, Helsinki, New York, and Sydney. Cities, 74(1), 156–168.

REFERENCES

Baum, S., O'Connor, K., & Yigitcanlar, T. (2009). The implications of creative industries for regional outcomes. *International Journal of Foresight and Innovation Policy*, 5(1), 44–64. DOI:10.1504/IJFIP.2009.022098

Benneworth, P., & Ratinho, T. (2014). Reframing the role of knowledge parks and science cities in knowledge-based urban development. *Environment and Planning*, 32(5), 784–808. DOI:10.1068/c1266r

Bereitschaft, B., & Cammack, R. (2015). Neighborhood diversity and the creative class in Chicago. *Applied Geography*, 63(1), 166–183. DOI:10.1016/j.apgeog.2015.06.020

Bowles, J. (2012). *New tech city*. Centre for an Urban Future, New York.

Bontje, M., & Musterd, S. (2009). Creative industries, creative class and competitiveness. *Geoforum*, 40(5), 843–852. DOI:10.1016/j.geoforum.2009.07.001

Borén, T., & Young, C. (2013). Getting Creative with the 'Creative City'? Towards New Perspectives on Creativity in Urban Policy. *International Journal of Urban and Regional Research*, 37(5), 1799–1815.

Brocker, J., Dohse, D., & Soltwedel, R. (Eds.). (2012). *Innovation clusters and interregional competition*. Springer, Berlin. DOI: 10.1007/978-3-540-24760-9

Brown, J., & Mczyski, M. (2009). Complex cities: locational choices of creative knowledge workers. *Built Environment*, 35(2), 238–252. DOI:10.2148/benv.35.2.238

Camay, S., Zielinski, E., & Zaranko, A. (2012). New York City's east river ferry. *Transportation Research Record*, 2274(1), 192–200. DOI:10.3141/2274-21

Carrillo, F.J., Yigitcanlar, T., García, B., & Lönnqvist, A. (2014). *Knowledge and the city*. Routledge, New York. DOI:10.4324/9781315856650

Carvalho, L., & Van Winden, W. (2017). Planned knowledge locations in cities. *International Journal of Knowledge-Based Development*, 8(1), 47–67. DOI:10.1504/IJKBD.2017.082429

Clifton, N., & Cooke, P. (2009) Creative knowledge workers and location in Europe and North America. *Creative Industries Journal*, 2(1), 73–89. DOI:10.1386/cij.2.1.73/1

Cooke, P. (2001). From technopoles to regional innovation systems. *Canadian Journal of Regional Science*, 24(1), 21–40.

Cooke, P. (2017). Complex spaces. *Journal of Open Innovation*, 3(2), 9. DOI: 10.1186/s40852-017-0060-5

Da Cunha, I.V., & Selada, C. (2009). Creative urban regeneration. *International Journal of Innovation and Regional Development*, 1(4), 371–386. DOI:10.1504/IJIRD.2009.022728

Darchen, S., & Tremblay, D.G. (2010). What attracts and retains knowledge workers/students. *Cities*, 27(4), 225–233. DOI:10.1016/j.cities.2009.12.009

Davis, C. H., Creutzberg, T., & Arthurs, D. (2009). Applying an innovation cluster framework to creative industry. *Innovation, Management, Policy & Practice*, 11(2), 201–214. DOI:10.5172/impp.11.2.201

D'Mello, M., & Sahay, S. (2007) I am kind of a nomad where I have to go places and places. *Information and Organization*, 17(3), 162–192. DOI:10.1016/j.infoandorg.2007.04.001

Durmaz, S.B. (2015). Analyzing the quality of place. *Journal of Urban Design*, 20(1), 93–124. DOI:10.1080/13574809.2014.972348

Eisenhardt, K.M. (1989). Building theories from case study research. *Academy of Management Review*, 14(4), 532–550. DOI: 10.2307/258557

Emmanuelle, N., & Isabelle M. (2013). A lexical analysis of voluntary press releases issued during takeover bids. *Comptabilité-Contrôle*, 2(19), 71–100. DOI: 10.3917/cca.192.0071

Esmaeilpoorarabi, N., Yigitcanlar, T., & Guaralda, M. (2016a). Place quality and urban competitiveness symbiosis? *International Journal of Knowledge-Based Development*, 7(1), 4–21. DOI:10.1504/IJKBD.2016.075444

Esmaeilpoorarabi, N., Yigitcanlar, T., & Guaralda, M. (2016b). Towards an urban quality framework. *International Journal of Knowledge-Based Development*, 7(3), 290–312. DOI:10.1504/IJKBD.2016.078556

Esmaeilpoorarabi, N., Yigitcanlar, T., & Guaralda, M. (2018). Place quality in innovation clusters: an empirical analysis of global best practices from Singapore, Helsinki, New York, and Sydney. *Cities*, 74(1), 156–168. DOI: 10.1016/j.cities.2017.11.017

Evers, H.D. (2008). Knowledge hubs and knowledge clusters. *ZEF Working Paper Series*, 27(1), 1–25. DOI:10.1142/9789814343688_0002

Flew, T. (2012). Creative suburbia. *International Journal of Cultural Studies*, 15(3), 231–246. DOI:10.1177/1367877911433746

Florida, R. (2005). *Cities and the creative class*. Routledge, New York. DOI:10.4324/9780203997673

Frenkel, A., Bendit, E., & Kaplan, S. (2013a). Residential location choice of knowledge-workers. *Cities*, 35(1), 33–41. DOI:10.1016/j.cities.2013.06.005

Frenkel, A., Bendit, E., & Kaplan, S. (2013b). The linkage between the lifestyle of knowledge-workers and their intra-metropolitan residential choice. *Computers, Environment and Urban Systems*, 39(1), 151–161. DOI:10.1016/j.compenvurbsys.2012.09.001

Glaser, B.G., & Strauss, A.L. (1967). *The discovery of grounded theory*. Aldine, Chicago.

Grant, J.L., & Buckwold, B. (2013). Precarious creativity: immigrant cultural workers. *Cambridge Journal of Regions, Economy and Society*, 6(1), 1–17. DOI:10.1093/cjres/rss008

Hackworth, J. (2002). Postrecession gentrification in New York City. *Urban Affairs Review*, 37(6), 815–843. DOI:10.1177/107874037006003

Hafeez, K., Foroudi, P., Dinnie, K., Nguyen, B., & Parahoo, S.K. (2016). The role of place branding and image in the development of sectoral clusters. *Journal of Brand Management*, 23(4), 383–402. DOI:10.1057/bm.2016.18

Hall, P. (1996). Revisiting the non-place urban realm: have we come full circle? *International Planning Studies*, 1(1), 7–15. DOI:10.1080/13563479608721640

Harvey, N., Jaynes, C., Khuong, Y., & Riscica, V. (2013). Framework for innovative public spaces. *Transportation Research Record*, 2350(1), 119–127. DOI:10.3141/2350-14

Hazelkorn, E., & Murphy, C. (2002). The cultural economy of Dublin. *Sociological Chronicles*, 3(1), 1–10.

He, J.L., & Gebhardt, H. (2014). Space of creative industries. *European Planning Studies*, 22(11), 2351–2368. DOI:10.1080/09654313.2013.837430

Heebels, B., & Van Aalst, I. (2010). Creative clusters in Berlin. *Geografiska Annaler*, 92(4), 347–363. DOI:10.1111/j.1468-0467.2010.00357.x

Homan, S. (2014). Liveability and creativity. *City, Culture and Society*, 5(3), 149–155. DOI:10.1016/j.ccs.2014.06.002

Kloosterman, R. C., & Trip, J.J. (2011). Planning for quality? *Journal of Urban Design*, 16(4), 455–470. DOI:10.1080/13574809.2011.585863

Lawton, P., Murphy, E., & Redmond, D. (2013). Residential preferences of the creative class? *Cities*, 31(1), 47–56. DOI:10.1016/j.cities.2012.04.002

Lee, J.H., Hancock, M.G., & Hu, M.C. (2014). Towards an effective framework for building smart cities. *Technological Forecasting and Social Change*, 89(1), 80–99. DOI:10.1016/j.techfore.2013.08.033

Lonnqvist, A., Kapyla, J., Salonius, H., & Yigitcanlar, T. (2014). Knowledge that matters. *European Planning Studies*, 22(10), 2011–2029. DOI:10.1068/a1302

Loures, L., Santos, R., & Panagopoulos, T. (2007). Urban parks and sustainable city planning. *Transactions on Environment and Development*, 3(10), 171–180. DOI:10.1016/j.proenv.2016.04.042

Makkonen, T., & Inkinen, T. (2014). Geographical and temporal variation of regional development and innovation in Finland. *Fennia-International Journal of Geography*, 193(1), 134–147.

McCann, E.J. (2008). Livable city/unequal city. *Revue Interventions Économiques*, 37(1), 1–18. DOI: 10.4000/interventionseconomiques.489

McNeill, D., Dowling, R., & Fagan, B. (2005). Sydney/global/city. *International Journal of Urban and Regional Research*, 29(4), 935–944. DOI:10.1111/j.1468-2427.2005.00629.x

Mendez, R., & Moral, S.S. (2011). Spanish cities in the knowledge economy. *European Urban and Regional Studies*, 18(2), 136–155. DOI:10.1177/0969776410381039

Mian, S., Fayolle, A., & Lamine, W. (2012). Building sustainable regional platforms for incubating science and technology businesses. *The International Journal of Entrepreneurship and Innovation*, 13(4), 235–247. DOI:10.5367/ijei.2012.0100

Millar, C.C., & Ju-Choi, C. (2010). Development and knowledge resources. *Journal of Knowledge Management*, 14(5), 759–776. DOI:10.1108/13673271011074881

Murphy, E., Fox-Rogers, L., & Redmond, D. (2015). Location decision making of creative industries. *Growth and Change*, 46(1), 97–113. DOI:10.1111/grow.12086

Musterd, S., & Gritsai, O. (2013). The creative knowledge city in Europe. *European Urban and Regional Studies*, 20(3), 343–359. DOI:10.1177/0969776412439199

Oldenburg, R. (1999). *The great good place.* Paragon House, New York.

Pancholi, S., Yigitcanlar, T., & Guaralda, M. (2014). Urban knowledge and innovation spaces. *Asia Pacific Journal of Innovation and Entrepreneurship*, 8(1), 15–38. DOI:10.1080/10630732.2016.1164443

Pancholi, S., Yigitcanlar, T., & Guaralda, M. (2015). Place making facilitators of knowledge and innovation spaces. *International Journal of Knowledge-Based Development*, 6(3), 215–240. DOI:10.1504/IJKBD.2015.072823

Pancholi, S., Yigitcanlar, T., & Guaralda, M. (2017). Governance that matters. *Journal of Place Management and Development*, 10(1), 73–87. DOI:10.1108/JPMD-08-2016-0053

Pancholi, S., Yigitcanlar, T., & Guaralda, M. (2019). Place making for innovation and knowledge-intensive activities. *Technological Forecasting & Social Change*, 146(Sep), 616–625. DOI:10.1016/j.techfore.2017.09.014.

Parker, R. (2010). Evolution and change in industrial clusters. *European Planning Studies*, 17(3), 245–260. DOI:10.1177/0969776409358244

Pelkonen, A. (2005). State restructuring, urban competitiveness policies and technopole building in Finland. *European Planning Studies*, 13(5), 685–705. DOI:10.1080/09654310500139319

Porter, M.E. (1990). The competitive advantage of nations. *Harvard Business Review*, 68(2), 73–93.

Porter, M.E. (1998). Location, clusters, and the new microeconomics of competition. *Business Economics*, 33(1), 7–13.

Sarimin, M., & Yigitcanlar, T. (2012). Towards a comprehensive and integrated knowledge-based urban development model. *International Journal of Knowledge-Based Development*, 3(2), 175–192. DOI:10.1504/IJKBD.2012.047035

Scott, A.J. (2006). Entrepreneurship, innovation and industrial development. *Small Business Economics*, 26(1), 1–24. DOI:10.1007/s11187-004-6493-9

Scott, A.J. (2010) Cultural economy and the creative field of the city. *Geografiska Annaler*, 92(2), 115–130. DOI:10.1111/j.1468-0467.2010.00337.x

Sepe, M. (2010). Liveability, quality and place identity in the contemporary city. *Journal of Place Management and Development*, 3(3), 221–246. DOI:10.1108/17538331011083952

Sepe, M., & Pitt, M. (2014). The characters of place in urban design. *Urban Design International*, 19(3), 215–227. DOI:10.1057/udi.2013.32

Smith, T., Nelischer, M., & Perkins, N. (1997). A framework for understanding the relationship between quality and physical form. *Landscape and Urban Planning*, 39(1), 229–241.

Sohn, D.W., & Kenney, M. (2007). Universities, clusters, and innovation systems. *World Development*, 35(6), 991–1004. DOI:10.1016/j.worlddev.2006.05.008

Storper, M., & Venables, A.J. (2004). Buzz: face-to-face contact and the urban economy. *Journal of Economic Geography*, 4(4), 351–370. DOI:10.1093/jnlecg/lbh027

Trip, J.J. (2007). Assessing quality of place. *Journal of Urban Affairs*, 29(5), 501–517. DOI: 10.1111/j.1467-9906.2007.00362.x

Van Winden, W., Van den Berg, L., & Pol, P. (2007). European cities in the knowledge economy. *Urban Studies*, 44(3), 525–549. DOI:10.1080/00420980601131886

Van Winden, W., De Carvalho, L., Van Tuijl, E., Van Haaren, J., & Van den Berg, L. (2013). *Creating knowledge locations in cities*. Routledge, London. DOI: 10.4324/9780203127162

Webber, M. (1964). *Explorations into urban structure*. University of Pennsylvania Press, Philadelphia.

Yigitcanlar, T., Baum, S., & Horton, S. (2007). Attracting and retaining knowledge workers in knowledge cities. *Journal of Knowledge Management*, 11(5), 6–17. DOI:10.1108/13673270710819762

Yigitcanlar, T., & Velibeyoglu, K. (2008). Knowledge-based urban development. *Local Economy*, 23(3), 195–207. DOI:10.1080/02690940802197358

Yigitcanlar, T., O'Connor, K., & Westerman, C. (2008a). The making of knowledge cities. *Cities*, 25(2), 63–72. DOI:10.1016/j.cities.2008.01.001

Yigitcanlar, T., Velibeyoglu, K., & Martinez-Fernandez, C. (2008b). Rising knowledge cities. *Journal of Knowledge Management*, 12(5), 8–20. DOI:10.1108/13673270810902902

Yigitcanlar, T. (2010). Making space and place for the knowledge economy. *European Planning Studies*, 18(11), 1769–1786. DOI:10.1080/09654313.2010.512163

Yigitcanlar, T., & Dur, F. (2013). Making space and place for knowledge communities. *Australasian Journal of Regional Studies*, 19(1), 36–63.

Yigitcanlar, T., & Lönnqvist, A. (2013). Benchmarking knowledge-based urban development performance. *Cities*, 31(1), 357–369. DOI:10.1016/j.cities.2012.11.005

Yigitcanlar, T., & Bulu, M. (2015). Dubaization of Istanbul. *Environment and Planning A*, 47(1), 89–107. DOI: 10.1068/a130209p

Yigitcanlar, T., & Sarimin, M. (2015). Knowledge-based urban development lessons from an emerging economy. *VINE*, 45(1), 126–147.

Yigitcanlar, T. (2016). *Technology and the city*. Routledge, New York.

Yigitcanlar, T., Edvardsson, I., Johannesson, H., Kamruzzaman, M., Ioppolo, G., & Pancholi, S. (2017). Knowledge-based development dynamics in less favoured regions. *European Planning Studies*, 25(12), 2272–2292. DOI:10.1080/09654313.2017.1358699

Yin, R.K. (2011). *Qualitative research from start to finish*. Guilford Press, London.

Youtie, J., & Shapira, P. (2008). Building an innovation hub. *Research Policy*, 37(8), 1188–1204. DOI:10.1016/j.respol.2008.04.012

Zaheer, S., & Nachum, L. (2011). Sense of place. *Global Strategy Journal*, 1(1–2), 96–108. DOI:10.1002/gsj.2

Zainal, Z. (2007). Case study as a research method. *Journal Kemanusiaan*, 9(1), 1–6.

CHAPTER 6

Place Quality Assessment of Innovation Districts

6.1 INTRODUCTION

Land-use policy plays a critical role to empower local socioeconomic conditions in cities and regions, and thereby helps in gaining a position in the global competitive markets. Today, the global economy is in transition from the neoclassic industrial paradigm to a knowledge-based one—focusing on innovation, creativity, and symbolic values (Cooke, 2001; Bontje & Musterd, 2009). This transition has pushed cities and regions to embrace new land-use types and structural changes, to foster their standing in the knowledge economy (Boddy, 1999; Sarimin & Yigitcanlar, 2012). Cities and regions implemented strategies to upgrade their infrastructures, improve the quality of life, and create an attractive environment for the new knowledge sectors (Anttiroiko, 2015; Yigitcanlar & Bulu, 2015).

In this context, a new type of land use—namely 'innovation districts'—has emerged, and has become the nucleus of knowledge-based activities in cities and regions (Yigitcanlar *et al.*, 2008a, 2008b). Innovation districts represent neighbourhood-scale spaces, such as science and technology parks, research hubs, industrial precincts, and creative clusters, where knowledge- and innovation-based activities beneficially agglomerate together (Pancholi *et al.*, 2014, 2015). The nature of such activities and their products are mostly intangible and rely on innovative ideas generated by educated and talented workforces; such workforces are referred to as knowledge workers, and are considered as a key asset for stimulating economic growth (Clifton, 2008; Pancholi *et al.*, 2017).

DOI: 10.1201/9781032657431-9

Consequently, in recent years, the policy focus has extended from attracting investments and industries in urban localities to attracting human assets as well (Porter, 1998; Glaeser, 2005; Pancholi *et al.*, 2019).

This transition encouraged scholars to investigate the place attributes that foster, attract, and retain knowledge workers and industries (Arora *et al.*, 2000; Florida, 2005; Yigitcanlar *et al.*, 2007; Asheim, 2009; Clifton & Cooke, 2009; Frenkel *et al.*, 2013a, 2013b; Brown, 2015). These attributes are often labelled based on spatial scale (e.g., region, city, cluster) investigation. For example, capital system, quality of life, and quality of place attributes are respectively assigned to denote the quality of region, city, and cluster (Carrillo, 2004, Craglia *et al.*, 2004; McCann, 2004, Heebels & Van Aalst, 2010; Durmaz, 2015; Esmaeilpoorarabi *et al.*, 2016b).

Nevertheless, previous studies mostly focused on regional and city scales, and overlooked the role that place quality can play in innovation districts—location of the day-to-day live-work-play activities—for attracting knowledge workers and industries (Evers, 2008; Kloosterman & Trip, 2011; Gu, 2014). More importantly, most of the studies ignored the synergies across different scales—e.g., whether the quality of life can be traded with place quality. Equally so, although scholars agreed that place quality is a multidimensional construct, the literature lacks a sound framework for the evaluation of place quality in innovation districts. This limits our understanding of whether place quality is a local-level phenomenon and bounded by the features within a cluster or what attributes of a region/city can affect the place quality of a cluster.

This chapter aims to develop a framework that can validly measure the place quality of innovation districts. The methodological approach adopts a Delphic hierarchy process approach. The study identifies a large pool of indicators through a comprehensive review of the literature, and then evaluates their suitability through an international Delphi survey involving 43 experts. By doing so, it classifies suitable indicators, and hence establishes a multidimensional and multiscalar framework of place quality in innovation districts. The framework, as an evaluation tool, has the potential to inform the planning, design, development, and management processes of innovation districts.

6.2 LITERATURE BACKGROUND

Combining a thick pool of knowledge workers with world-class infrastructure, amenities, and global markets is a current strategy to nourish the growth of innovation districts (Esmaeilpoorarabi *et al.*, 2016a). Thus, to systematically foster, attract, and retain knowledge industries and workers, one needs to realise where they choose to locate and why. With respect to the importance of economic factors and work conditions, knowledge workers no longer tend to locate solely where the jobs are (Yigitcanlar *et al.*, 2007). They chose their locations based on specific place qualities and lifestyle options that fulfil their sophisticated needs and fit their creative identities—i.e., residential amenities, authenticity, quality of meeting places for business and leisure purposes (McCann, 2004; Kloosterman & Trip, 2011).

Place quality does not solely relate to the built environment quality. Natural environmental qualities are fundamental assets at the regional, city, and cluster scales, and through effective sustainable urban development practices industrial and urban symbiosis are in the process of merging in many urban locations (Arbolino *et al.*, 2017; Taddeo *et al.*, 2017; Aquilani *et al.*, 2018). Indirectly, high-level place quality also influences where knowledge industries locate as they seek places with the highest concentration of talent (Florida, 2005). Therefore, enhancing place quality is an effective approach to attract knowledge workers and industries, gathering agglomeration of talents, and forming an attractive socioeconomic platform.

Number of studies have highlighted these characteristics of urban environments as 'quality of place', and emphasised that place quality mostly relies on the intangible conditions of a place (or in other words soft factors)—e.g., urban ambiance, sociocultural characteristics, level of diversity, tolerance and openness of the population (Florida, 2005; Bereitschaft & Cammack, 2015). However, the importance of soft factors does not totally detract from the relevance of hard factors, as reported in many studies, such as investment availability, job opportunity, and cost of living (Storper & Manville, 2006; Storper & Scott, 2009; Darchen & Tremblay, 2010; Scott, 2010; Alfken *et al.*, 2015).

There is also a third approach, which claims an attractive place should be capable of offering ideal conditions both in terms of hard labour market and economic factors, as well as soft quality-based factors (Brown & Mczyski, 2009; Yigitcanlar & Dur, 2013). Recently, this combined approach has gained popularity (Bontje & Musterd, 2009; Boren & Young, 2013; Durmaz, 2015). Such a balanced viewpoint is also important to support place quality in innovation districts. This research, hence, suggests a multidimensional set of combined hard and soft indicators for investigating place quality in innovation districts.

There are also fundamental relations between different geographic scales that need to be considered in evaluating place quality at the cluster scale (Kitson *et al.*, 2004). Place quality is not an isolated phenomenon; it is dramatically influenced by the performance of the city and region that host the cluster. At the regional scale, place quality is usually studied through a 'general capital system' (Carrillo, 2004; Carrillo *et al.*, 2014). The capital system focuses on the assets of urban spaces as a whole (Sotarauta & Linnamaa, 2001). It is a sustainable combination of visible—e.g., physical structures and environment—and invisible assets—e.g., human and cultural values—of a region.

These regional assets aim to provide macro-scale requirements of knowledge-based growth, and empower the performance of cities and their innovation districts (Yigitcanlar & Dur, 2013). At the city scale, place quality is strongly related to quantity and standard of amenities provided, which are known as 'quality of life' (Ballas, 2013). In comparison with the regional scale, cities are more prone to subjective factors and personal perception of place—e.g., life and job satisfaction, wellbeing, happiness (Malkina-Pykh & Pykh, 2008). These characteristics of cities play a critical role in shaping their attractiveness (Marans, 2003); consequently, improving the quality of life in turn enhances the ability of cities to attract a thick pool of talents and investments, which in return empowers innovation districts (Santos & Martins, 2007; Morais *et al.*, 2013).

At the cluster level, knowledge workers resonate with certain characteristics of the day-to-day urban environment; this touchable scale of the place is where personal and social life happen and is the soul of the dynamic interaction between people and their routine/environment (Clifton, 2008; Heebels & Van Aalst, 2010). Place quality at the cluster scale can be assumed as personal experience at street level, where knowledge workers enjoy different types of entertainment—e.g., music, food—as well as different ways of socialising (Darchen & Tremblay, 2010). The cluster scale fosters the connection of firms and individuals and the spillover of knowledge (Arora *et al.*, 2000).

This research proposes a multidimensional and multiscale conceptual framework for measuring place quality in innovation districts, which considers the interactive connotation of various geographic scales. Figure 6.1, the conceptual framework, represents the distinctive nature of place quality elements at various geographic scales, which contribute to creating attractive innovation districts. Dimensions of the general capital system, quality of life, and place quality are retrieved respectively from

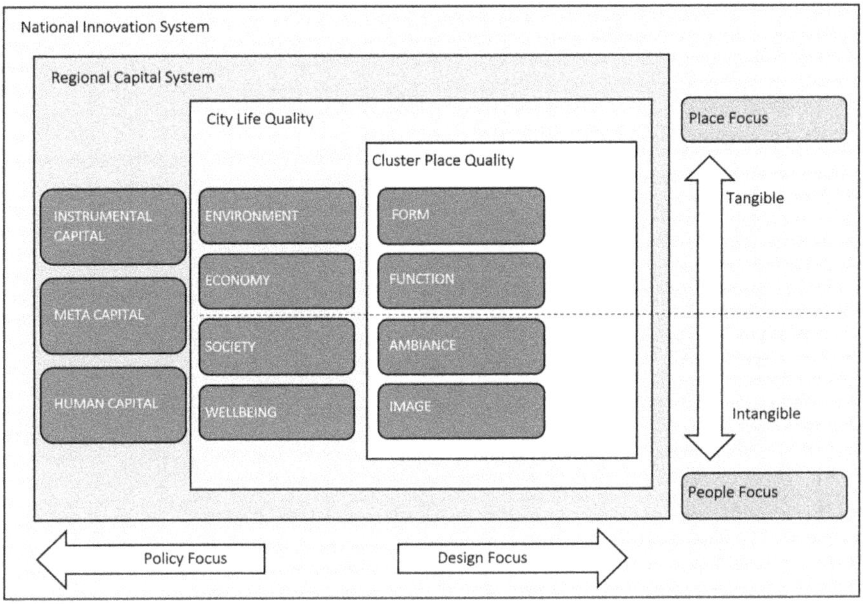

FIGURE 6.1 Conceptual framework of innovation district place quality, by authors.

Carrillo (2004), Ballas (2013), and the previous chapter (Chapter 5). These elements are expanded and examined in detail in the following sections.

6.3 METHODOLOGY

6.3.1 Research Design

After reviewing the interdisciplinary literature and best practices, the conceptual framework (Figure 6.1) is expanded into a detailed framework that contains the factors or indicators of place quality at regional, city, and cluster levels (Table 6.1). A Delphic hierarchy process approach—using the Delphi method and analytical hierarchy process (Khorramshahgol & Moustakis, 1988)—is adopted as the methodological approach. Firstly, the framework is validated through an international Delphi survey by eminent experts of the field to determine adequacy and accuracy of the indicators. In consequence, the study developed an evaluation model, which can be used effectively in evaluating place quality in innovation districts. Secondly, the methodological approach is advanced through retrieving weights by employing an analytic hierarchy process (AHP).

TABLE 6.1 General Framework of Innovation District Place Quality

Scale	Theme	Category	Indicator	Reference
Region	**Instrumental capital**	Tangible	Natural	Deas & Giordano, 2001; Carrillo, 2004; Carrillo *et al.*, 2014
			Artificial	Carrillo, 2004; Metaxiotis *et al.*, 2010; Carrillo *et al.*, 2014
		Intangible/symbolic	Physical and symbolic system	Carrillo, 2004; Yigitcanlar *et al.*, 2012; Carrillo *et al.*, 2014
			Digital system	Carrillo, 2004; Carrillo *et al.*, 2014
	Mata capital	Articulation	Financial	Bontis, 2004; Carrillo, 2004; Bontje & Musterd, 2009; Storper & Scott, 2009; Carrillo *et al.*, 2014
			Relational	Deas & Giordano, 2001; Carrillo, 2004; Storper & Scott, 2009; Carrillo *et al.*, 2014
		Referential	Identity	Carrillo, 2004; Carrillo *et al.*, 2014
			Intelligence	Carrillo, 2004; Bontje & Musterd, 2009; Millar & Ju-Choi, 2010; Yigitcanlar, 2010; Carrillo *et al.*, 2014
	Human capital	Collectives	Cultural liveliness	Carrillo, 2004; Yigitcanlar, 2010; Carrillo *et al.*, 2014
			Evaluative capacity	Carrillo, 2004; Carrillo *et al.*, 2014
		Individuals	Health	Bontis, 2004; Carrillo, 2004; Carrillo *et al.*, 2014
			Education	Bontis, 2004; Carrillo, 2004; Carrillo *et al.*, 2014
City	**Environment**	Natural environment	Physical environment	Brown & Mczyski, 2009; Ballas, 2013
			Spatial conditions	Van Winden *et al.*, 2007, 2013
		Built environment	Physical urban development	Darchen, & Tremblay, 2010; Mendez & Moral, 2011
			City image and brand	McNeill *et al.*, 2005; Musterd & Gritsai, 2013; Homan, 2014;

	Infrastructure and facilities	Infrastructure	Lawton *et al.*, 2013; Lee *et al.*, 2014; Brown, 2015
		Public facilities	Brown & Mczyski, 2009; Van Winden *et al.*, 2007, 2013
Economy	Affordability	Properties	Alawadi & Dooling, 2016; Lawton *et al.*, 2013
		Service and amenities	Darchen, & Tremblay, 2010; Lee *et al.*, 2014
	Economic development	Investment and production	Hazelkorn & Murphy, 2002; McCann, 2004; Scott, 2010; Lee *et al.*, 2014
		Employment and income	Scott, 2010; Brown & Mczyski, 2009; Lawton *et al.*, 2013; Brown, 2015
	Internationalisation	Connection	Brown, 2015; Lee *et al.*, 2014
		Tourism	Yigitcanlar & Velibeyoglu, 2008; Lee, *et al.*, 2014
Society	Community structure	Population structure	Van Winden *et al.*, 2007, 2013; Bereitschaft & Cammack, 2015
		Lifestyle opportunities	Brown & Mczyski, 2009; Darchen, & Tremblay, 2010; Bereitschaft & Cammack, 2015
	Culture	Cultural facilities	Brown & Mczyski, 2009; Musterd & Gritsai, 2013
		Cultural dynamics	Darchen, & Tremblay, 2010; Sepe & Pitt, 2014
	Social life	Social services and activities	Darchen, & Tremblay, 2010
		Social participation	Nurick & Johnson, 1998; Lee *et al.*, 2014
Wellbeing	Health	Physical	Morais *et al.*, 2013; Viitanen & Kingston, 2014
		Mental	Viitanen & Kingston, 2014
	Personal development	Education	D'Mello & Sahay, 2007; Van Winden *et al.*, 2007, 2013; Musterd & Gritsai, 2013
		Individual growth	D'Mello & Sahay, 2007
	Safety & security	Personal safety	Nurick & Johnson, 1998; Santos & Martins, 2007
		Personal security	Brown & Mczyski, 2009; Darchen, & Tremblay, 2010

(Continued)

TABLE 6.1 *(Continued)* General Framework of Innovation District Place Quality

Scale	**Theme**	**Category**	**Indicator**	**Reference**
Cluster	**Form**	Location	Centrality	Trip, 2007; Kloosterman & Trip, 2011; Durmaz, 2015
			Unique surrounding	Pelkonen, 2005; Youtie, & Shapira, 2008; Da Cunha & Selada, 2009
		Urban form and structure	Urban form	Smith *et al.*, 1997; Da Cunha & Selada, 2009; Kloosterman & Trip, 2011
			Urban spatial structure	Smith *et al.*, 1997; He & Gebhardt, 2014; Durmaz, 2015
		Design	Urban design	Smith *et al.*, 1997; Loures *et al.*, 2007; Da Cunha & Selada, 2009
			Architectural design	Smith *et al.*, 1997; Loures *et al.*, 2007; Kloosterman & Trip, 2011; Durmaz, 2015
		Amenities	Essential amenities	Loures *et al.*, 2007; Da Cunha & Selada, 2009; He & Gebhardt, 2014
			Advanced amenities	Loures *et al.*, 2007; Heebels & Van Aalst, 2010; Frenkel *et al.*, 2013a, 2013b; Murphy *et al.*, 2015
	Function	Services	Urban mobility	Benneworth & Ratinho, 2014; He & Gebhardt, 2014
			Management	Da Cunha & Selada, 2009; Yigitcanlar *et al.*, 2017
		Land use	Mixed-use development	Loures *et al.*, 2007; Kloosterman & Trip, 2011; Frenkel *et al.*, 2013a, 2013b; He & Gebhardt, 2014
			Property availability and value	Frenkel *et al.*, 2013a, 2013b; He & Gebhardt, 2014
		Work condition	Thick labour market	Sohn & Kenney, 2007; Murphy *et al.*, 2015
			Professional networks	Evers, 2008; Messeni-Petruzzelli *et al.*, 2010; Frenkel *et al.*, 2013a, 2013b; He & Gebhardt, 2014; Murphy *et al.*, 2015

	Technology	Technology adoption	Sohn & Kenney, 2007; Trip, 2007; Benneworth & Ratinho, 2014
		Knowledge/creative industries	Sohn & Kenney, 2007; Trip, 2007; Baum *et al.*, 2009; Parker, 2010; Benneworth & Ratinho, 2014
Ambiance	Cultural milieu	Public and cultural spaces	Da Cunha & Selada, 2009; Flew, 2012
		Cultural events	Trip, 2007; Durmaz, 2015
	Social interaction	Work climate	Frenkel *et al.*, 2013a & b; Murphy *et al.*, 2015
		Social connections	Da Cunha & Selada, 2009; Heebels & Van Aalst, 2010; Grant & Buckwold, 2013
	Diversity	Diverse communities	Florida, 2005; Trip, 2007; Flew, 2012
		Tolerance	Florida, 2005; Sepe, 2010; Durmaz, 2015
	Creativity	Creative communities	Florida, 2005; Trip, 2007; Durmaz, 2015
		Openness	Florida, 2005; Sepe, 2010; Flew, 2012
Image	Lifestyle	Pace of life	Kloosterman & Trip, 2011; Flew, 2012
		Variety of lifestyle	Sepe, 2010; Frenkel *et al.*, 2013a, 2013b
	Personal safety	Social control	Durmaz, 2015; Smith *et al.*, 1997
		Safety and security	Trip, 2007; Kloosterman & Trip, 2011; Sepe, 2010
	Sense of place	Place attachment	Durmaz, 2015; Trip, 2007; Sepe, 2010
		Social cohesion	Heebels & Van Aalst, 2010
	Identity	Authenticity	Florida, 2005; Heebels & Van Aalst, 2010; Sepe, 2010; Trip, 2007
		Brand	Sepe, 2010; Hafeez *et al.*, 2016

6.3.2 Selection of the Delphic Hierarchy Process Approach

Rand Corporation introduced the Delphi method in 1950, and since then it has been widely used in research (Dalkey & Helmer, 1963; Hsu & Sandford, 2007; He *et al.*, 2016). The method, as a group communication process, aims to deal with complex problems by exploring expert views, facilitating problem solving, and suggesting alternative policies (Alawadi & Dooling, 2016). The Delphi method is mainly a reiterative systematic policymaking process that utilises a series of anonymous questionnaires to collect expert opinions (Chan *et al.*, 2014; Julsrud & Priya-Uteng, 2015; Norouzian-Maleki *et al.*, 2015; Ruppert & Duncan, 2017; Rust, 2017). In terms of number of required rounds, several studies have stated that the most effective number of rounds to reach a consensus is usually two (Chiang & Lei, 2016; Gigovic *et al.*, 2016; Soria-Lara & Banister, 2017) or three (Holey *et al.*, 2007; Jordan & Javernick-Will, 2013; Singhal *et al.*, 2013). However, some suggest that occasionally it may take up to four or even five rounds until a consensus is reached (Ruppert & Duncan, 2017).

The Delphi method was applied in this study as: (a) the literature acknowledges that place quality in innovation districts is impacted by diverse factors at different geographic scales (Trip, 2007; Kloosterman & Trip, 2011; Esmaeilpoorarabi *et al.*, 2016b). However, there is no clear agreement on the indicators of place quality at each scale, and their level of impact on the cluster scale. Delphi is particularly suited in such situations, where there is a lack of actual resources and documents (Ruppert & Duncan, 2017). Hence, this study engages the Delphi method to investigate indicators and measures of the framework. (b) The Delphi is widely utilised for strategic urban planning and land-use studies (Singhal *et al.*, 2013; Gigovic *et al.*, 2016). Therefore, the chapter employs Delphi to establish a set of recommendations and policies for planners and policymakers. (c) Delphi has unique anonymity, which enables participants to provide their judgement freely, and guarantees more truthful responses (Huang & Lee, 2016). (d) Delphi's feedback process encourages interaction between experts. These feedbacks are considered an advantage for this research in order to achieve reliable results (Julsrud & Priya-Uteng, 2015).

While Delphi helps in defining indicators of place quality, establishing the relationships between such indicators is the most critical part of this process. The research, thus, employed the AHP calculation method to investigate the relationship between the finalised indicators, and determine their importance. AHP is all about how people process their decision-making procedure, which is useful in solving complex policy-making problems with multicriteria conditions (Hsueh *et al.*, 2013; Kim *et al.*, 2016). A complex issue can be better addressed by dividing it into separated smaller parts, creating a hierarchy structure, and analysing each part independently (Saaty, 1980). AHP eases the assessment of relationships among criteria by requesting only simple comparisons by experts, and then derives a weight for each criterion and sub-criteria (Yang *et al.*, 2016).

AHP is used in this study as: (a) place quality in innovation districts is a complex phenomenon, which contains various multidimensional and multiscalar factors. AHP divides these factors into comparable categories, and simplifies the assessment process; (b) instead of an equal weighting system, AHP calculates the weights of sub-criteria based on the importance of their scales and heading criteria; and (c) AHP establishes an effective policymaking model—contains both quantities and qualitative results of the study—that provides a holistic understanding of the place quality in innovation districts.

6.3.3 Identification of Place Quality Indicators

An indicator is a parameter, or a value derived from parameters that can reflect or describe the state of a phenomenon (Dur & Yigitcanlar, 2015). There are different ways to approach the initial phase of the Delphi survey for indicator selection. Some studies begin with a series of open-ended questions to collect ideas from experts (Yang *et al.*, 2016; Chiang & Lei, 2016; Yoon & Kim, 2016). However, most studies begin with a review of the literature to provide an initial set of indicators that can be edited and assessed by the Delphi panel (Schuckmann *et al.*, 2012; Jozi *et al.*, 2013; Ameyaw & Chan, 2015; Kiba-Janiak, 2016). Hence, it is common practice to use a structured questionnaire along with a series of open-ended questions in the first round; this strategy not only evaluates the validity, suitability, and importance of proposed indicators, but also retrieves other disregarded indicators through open-ended questions (Norouzian-Maleki *et al.*, 2015).

This study reviewed over 50 scholarly sources to identify the most frequently cited indicators (Table 6.1), which were grouped according to the conceptual framework (Figure 6.1). The review identified a total of 68 relevant indicators. The indicators, their thematic classes, and the sources are presented in the proposed framework (Table 6.1). The Delphi survey also provided opportunities to include new indicators deemed important by the experts.

6.3.4 Experts of the Delphi Study

The Delphi study followed three main principles to identify experts: (a) experts were selected from academic and professional sectors to acquire both theory- and practice-oriented views; (b) as opinions vary among the experts with different backgrounds, the panel was grouped from different geographical locations for wider validity—i.e., Europe, North America, Latin America, Pacific Region, Asiatic Region, Middle East—(Ameyaw & Chan, 2015; Ruppert & Duncan, 2017); (c) experts were sourced from different disciplines to provide an heterogeneous landscape to the research (Schuckmann *et al.*, 2012; Meijering *et al.*, 2015). These disciplinary areas include business, management and accounting; economics, finance, and econometrics; geography, planning, and development; architecture and urban design; and sociology and political science.

Academic members of the Delphi panel are selected among the individuals with an academic position, and published on the innovation district topic in international peer-reviewed journals during the last five years (Meijering *et al.*, 2015). Selected experts from the professional sector are required to hold a position at a public or private professional organisation, and actively involved in the planning, design, development, or management of an innovation district during the last five years. This selection protocol guaranteed the quality of the sample pool and the reliability of expert inputs. While the quality of the expert sample is important, it is also vital to engage a reasonable number of experts in the study. Some studies suggested a panel of 10–15 experts for homogenous samples is adequate (Mafi *et al.*, 2015, Norouzian-Maleki *et al.*, 2015). However, for heterogeneous samples, a panel of 30–50 practitioners is large enough to ensure diversity of opinions, and small enough to establish a consensus (Ameyaw & Chan, 2015; Alawadi & Dooling, 2016). To reach the target number of 30–50 participants, 200 experts from all across the globe were contacted.

An invitation email was sent to the identified 200 experts along with an information sheet and a link to the online survey. Two weeks after sending the request, reminders were sent out; this process was repeated three times, fortnightly. By the end of the eighth week, 43 experts completed the survey, representing a response rate of 22%. As the sample size fulfilled the expectation of this research—30–50 participants—Round 1 was closed. Then, an email was sent to the 43 experts who participated in Round 1, to invite them to Round 2. After three reminder emails, 34 experts participated in the Round 2 survey, yielding an 80% response rate. This response rate was found acceptable in association with the previous studies on Delphi (Jozi *et al.*, 2013; Colagiuri *et al.*, 2015; Chiang & Lei, 2016).

6.3.5 Delphi Survey

A Delphi survey questionnaire was developed based on the indicators retrieved from the literature (Table 6.1). The questionnaire contained both Likert-scale and open-ended questions to collect qualitative and quantitative data. Open-ended questions were added in order to clarify some specific aspects of the topic. The Likert-scale questions were used to: (a) measure the suitability of the recommended indicators, categories, and scales; (b) assess the level of consensus; and (c) generate the weighting system.

This study employed an 11-point scale (from 0–not important at all to 10–extremely important), which provides a more accurate data for the calculation of indicator weightings using the AHP method than a 5-point scale (Schuckmann *et al.*, 2012; Jozi *et al.*, 2013; Hassanzadeh *et al.*, 2014; Alshehri *et al.*, 2015; Julsrud & Priya-Uteng, 2015; Kiba-Janiak, 2016; Kim *et al.*, 2016; Perveen *et al.*, 2017; Soria-Lara & Banister, 2017).

Before starting the Delphi study, the questionnaire was reviewed for consistency and completeness by three local experts through a pilot study. Some subsequent changes were made based on the pilot outcomes, rephrasing of indicators, modifying definitions, and improving the arrangement and scoring system. For example, 'creativity', 'control', and 'technology' terms were respectively rephrased as 'creative community', 'social control', and 'technology adaptation'; 'authenticity' was moved to the category of the 'identity'. The scoring system was also changed from a 7-point scale to an 11-point scale. Finally, a two-round online Delphi survey was conducted between June and September 2017.

Round 1 (June and July 2017): The Key Survey tool—an enterprise survey platform—was used to distribute the questionnaire. Due to participants' different backgrounds, they were provided with a brief description of the project, research aims, objectives, and keywords. The conceptual framework (Figure 6.1) and the general framework (Table 6.1) were introduced. Descriptions of indicators were also included to reduce the risk of biased interpretations and misunderstandings. Experts were requested to score the importance of the 3 scales, 11 themes, and 68 indicators in fostering the place quality of innovation districts. The three scales and their themes—those are categorised under each scale—were questioned in a comparative way following the procedure as outlined in the AHP literature (Saaty, 1980; Yang *et al.*, 2016). For example, experts were asked to score: 'Between region and city, which scale has more impact on shaping place quality in innovation districts'. Five open-ended questions were placed to collect experts' opinions around adequacy of indicators, the validity of the classifications, and accuracy of provided descriptions for indicators. By the end of Round 1, both qualitative and quantitative data were analysed and used to modify the initial framework. However, a second round was still necessary to validate the rating scores for AHP and weighting purposes.

Round 2 (August and September 2017): Round 2 was carried out among the panellists who participated in Round 1. The structure of the questionnaire was like Round 1. The modified framework was provided to the experts along with a summary of statistics (average value) from the previous round. Additionally, the questionnaires were personalised for participants to include their previous responses to each question. This feedback process was performed for encouraging panellists to reconsider their original ratings, and to reach a group consensus. Experts were requested to score the impact of the 3 scales, 11 themes, and 68 indicators through Likert-scale questions. Only one open-ended question was included to collect general comments. A consensus was established in Round 2; hence, a further round was not needed.

6.3.6 Analysis of the Delphi Survey

In the Delphi method, there is no definite technique for analysing the data and evaluating the consensus level (Holey *et al.*, 2007). Nevertheless, the most common analytical statistics used in Delphi studies are central tendency measures (i.e., mean, median, mode) and dispersion measures (i.e., standard deviation, inter-quartile range). The most common definition for consensus is the percentage of agreement (Diamond *et al.*, 2014). This study used the mean values to show the group aggregate ranks, which suits feedback reports and weight calculations (Holey *et al.*, 2007; Jordan & Javernick-Will, 2013; Singhal *et al.*, 2013). Standard deviation (SD) is calculated to show the distribution of the experts' responses around the mean values. A lower level of SD occurs when responses are clustered more closely around the mean values, which means a stronger agreement. Taking the literature's explanation into account, the SD below 1 in a 5-point Likert-scale questionnaire is mainly accepted as a low dispersion level in the rating (Julsrud & Priya-Uteng, 2015; Perveen *et al.*, 2017). Accordingly, it is reasonable to define the SD below 2 for the current 11-point Likert-scale questionnaire.

Additionally, Cronbach's α is measured to examine the reliability of the questionnaire and the internal homogeneity and consistency of opinion among the panellists. Cronbach's α typically takes on values between 0 and 1. Lower values demonstrate that the ratings of the experts are completely unrelated to each other, while values above 0.7 mean that the ratings are strongly associated (Hassanzadeh *et al.*, 2014; Mafi *et al.*, 2015). To ensure the reliability of the data for further analysis, the study initially measured the validity of Cronbach's α and SD, and then employed the data to measure the level of agreement in each round.

Two types of agreement levels are calculated in each round: (a) the overall agreement, which shows the overall percentage of votes for 'high impact' and 'extremely high impact' rates, together (Ruppert & Duncan, 2017; Sutterluty *et al.*, 2017); and (b) the specific agreement, which shows the percentage of votes for 'high impact' and 'extremely high impact' rates, separately—followed by the previous studies (Schuckmann *et al.*, 2012; Von der Gracht, 2012; Meijering *et al.*, 2015; Zeeman *et al.*, 2016; Perveen *et al.*, 2017). If most votes (>50%) show an overall agreement on indicators, then it means the indicator has a notable impact in shaping place quality and should be retained in the evaluation framework.

On the other hand, the specific agreement level is evaluated to consider priorities in each category. If an indicator receives more than 50% of votes on 'extremely high impact', then it should be marked as high priority in the related category. Finally, the research tested the stability of responses between rounds by calculating Kendall's coefficient (Ameyaw & Chan, 2015; He *et al.*, 2016; Kiba-Janiak, 2016). Kendall's W is calculated for both mean values, and the percentage of overall agreement among continuous rounds—and not for the level of agreement between participants. Kendall's W greater than 0.5 indicates the stability of mean scores and the consistency of expert judgements among the Delphi rounds. This means further rounds would not produce any extra convergence of opinions (He *et al.*, 2016).

After finalising indicators and the general framework, and confirming the reliability of data, quantitative statistics are converted to weighting systems; the mean values are employed to generate the equal and AHP-based weights

(Singhal *et al.*, 2013; Alshehri *et al.*, 2015). In both systems, the weight of each indicator is assigned between 0 and 1, and the sum of all indicators is equal to 1 (Jin *et al.*, 2016). This is also applied to the weights of themes and scales. The equal weighting is simply calculated by dividing the mean value of each indicator to the total mean values of all indicators:

$$Ew\ (n) = m\ (n)/t$$

Where: m (1, 2, … , n) = Mean of indicator (1, 2, … , n), t = Mean of indicator (1) + Mean of indicator (2) + … + Mean of indicator (n), Ew (1, 2, … , n) = Equal Weight (1, 2, … , n)

In contrast to the equal weighting, the AHP-based weighting approach assigns specific weights to each scale, theme, and indicator. Initially, the weightings are evaluated for the three scales. Then, these weightings are used to evaluate the weightings of themes through the following process:

$$WRT(n) = (MRT(n) * WR)/(MRT(1) + MRT(2) + \ldots + MRT(n))$$
$$WCIT(n) = (MCIT(n) * WCI)/(MCIT(1) + MCIT(2) + \ldots + MCIT(n))$$
$$WCLT(n) = (MCLT(n) * WCL)/(MCLT(1) + MCLT(2) + \ldots + MCLT(n))$$

Where: R = Region, CI = City, CL = Cluster, RT = Regional scale, CIT = City scale, CLT = Cluster scale, RT (1, 2, … , n) = Theme (1, 2, … , n) categorised under the Regional scale, CIT (1, 2, … , n) = Theme (1, 2, … , n) categorised under the City scale, CLT (1, 2, … , n) = Theme (1, 2, … , n) categorised under the Cluster scale, M = Mean, W = Weight

This process is repeated to calculate weights of indicators, which are categorised under each theme. The function of weights in this process is to express the importance of each indicator, theme, or scale relative to all others.

6.4 RESULTS

6.4.1 Response Rates and Expert Profiles

Participants' profiles were checked to ensure a balanced heterogeneous sample of experts. The prevalent groups of panellists were from 'business, management and accounting' backgrounds (30%), followed by 'geography, planning and development' sectors (26%). These two groups of expertise were completely familiar with different aspects of the topic and directly involved in the strategy making and development process of the innovation districts. On the other hand, the lower representation of 'economist' (16%), 'architect' (14%), and 'sociologist'

(14%) was since their expertise solely covered limited aspects of the innovation district topic. As for the geographical distribution, 35% of the panellists were from Europe, followed by Latin America (16%), Pacific Region (16%), Middle East (14%), North America (12%), and Asiatic Region (7%). These figures confirmed that a heterogeneous sample of experts was assembled; and ensured a fair range of opinions from all areas of expertise and all regions of the world was collected. The variety of ideas provided an invaluable input and critical insight in selecting the indicators and improving the framework.

6.4.2 Consensus Level and Selection of Indicators in Round 1

Round 1: The experts rated 68 indicators, 11 themes, and 3 scales, yielding a total of 82 responses from each expert. In addition, 29 comments were received through open-ended questions, which mainly focused on description of indicators and arrangement of the framework rather than suggesting additional indicators. For instance, a number of comments suggested that 'political state' and 'institutions' were neglected from the regional scale and that needed to be considered under the category of 'relational'; or for example, 'work climate' related to social factors and was more likely to be classified under the 'ambiance'. Taking the new ideas provided by the experts into account, the description and classification of indicators were updated and revised; even the title of some indicators was transformed to become adapted to their new descriptions and categories. Appendices 6.A–6.C present the framework based on the qualitative analysis of expert opinions in each scale. After the qualitative analysis, the reliability of the quantitative data was examined by evaluating Cronbach's α. It was found to be 0.978 and above the minimum value of 0.7, which indicates a high level of internal consistency and hence reliability.

The mean value, SD, and overall and specific agreements were calculated. Table 6.2 provides an aggregated summary of these results. SD was found to be lower than 2 for all indicators, themes, and scales—except 'digital system', indicating that the responses were all convergent and reliable. Further analysis demonstrated that the experts reached a consensus for all the indicators, themes, and scales with an overall agreement of more than 50% (average agreement level of 82.75%), suggesting the importance of shaping place quality in innovation districts. In this regard, the 'region' and 'instrumental capital' from the classification of 'scale' and 'theme', and the 'natural' and 'tourism' from the classification

TABLE 6.2 Round 1 Results

Scale	M	SD	OA	EH	H	Theme	M	SD	OA	EH	H	Category	Indicator	M	SD	OA	EH	H
Region	6.81	1.62	51.2	18.6	32.6	**Instrumental capital**	7.32	1.97	69.8	30.3	39.5	Tangible	Natural	6.81	1.93	62.8	21	41.8
													Artificial	7.55	1.58	69.8	27.9	41.9
												Intangible/ symbolic	Physical and symbolic system	7.20	1.83	67.4	46.6	20.8
													Digital system	7.67	2.11	75.6	41.8	33.8
						Mata capital	7.53	1.76	79.1	30.2	48.9	Articulation	Financial	7.55	1.58	81.4	32.6	48.8
													Relational	7.97	1.41	86	34.8	51.2
												Referential	Identity	7.88	1.40	88.4	32.6	55.8
													Intelligence	7.60	1.87	79.1	32.6	46.5
						Human capital	8.04	1.90	81.4	51.2	30.02	Collectives	Cultural liveliness	7.86	1.66	86	34.8	51.2
													Evaluative capacity	7.65	1.72	79.1	32.6	46.5
												Individuals	Health	7.55	1.67	76.8	27..9	48.7
													Education	8.51	1.48	95.3	53.5	41.8
City	8.30	1.06	95.3	41.8	53.5	**Environment**	8.04	1.50	81.4	37.2	44.2	Natural environment	Physical environment	7.79	1.36	83.7	28	55.7
													Spatial conditions	7.65	1.51	71.9	37.2	34.7
												Built environment	Physical urban development	7.97	1.47	83.7	41.8	41.9
													City image and brand	7.86	1.47	76.8	41.9	34.9
												Infrastructure and facilities	Infrastructure	8.20	1.68	86	51.2	34.8
													Public facilities	8.13	1.36	88.4	44.2	44.2
						Economy	8.27	1.32	93	39.5	53.5	Affordability	Properties	7.79	1.36	83.8	32.6	51.2
													Service and amenities	8.32	1.29	95.3	46.5	48.8

												Economic development	Investment and production	7.90	1.41	83.8	41.6	41.9
													Employment and income	8.41	1.40	90.8	58.2	32.6
												Internationalisation	Connection	8.69	1.12	97.7	62.8	34.9
													Tourism	7.09	1.61	62.8	21	41.8
						Society	8.39	1.35	88.4	58.2	30.2	Community structure	Population structure	7.86	1.35	83.8	32.6	51.2
													Lifestyle opportunities	8.09	1.36	88.4	34.9	53.5
												Culture	Cultural facilities	7.95	1.25	86	37.2	48.8
													Cultural dynamics	8.00	1.13	90.8	39.6	51.2
												Social life	Social services and activities	7.88	1.38	81.4	32.6	48.8
													Social participation	7.62	1.53	72.1	30.2	41.9
						Wellbeing	8.32	1.29	90.9	51.2	39.5	Health	Physical	7.86	1.70	81.4	37.2	44.2
													Mental	7.88	1.78	79.1	39.6	39.5
												Personal development	Education	8.90	1.41	88.4	74.4	14
													Individual growth	8.69	1.44	90.7	62.8	27.9
												Safety and security	Personal safety	8.44	1.33	93	51.2	41.8
													Personal security	8.44	1.33	93	51.2	41.8
Cluster	8.44	1.61	86	53.4	32.6	**Form**	7.74	1.53	79.1	32.5	46.6	Location	Centrality	8.09	1.46	83.8	41.9	41.9
													Unique surrounding	7.34	1.84	72.1	27.9	44.2
												Urban form and structure	Urban form	7.39	1.64	72.1	30.2	41.9
													Urban spatial structure	7.41	1.75	78.8	25.3	53.5
												Design	Urban design	7.79	1.42	83.7	37.2	46.5

(Continued)

TABLE 6.2 *(Continued)* Round 1 Results

Scale	M	SD	OA	EH	H	Theme	M	SD	OA	EH	H	Category	Indicator	M	SD	OA	EH	H
													Architectural design	7.39	1.77	74.4	30.2	44.2
												Amenities	Essential amenities	8.06	1.47	86	39.5	46.5
													Advanced amenities	7.83	1.80	79.1	34.9	44.2
						Function	8.23	1.32	86	44.2	41.8	Services	Urban mobility	8.62	1.22	93	55.8	37.2
													Management	8.09	1.50	88.4	41.9	46.5
												Land use	Mixed use development	7.86	1.25	90.7	27.9	62.8
													Property availability and value	8.11	1.28	90.7	39.5	51.2
												Work condition	Thick labour market	8.39	1.48	88.4	58.2	30.2
													Professional networks	8.46	1.47	86	60.5	25.5
												Company profile	Technology adoption	8.25	1.72	79.1	53.5	25.6
													Knowledge/ creative industry	8.55	1.60	88.4	58.2	30.2
						Ambiance	8.51	1.37	93	53.5	39.5	Cultural milieu	Public and cultural spaces	7.97	1.50	83.8	41.9	41.9
													Cultural events	7.62	1.51	79.1	37.3	41.8
												Social context	Work climate	8.39	1.40	90.7	53.5	37.2
													Social interactions	8.27	1.42	83.8	44.2	39.6
												Diversity	Diverse communities	7.93	1.62	83.8	41.9	41.9

							Tolerance	8.34	1.63	83.8	55.9	27.9
						Creativity	Creative communities	8.34	1.77	86	55.8	30.2
							Openness	8.51	1.47	88.4	62.8	25.6
Image	8.30	1.26	95.4	46.5	48.9	Lifestyle	Pace of life	7.37	1.56	69.8	30.2	39.6
							Variety of lifestyle	7.81	1.65	79.1	37.3	41.8
						Personal safety	Informal social control	7.09	1.95	69.8	23.3	46.6
							Safety and security	8.16	1.98	81.4	48.9	32.5
						Sense of place	Place attachment	7.44	1.74	81.4	23.3	58.1
							Social cohesion	7.95	1.60	86	37.2	48.8
						Identity	Authenticity	7.88	1.68	76.7	44.8	34.9
							Brand	7.79	1.92	76.7	46.5	30.2

of 'indicators' reached the lowest agreements with 51.2%, 69.8%, and 62.8%, respectively. The highest agreement was in the 'city' scale, 'image' theme, and 'education' indicator, with 95.3%. However, only 33 out of 82 questioned scales, themes, and indicators gained a specific agreement in Round 1. This suggests a weak consensus among the experts. Round 2, hence, mostly focused on improving the overall level of agreement, and achieving a consensus on the scales, themes, and indicators.

6.4.3 Consensus Level and Identification of Priorities in Round 2

Round 2: Table 6.3 displays the results obtained from Round 2. SD indicated a significant decline in all classifications, and accordingly, an improvement in the level of consensus. The average SD reduced 0.41 points between rounds (R1-R2: 1.54–1.13). Also, a slight improvement was found in the level of overall agreement in all classifications (improved by 10% on average). However, the most significant improvement was achieved in the level of specific agreements where 77 out of 82 scales, themes, and indicators reached a consensus. One scale, 5 themes, and 20 indicators were marked as 'extremely high impact' and 'high priority', which are highlighted in Table 6.3. Kendall's W showed the value of 0.81 and 0.91 for the mean values and the percentage of overall agreement among rounds. These numbers are greater than the cut value of 0.5—Kendall's W value ranges from 0 (no agreement) to 1 (complete agreement).

The authors decided to stop the Delphi rounds as: (a) generally, the 15% or lower changes in the level of agreement was considered a stable situation (Von der Gracht, 2012). As the overall level of agreement was changed only 10%, it can be assumed that in the majority of classifications the experts upheld their earlier opinions. (b) Most of the criteria (94%) reached the specific agreement on their level of impact, which was the main purpose of the Delphi survey. (c) Kendall's W demonstrated great stability between rounds, which means repeating Delphi rounds would not provide any extra contribution to the research. By the end of the Delphi study and after ensuring the stability of the results, the mean values—collected in Round 2—were employed to calculate weights of the framework.

6.5 FINDINGS

The evaluation framework, along with weights and priorities, is the most significant outcome of this Delphic hierarchy process study. Table 6.4

TABLE 6.3 Round 2 Results

Scale	M	SD	OA	EH	H	Theme	M	SD	OA	EH	H	Category	Indicator	M	SD	OA	EH	H
Region	7.41	1.28	76	17	59	**Instrumental capital**	7.50	1.35	76	30	46	Tangible	Natural	7.06	1.37	76	12	64
													Artificial	7.76	1.21	85	30	55
												Intangible/ symbolic	Physical and symbolic system	7.44	1.33	82	18	64
													Digital system	7.74	1.62	91	20	71
						Mata capital	7.94	0.92	94	20	74	Articulation	Financial	7.29	1.54	79	15	64
													Relational	8.03	1.06	94	26	68
												Referential	Identity	8.17	0.97	97	29	68
													Intelligence	7.88	1.51	94	30	64
						Human capital	8.53	0.96	97	56	38	Collectives	Cultural liveliness	8.12	1.04	97	37	60
													Evaluative capacity	8.00	0.98	94	30	64
												Individuals	Health	8.06	1.10	91	35	56
													Education	8.56	0.96	100	53	47
City	8.53	0.83	100	44	56	**Environment**	8.41	1.09	94	50	44	Natural environment	Physical environment	8.15	1.13	94	34	60
													Spatial conditions	7.97	1.08	85	35	50
												Built environment	Physical urban development	8.32	0.84	97	42	55
													City image and brand	8.15	1.18	88	40	48
												Infrastructure and facilities	Infrastructure	8.32	1.27	97	53	44
													Public facilities	8.47	0.86	100	53	47
						Economy	8.38	0.74	100	41	59	Affordability	Properties	7.94	0.78	97	17	80
													Service and amenities	8.62	0.85	100	50	50

(Continued)

TABLE 6.3 *(Continued)* Round 2 Results

Scale	M	SD	OA	EH	H	Theme	M	SD	OA	EH	H	Category	Indicator	M	SD	OA	EH	H
												Economic development	Investment and production	8.21	0.91	97	32	65
													Employment and income	8.62	0.95	97	67	30
												Internationalis-ation	Connection	8.91	0.79	100	70	30
													Tourism	7.17	1.56	82	12	70
						Society	8.56	0.89	97	50	47	Community structure	Population structure	8.00	1.15	91	24	67
													Lifestyle opportunities	8.15	1.13	94	32	62
												Culture	Cultural facilities	8.18	0.83	97	30	67
													Cultural dynamics	8.18	0.87	96	32	65
												Social life	Social services and activities	8.03	1.06	91	30	61
													Social participation	8.09	1.24	88	32	56
						Wellbeing	8.56	0.93	97	53	44	Health	Physical	8.09	1.19	88	38	50
													Mental	8.06	1.23	88	35	53
												Personal development	Education	8.97	0.99	94	91	3
													Individual growth	8.79	1.07	94	80	14
												Safety and security	Personal safety	8.74	1.26	94	60	34
													Personal security	8.74	1.31	94	60	34
Cluster	8.82	0.99	97	68	29	**Form**	7.74	1.50	94	29	65	Location	Centrality	8.21	0.91	91	38	53
													Unique surrounding	7.79	1.09	88	27	61
													Urban form	7.50	1.40	88	18	70

						Urban form and structure	Urban spatial structure	7.71	1.34	94	18	76
						Design	Urban design	7.97	1.49	91	44	47
							Architectural design	7.71	1.47	88	30	58
						Amenities	Essential amenities	8.35	0.92	97	30	67
							Advanced amenities	8.15	0.96	94	30	64
Function	8.41	1.10	94	47	47	Services	Urban mobility	8.79	0.88	97	65	32
							Management	8.21	1.09	94	35	62
						Land use	Mixed use development	8.21	1.04	97	32	65
							Property availability and value	8.44	0.99	97	41	56
						Work condition	Thick labour market	8.79	0.77	100	70	30
							Professional networks	8.79	0.84	97	65	32
						Company profile	Technology adoption	8.53	1.16	97	53	44
							Knowledge/ creative industry	8.76	0.99	97	60	37
Ambiance	8.64	0.92	100	59	41	Cultural milieu	Public and cultural spaces	8.18	0.99	97	32	65
							Cultural events	7.88	1.23	94	30	64
						Social context	Work climate	8.68	0.91	100	70	30
							Social interactions	8.65	1.01	97	67	32
						Diversity	Diverse communities	8.35	1.04	97	37	60
							Tolerance	8.79	0.98	97	62	35

(Continued)

TABLE 6.3 *(Continued)* Round 2 Results

Scale	M	SD	OA	EH	H	Theme	M	SD	OA	EH	H	Category	Indicator	M	SD	OA	EH	H
												Creativity	Creative communities	8.74	1.02	100	62	38
													Openness	8.74	1.14	97	65	32
						Image	8.21	0.88	94	38	56	Lifestyle	Pace of life	7.53	1.35	82	20	62
													Variety of lifestyle	8.09	0.96	94	32	62
												Personal safety	Informal social control	7.15	1.64	74	18	56
													Safety and security	8.41	1.83	88	58	30
												Sense of place	Place attachment	7.53	1.76	88	26	62
													Social cohesion	8.32	1.32	97	47	50
												Identity	Authenticity	8.00	1.39	88	32	56
													Brand	7.76	1.81	82	35	47

TABLE 6.4 Innovation District Place Quality Evaluation Framework

Scale	EW	RA	Theme	EW	RA	AW	Category	Indicator	EW	RA	AW
Region	0.30	3	**Instrumental capital**	0.083	8	0.094	Tangible	Natural	0.013	4	0.022
								Artificial	0.014	3	0.024
							Intangible/symbolic	Physical and symbolic system	0.013	4	0.023
								Digital system	0.014	3	0.024
			Mata capital	0.088	6	0.100	Articulation	Financial	0.013	4	0.023
								Relational	0.014	3	0.025
							Referential	Identity	0.015	2	0.026
								Intelligence	0.014	3	0.025
			Human capital	0.094	2	0.106	Collectives	Cultural liveliness	0.015	2	0.026
								Evaluative capacity	0.014	3	0.026
							Individuals	Health	0.014	3	0.026
								Education	0.015	2	0.028
City	0.34	2	**Environment**	0.093	3	0.085	Natural environment	Physical environment	0.015	2	0.014
								Spatial conditions	0.014	3	0.014
							Built environment	Physical urban development	0.015	2	0.014
								City image and brand	0.015	2	0.014
							Infrastructure and facilities	Infrastructure	0.015	2	0.014
								Public facilities	0.015	2	0.015
			Economy	0.092	4	0.085	Affordability	Properties	0.014	3	0.014
								Service and amenities	0.016	1	0.015
							Economic development	Investment and production	0.015	2	0.014
								Employment and income	0.016	1	0.015
							Internationalisation	Connection	0.016	1	0.015
								Tourism	0.013	4	0.012
			Society	0.094	2	0.087	Community structure	Population structure	0.015	3	0.014
								Lifestyle opportunities	0.015	2	0.015

(Continued)

TABLE 6.4 *(Continued)* Innovation District Place Quality Evaluation Framework

Scale	EW	RA	Theme	EW	RA	AW	Category	Indicator	EW	RA	AW
							Culture	Cultural facilities	0.015	2	0.015
								Cultural dynamics	0.015	2	0.015
							Social life	Social services and activities	0.014	3	0.014
								Social participation	0.014	3	0.014
			Wellbeing	0.094	2	0.087	Health	Physical	0.015	2	0.014
								Mental	0.014	3	0.014
							Personal development	Education	0.016	1	0.015
								Individual growth	0.016	1	0.015
							Safety and security	Personal safety	0.016	1	0.015
								Personal security	0.016	1	0.015
Cluster	0.36	1	**Form**	0.085	7	0.084	Location	Centrality	0.015	2	0.011
								Unique surrounding	0.014	3	0.010
							Urban form and structure	Urban form	0.014	3	0.010
								Urban spatial structure	0.014	3	0.010
							Design	Urban design	0.014	3	0.011
								Architectural design	0.014	3	0.010
							Amenities	Essential amenities	0.015	2	0.011
								Advanced amenities	0.015	2	0.011
			Function	0.093	3	0.091	Services	Urban mobility	0.016	1	0.012
								Management	0.015	2	0.011
							Land use	Mixed-use development	0.015	2	0.011
								Property availability and value	0.015	2	0.011
							Work condition	Thick labour market	0.016	1	0.012
								Professional networks	0.016	1	0.012
							Company profile	Technology adoption	0.015	1	0.011
								Knowledge/creative industry	0.016	1	0.012

		Ambiance	0.095	1	0.093	Cultural milieu	Public and cultural spaces	0.015	2	0.011
							Cultural events	0.014	3	0.011
						Social context	Work climate	0.016	1	0.012
							Social interactions	0.016	1	0.012
						Diversity	Diverse communities	0.015	2	0.011
							Tolerance	0.016	1	0.012
						Creativity	Creative communities	0.016	1	0.012
							Openness	0.016	1	0.012
		Image	0.090	5	0.089	Lifestyle	Pace of life	0.014	3	0.011
							Variety of lifestyle	0.015	2	0.011
						Personal safety	Informal social control	0.013	4	0.010
							Safety and security	0.015	2	0.012
						Sense of place	Place attachment	0.014	3	0.011
							Social cohesion	0.015	2	0.012
						Identity	Authenticity	0.014	3	0.011
							Brand	0.014	3	0.011
Total	1.00		1.00		1.00			1.00		1.00

not only provides the equal and AHP-based weights, but also highlights the priorities selected by the experts. This systematic framework collapses a complicated issue into a simplified classification and provides a thorough process for data collection, analysis, and policymaking. While the framework is applicable in both qualitative and quantitative types of research, weighting systems are more effective in studies with quantitative data.

On the one hand, the equal weighting approach evaluates the performance of an innovation district considering every indicator is critical. This approach is useful to determine weak and strong attributes of the innovation district. On the other hand, the AHP-based weighting approach is more suitable for evaluating the general place quality performance of an innovation district. This provides a holistic perspective. While equal weights associate with the process of decision- and strategy-making around an individual innovation district, the AHP-based weights are more likely to be used in comparative analysis of different innovation districts from various contexts.

The experts exerted the highest importance of the cluster scale (by margin only) to foster place quality in innovation districts, and therefore, this scale needs to be considered as a high priority in planning, designing, developing, and managing processes. Within this cluster scale, the 'ambiance' theme received the highest importance from the experts—other significant ones were human capital, environment, society, and wellbeing. The ambiance theme consists of eight indicators, of which five indicators stand out: 'work climate', 'social interaction', 'tolerance', 'creative communities', 'openness'. They all represent the critical impact of soft factors on creating an attractive environment in innovation districts.

In general, the panellists did not admit the importance and priority of the 'Function' theme as a representative of hard factors. However, they assessed five indicators under this category as 'high priority', which indirectly confirms the importance of traditional hard factors: 'urban mobility', 'thick labour market', 'professional networks', 'technology adoption', 'knowledge and creative industry'. The 'safety and security' indicator is also found as a significant factor in providing a liveable place quality in innovation districts. In sum, both soft and hard factors are found equally important in the cluster scale.

While the experts only identified the cluster scale as 'high priority', the regional (30%) and city (34%) impacts were also scored closely to the cluster impact (36%). These close scores clarify the fact that many themes and indicators, under the classification of city and regional scales, are marked as 'high priority'. For instance, the 'human capital' theme and 'education' indicator—under the classification of the regional scale—were marked as 'high priority'; this clarifies that fostering high-educated and talented workforces is a macro-scale issue, and is not limited to the presence of them in the cluster scale.

Among the city scale's themes, 'environment', 'society', and 'wellbeing' were evaluated as 'high priority'. The 'economy', which contained hard factors, was not selected by experts as a priority; in spite of this fact, three indicators under this category were labelled as 'high priority': 'service and amenities', 'employment and income', 'connection'. Under the classification of 'environment', only two hard factors of 'infrastructure' and 'public facilities' were found 'extremely high important'.

On the other hand, though 'society' as a soft factor was considered a priority, none of its indicators were scored high enough to become a priority. However, this situation might be caused by the fact that all indicators in this classification were rated approximately high, which in general makes the 'society' theme important. Totally, only four soft factors, under the 'wellbeing' theme, were marked as high priorities: 'education', 'individual growth', 'personal safety', 'personal security'. In sum, hard factors play a significant role in transforming a city into an attractive context and a proper host for innovation districts.

6.6 DISCUSSION AND CONCLUSION

The findings, in general, have provided invaluable insights for land-use policymakers, who aim to enhance the competitive advantage of their cities or regions. In this regard, the findings support the initial hypothesis that place quality in innovation districts not only is a complex multidimensional phenomenon, but also is impacted by the quality of clusters' context—i.e., city and region. The findings also confirm the important contribution of both hard factors and soft factors in creating attractive innovation districts.

The emergence of soft factors alongside with hard factors shows the demand for adapting with requirements of the knowledge economy. For instance, at the regional scale, the priority of 'human capital' and 'education'—rather than 'financial' factors—demonstrates the importance of highly educated and talented workers as driving forces of the new economy. Such a tendency is caused by the nature of knowledge-based activities, where economic success is relied on translating new ideas and creativity into economic return (Petrov, 2008; Alfken *et al.*, 2015); and with no doubt, such ideas are generated by the right kind of human capital, who are educated and known by the label of knowledge workers (Florida, 2005; Yigitcanlar *et al.*, 2007).

The findings resonate with the 'human capital theory', which claims that knowledge economy growth occurs in places that have a higher concentration of knowledge workers (Glaeser, 2005; Laroche *et al.*, 1999; Gabe *et al.*, 2012). The findings at the regional scale support Florida's (2005) 'creative class thesis', which emphasises that fostering and attracting knowledge workers are prior to focusing on knowledge industries. Similar to the regional scale, the priority of 'education' and 'individual growth' are identified at the city level. However, the 'population structure' is not scored the same importance; this indicates that in the era of knowledge economy, the quality of workforces outreaches the quantity of them.

At the cluster scale, similarly, a strong presence of 'creative communities'—i.e., knowledge workers—is found as a priority. However, it is hard to claim if indicators such as 'human capital', 'education', 'individual growth', and 'creative communities' could be exactly categorised as soft factors at all scales. While these indicators mean the same at different scales, they become more tangible at the regional and city levels—for instance, the existence of universities and research centres. Consequently, these indicators mainly utilise soft factors to attract knowledge workers at the cluster level. In contrast, they explain the responsibility of city and regional scales for providing facilities—through hard factors—and fostering such talent force.

Like the above-mentioned indicators, many elements of place quality are relevant to more than one scale, and the difference between various scales is gradual (Trip, 2007). The 'safety and security' issue is a good example of such indicators that represent various aspects of the same concept in different scales. From the perspective of the cluster level, safety and security relate to day-to-day issues such as petty crimes or street safety for pedestrians. However, perceived safety is also related to the overall crime level in the city, which may in turn influence the image of an innovation district. It is a priority to provide safety and security at both cluster and city levels, to create an attractive and liveable image for innovation districts.

In contrast, some indicators are more likely to prevail solely on one scale. For instance, the 'environment' theme is only selected as a priority at the city level, which indicates the regional scale is too broad, and the cluster scale is too limited to reflect the environmental characteristic of a location. These findings support previous studies that considered natural and built environment as a significant tool for attracting knowledge workers to cities (Brown & Mczyski, 2009; Darchen & Tremblay, 2010; Alfken *et al.*, 2015).

The results also identified significant dissimilarities around the role of hard and soft factors at each geographical scale. Florida (2005) claimed that knowledge workers choose locations based on non-economic factors—such as tolerance, openness, diversity, and authenticity—where many scholars have criticised this view. These studies insist on the importance of economic factors, infrastructure, services, and employment opportunities (Storper & Manville, 2006; Storper & Scott, 2009; Darchen & Tremblay, 2010; Scott, 2010). However, our findings revealed that neither Florida's thesis nor its anti-thesis are incorrect. The solution is as simple as choosing the right scale for the application of such hard and soft factors. Florida's anti-thesis is more appropriate for the city scale, where essential amenities need to be provided for people.

The city level is responsible to offer a safe and successful economic environment, and a thick pool of work opportunities. It is also responsible for providing appropriate infrastructures and adequate services and amenities, which guarantee a high level of life quality and fulfil the basic expectation of knowledge workers. However, when it comes to the cluster scale, the competition is not about such amenities anymore; as they are already provided by the city. What is vital at the cluster level is to offer an extra attractive ambiance—e.g., providing a dynamic social or professional network, and supporting diverse communities and various lifestyles. Offering an appealing ambiance will empower competitive advantages for attracting and retaining knowledge workers and industries.

While the above-mentioned discussion reveals some theoretical findings, the evaluation framework represents the practical outcome of the study. The framework offers guidance to urban planners and policymakers to evaluate place quality in innovation districts, compare different clusters, and reveal their exact weaknesses and strength points. It also helps in making reliable strategies for developing and managing innovation districts.

However, the following limitations of this study should be taken into consideration: (a) The expert's comments were analysed in a workshop to reduce the risk of misunderstanding or biased interpretation; however, there still might be some biased in analysing the qualitative data. (b) There might be a potential risk of data reliability and validity issues due to the way expertise area categories and geographic locations are determined, and experts from these categories and locations are invited. While in general statistical analysis and literature supports the findings, different experts might have provided different inputs in the study. (c) The study only considered generic indicators, which are applicable to all innovation districts. While characteristics of innovation districts are place-based and difficult to generalise, specific indicators of a local context might need to be considered. (d) The evaluation model has the potential to become an effective tool for investigating place quality in innovation districts. However, before applying the model, the weighting system may need to be modified based on the unique characteristics of each innovation district.

This chapter explored a practical approach to understand how place quality contributes to the attractiveness of innovation districts by determining some indicators, and developing a framework. However, the findings of this study have potential implications. For example, the insights generated in this research help city administrators, planners, and urban designers, firstly, to better understand the critical place quality issues in their innovation districts. Then the proposed framework can be used as a tool to assess the strengths and weaknesses of their innovation districts. Lastly, the outcomes of the assessment can help them deliver high-quality and performance innovation districts.

The proposed framework is only a starting point for future studies aiming to evaluate place quality in innovation districts. Our prospective studies will focus on testing and applying the framework in Australian innovation districts. Evaluating place quality in real-world innovation districts will reveal detailed insights into their strengths, weaknesses, and potentials. However, further investigations are required to identify appropriate methods for applying the model, and converting the result into practical guidelines (Esmaeilpoorarabi *et al.*, 2018).

ACKNOWLEDGEMENTS

This chapter, with permission from the copyright holder, is a reproduced version of the following journal article: Esmaeilpoorarabi, N., Yigitcanlar,

T., Guaralda, M., & Kamruzzaman, M., (2018). Evaluating place quality in innovation districts: a Delphic hierarchy process approach. Land Use Policy, 76(1), 471-486.

REFERENCES

Alawadi, K., & Dooling, S. (2016). Challenges and opportunities for participatory planning approaches within Dubai's urban context. *Journal of Urbanism*, 9(3), 276–301. DOI:10.1080/17549175.2015.1045924

Arbolino, R., Carlucci, F., Cirà, A., Ioppolo, G., & Yigitcanlar, T. (2017). Efficiency of the EU regulation on greenhouse gas emissions in Italy. *Ecological Indicators*, 81, 115–123. DOI:10.1016/j.ecolind.2017.05.053

Alfken, C., Broekel, T., & Sternberg, R. (2015) Factors explaining the spatial agglomeration of the creative class. *European Planning Studies*, 23(12), 2438–2463. DOI:10.1080/09654313.2014.979767

Alshehri, S.A., Rezgui, Y., & Li, H. (2015). Disaster community resilience assessment method. *Natural Hazards*, 78, 395–416. DOI:10.1007/s11069-015-1719-5

Ameyaw, E.E., & Chan, A.P. (2015). Evaluating key risk factors for PPP water projects in Ghana. *Journal of Facilities Management*, 13(2), 133–155. DOI:10.1108/JFM-10-2013-0051

Anttiroiko, A.V. (2015). City branding as a response to global intercity competition. *Growth and Change*, 46(2), 233–252.

Aquilani, B., Silvestri, C., Ioppolo, G., & Ruggieri, A. (2018). The challenging transition to bio-economies. *Journal of Cleaner Production*, 172, 4001–4009. DOI:10.1016/j.jclepro.2017.03.153

Arora, A., Florida, R., Gates, G.J., & Kamlet, M. (2000). *Human capital, quality of place, and location*. Heinz Endowments and Sustainable Pittsburgh, Pittsburgh.

Asheim, B., & Hansen, H.K. (2009). Knowledge bases, talents, and contexts. *Economic Geography*, 85(4), 425–442. DOI: 10.1111/j.1944-8287.2009.01051.x

Ballas, D. (2013). What makes a 'happy city'? *Cities*, 32(1), 39–50. DOI:10.1016/j.cities.2013.04.009

Baum, S., O'Connor, K., & Yigitcanlar, T. (2009). The implications of creative industries for regional outcomes. *International Journal of Foresight and Innovation Policy*, 5, 44–64. DOI:10.1504/IJFIP.2009.022098

Benneworth, P., & Ratinho, T. (2014). Reframing the role of knowledge parks and science cities in knowledge-based urban development. *Environment and Planning*, 32(5), 784–808. DOI:10.1068/c1266r

Bereitschaft, B., & Cammack, R. (2015). Neighborhood diversity and the creative class in Chicago. *Applied Geography*, 63, 166–183. DOI:10.1016/j.apgeog.2015.06.020

Boddy, M. (1999). Geographical economics and urban competitiveness. *Urban Studies*, 36(5-6), 811–842. DOI: 10.1080/0042098993

Bontis, N. (2004). National intellectual capital index. *Journal of Intellectual Capital*, 5(1), 13–39. DOI:10.1108/14691930410512905

Bontje, M., & Musterd, S. (2009). Creative industries, creative class and competitiveness. *Geoforum*, 40(5), 843–852. DOI:10.1016/j.geoforum.2009.07.001

Boren, T., Young, C., Kulturgeografiska, I., Stockholms, U., & Samhallsvetenskapliga, F. (2013). Getting creative with the creative city? *International Journal of Urban and Regional Research*, 37, 1799–1815. DOI:10.1111/j.1468-2427.2012.01132.x

Brown, J., & Mczyski, M. (2009). Complexcities: locational choices of creative knowledge workers. *Built Environment*, 35(2), 238–252. DOI:10.2148/benv.35.2.238

Brown, J. (2015). Home from home? *European Planning Studies*, 23(12), 2336–2355. DOI:10.1080/09654313.2014.988012

Carrillo, F.J. (2004). Capital cities. *Journal of Knowledge Management*, 8(5), 28–46. DOI:10.1108/1367327041058738

Carrillo, F.J., Yigitcanlar, T., García, B., & Lönnqvist, A. (2014). *Knowledge and the city*. Routledge, New York. DOI:10.4324/9781315856650

Chan, S.L., Wey, W.M., & Chang, P.H. (2014). Establishing disaster resilience indicators for Tan-sui river basin in Taiwan. *Social Indicators Research*, 115(1), 387–418. DOI:10.1007/s11205-012-0225-3

Chiang, Y. C., & Lei, H. Y. (2016). Using expert decision-making to establish indicators of urban friendliness for walking environments. *International Journal of Health Geographics*, 15(1), 1–12. DOI:10.1186/s12942-016-0071-7

Clifton, N. (2008). The "creative class" in the UK. *Human Geography*, 90(1), 63–82. DOI:10.1111/j.1468-0467.2008.00276.x

Clifton, N., & Cooke, P. (2009) Creative knowledge workers and location in Europe and North America. *Creative Industries Journal*, 2(1), 73–89. DOI:10.1386/cij.2.1.73/1

Colagiuri, R., Boylan, S., & Morrice, E. (2015). Research Priorities for NCD prevention and climate change. *International Journal of Environmental Research and Public Health*, 12(10), 12941–12957. DOI:10.3390/ijerph121012941

Cooke, P. (2001). From technopoles to regional innovation systems. *Canadian Journal of Regional Science*, 24, 21–40.

Craglia, M., Leontidou, L., Nuvolati, G., & Schweikart, J. (2004). Towards the development of quality of life indicators in the digital city. *Environment and Planning B*, 31(1), 51–64. DOI:10.1068/b12918

Da Cunha, I.V., & Selada, C. (2009). Creative urban regeneration. *International Journal of Innovation and Regional Development*, 1(4), 371–386. DOI:10.1504/IJIRD.2009.022728

Dalkey, N. C., & Helmer, O. (1963). An experimental application of the Delphi method to the use of experts. *Management Science*, 9, 458–467.

Darchen, S., & Tremblay, D.G. (2010). What attracts and retains knowledge workers/students. *Cities*, 27(4), 225–233. DOI:10.1016/j.cities.2009.12.009

Deas, I., & Giordano, B. (2001). Conceptualising and measuring urban competitiveness in major English cities. *Environment and Planning A*, 33(8), 1411–1429. DOI:10.1068/a33142

Diamond, I.R., Grant, R.C., Feldman, B.M., Pencharz, P.B., Ling, S.C., Moore, A.M., & Wales, P. W. (2014). Defining consensus. *Journal of clinical epidemiology*, 67, 401–409. DOI:10.1016/j.jclinepi.2013.12.002

D'Mello, M., & Sahay, S. (2007). I am kind of a nomad where I have to go places and places. *Information and Organization*, 17(3), 162–192. DOI:10.1016/j.infoandorg.2007.04.001

Dur, F., & Yigitcanlar, T. (2015). Assessing land-use and transport integration via a spatial composite indexing model. *International Journal of Environmental Science and Technology*, 12(3), 803–816. DOI:10.1007/s13762-013-0476-9

Durmaz, S.B. (2015). Analyzing the quality of place. *Journal of Urban Design*, 20(1), 93–124. DOI:10.1080/13574809.2014.972348

Esmaeilpoorarabi, N., Yigitcanlar, T., & Guaralda, M. (2016a). Place quality and urban competitiveness symbiosis? *International Journal of Knowledge-Based Development*, 7(1), 4–21. DOI:10.1504/IJKBD.2016.075444

Esmaeilpoorarabi, N., Yigitcanlar, T., & Guaralda, M. (2016b). Towards an urban quality framework. *International Journal of Knowledge-Based Development*, 7(3), 290- 312. DOI:10.1504/IJKBD.2016.078556

Esmaeilpoorarabi, N., Yigitcanlar, T., Guaralda, M., & Kamruzzaman, M., (2018). Evaluating place quality in innovation districts: a Delphic hierarchy process approach. *Land Use Policy*, 76(1), 471–486. DOI: 10.1016/j.landusepol.2018.02.027

Evers, H.D. (2008). Knowledge hubs and knowledge clusters. *ZEF Working Paper Series*, 27(1), 1–25. DOI:10.1142/9789814343688_0002

Flew, T. (2012). Creative suburbia. *International Journal of Cultural Studies*, 15(3), 231–246. DOI:10.1177/1367877911433746

Florida, R. (2005). *Cities and the creative class*. Routledge, New York. DOI:10.4324/9780203997673

Frenkel, A., Bendit, E., & Kaplan, S. (2013a). Residential location choice of knowledge-workers. *Cities*, 35(1), 33–41. DOI:10.1016/j.cities.2013.06.005

Frenkel, A., Bendit, E., & Kaplan, S. (2013b). The linkage between the lifestyle of knowledge-workers and their intra-metropolitan residential choice. *Computers, Environment and Urban Systems*, 39(10), 151–161. DOI:10.1016/j.compenvurbsys.2012.09.001

Gabe, T., Florida, R., Mellander, C. (2012). The creative class and the crisis. *Cambridge Journal of Regions, Economy and Society*, 6(1), 37–53. DOI:10.1093/cjres/rss012

Gigovic, L., Pamucar, D., Lukic, D., & Markovic, S. (2016). GIS-Fuzzy DEMATEL MCDA model for the evaluation of the sites for ecotourism development. *Land Use Policy*, 58, 348–365. DOI:10.1016/j.landusepol.2616.07.030

Glaeser, E. (2005). Review of Richard Florida's the rise of the creative class. *Regional Science and Urban Economics*, 35(5), 593–596. DOI:10.1016/j.regsciurbeco.2005.01.005

Grant, J.L., & Buckwold, B. (2013). Precarious creativity: immigrant cultural workers. *Cambridge Journal of Regions, Economy and Society*, 6(1), 1–17. DOI:10.1093/cjres/rss008

Gu, X. (2014). Cultural industries and creative clusters in Shanghai. *City, Culture and Society*, 5(3), 123–130. DOI:10.1016/j.ccs.2014.07.005

Hafeez, K., Foroudi, P., Dinnie, K., Nguyen, B., & Parahoo, S.K. (2016). The role of place branding and image in the development of sectoral clusters. *Journal of Brand Management*, 23(4), 383–402. DOI:10.1057/bm.2016.18

Hassanzadeh, A., Namdarian, L., Elahi, S. B., & Majidpour, M. (2014). Impact of technology foresight on the policy-making process in Iran. *Science, Technology and Society*, 19(3), 275–304. DOI:10.1177/0971721814548112

Hazelkorn, E., & Murphy, C. (2002). The cultural economy of Dublin. *Sociological Chronicles*, 3(1), 1–10.

He, J.L., & Gebhardt, H. (2014). Space of creative industries: a case study of spatial characteristics of creative clusters in Shanghai. *European Planning Studies*, 22(11), 2351–2368. DOI:10.1080/09654313.2013.837430

He, S.Y., Kuo, Y.H., & Wu, D. (2016). Incorporating institutional and spatial factors in the selection of the optimal locations of public electric vehicle charging facilities. *Transportation Research Part C*, 67, 131–148. DOI:10.1016/j.trc.2016.02.003

Heebels, B., & Van Aalst, I. (2010). Creative clusters in Berlin. *Geografiska Annaler*, 92(4), 347–363. DOI:10.1111/j.1468-0467.2010.00357.x

Holey, E. A., Feeley, J. L., Dixon, J., & Whittaker, V. J. (2007). An exploration of the use of simple statistics to measure consensus and stability in Delphi studies. *BMC Medical Research Methodology*, 7(1), 52–62. DOI:10.1186/1471-2288-7-52

Homan, S. (2014). Liveability and creativity. *City, Culture and Society*, 5(3), 149–155. DOI:10.1016/j.ccs.2014.06.002

Huang, W. C., & Lee, Y. Y. (2016). Strategic planning for land use under extreme climate changes. *Sustainability*, 8(1), 1–17. DOI:10.3390/su8010053

Hsu, C.C., & Sandford, B.A. (2007). The Delphi technique. *Practical Assessment, Research & Evaluation*, 12, 1–8.

Hsueh, S.L., Lee, J.R., & Chen, Y.L. (2013). DFAHP multicriteria risk assessment model for redeveloping derelict public buildings. *International Journal of Strategic Property Management*, 17(4), 333–346. DOI:10.3846/1648715X.2013.852995

Jin, J., Yao, J., & Wang, J.X. (2016). Value of planning and design of Buddhist temples in Hebei Province. *Open House International*, 41(3), 44–49. DOI:10.1108/OHI-03-2016-B0007

Jordan, E., & Javernick-Will, A. (2013). Indicators of community recovery. *Natural Hazards Review*, 14(1), 21–28. DOI:10.1061/(ASCE)NH.1527-6996.0000087

Jozi, S.A., Sohrabi Balsini, M., & Jamshidi, A. (2013). Investigating the promotion possibility of no-hunting areas to the protected areas. *Journal of the Indian Society of Remote Sensing*, 41(3), 577–585. DOI:10.1007/s12524-012-0250-3

Julsrud, T.E., & Priya-Uteng, T. (2015). Technopolis, shared resources or controlled mobility? *European Journal of Futures Research*, 3(1), 1–13. DOI:10.1007/s40309-015-0069-6

Khorramshahgol, R., & Moustakis, V.S. (1988). Delphic hierarchy process. *European Journal of Operational Research*, 37(3), 347–354. DOI: 10.1016/0377-2217(88)90197-X

Kiba-Janiak, M. (2016). Key success factors for city logistics from the perspective of various groups of stakeholders. *Transportation Research Procedia*, 12, 557–569. DOI:10.1016/j.trpro.2016.02.011

Kim, H.Y., Choi, Y., Kim, H., & Oh, S.H. (2016). Planning for the suitable? *KSCE Journal of Civil Engineering*, 20, 359–366. DOI:10.1108/K-11-2021-1114

Kitson, M., Martin, R., & Tyler, P. (2004) Regional competitiveness. *Regional Studies*, 38(90), 991–999. DOI:10.1080/0034340042000320816

Kloosterman, R.C., & Trip, J.J. (2011). Planning for quality? *Journal of Urban Design*, 16(4), 455–470. DOI:10.1080/13574809.2011.585863

Laroche, M., Merette, M., Ruggeri, G.C. (1999) On the concept and dimensions of human capital in a knowledge-based economy context. *Canadian Public Policy*, 25(1), 87–100.

Lawton, P., Murphy, E., & Redmond, D. (2013). Residential preferences of the creative class? *Cities*, 31(1), 47–56. DOI:10.1016/j.cities.2012.04.002

Lee, J.H., Hancock, M.G., & Hu, M.C. (2014). Towards an effective framework for building smart cities. *Technological Forecasting and Social Change*, 89(1), 80–99. DOI:10.1016/j.techfore.2013.08.033

Loures, L., Santos, R., & Panagopoulos, T. (2007). Urban parks and sustainable city planning. *Transactios on Environment and Development*, 3(10), 171–180. DOI:10.1016/j.proenv.2016.04.042

Mafi, D., Feghhi, J., Danehkar, A., & Yarali, N. (2015). Classification and prioritization of negative factors affecting on mangrove forests using Delphi method. *Advances in Bioresearch*, 63, 78–92. DOI:10.15515/abr.0976-4585.6.3.7892

Malkina-Pykh, I.G., & Pykh, Y.A. (2008) Quality-of-life indicators at different scales. *Ecological Indicators*, 8(6), 854–862. DOI:10.1016/j.ecolind.2007.01.008

Marans, R.W. (2003). Understanding environmental quality through quality of life studies. *Landscape and Urban Planning*, 65(1-2), 73–83. DOI:10.1016/S0169-2046(02)00239-6

McCann, E.J. (2004). Best places. *Urban Studies*, 41(10), 1909–1929. DOI:10.1080/0042098042000256314

McNeill, D., Dowling, R., & Fagan, B. (2005). Sydney/global/city. *International Journal of Urban and Regional Research*, 29(4), 935–944. DOI:10.1111/j.1468-2427.2005.00629.x

Meijering, J.V., Tobi, H., Van den Brink, A., Morris, F., & Bruns, D. (2015). Exploring research priorities in landscape architecture. *Landscape and Urban Planning*, 137, 85–94. DOI: 10.1016/j.landurbplan.2015.01.002

Mendez, R., & Moral, S.S. (2011). Spanish cities in the knowledge economy. *European Urban and Regional Studies*, 18(2), 136–155. DOI:10.1177/0969776410381039

Messeni-Petruzzelli, A., Albino, V., Carbonara, N., & Rotolo, D. (2010). Leveraging learning behavior and network structure to improve knowledge gatekeepers' performance. *Journal of Knowledge Management*, 14(5), 635–658. DOI:10.1108/13673271011074818

Metaxiotis, K., Carrillo, F.J., & Yigitcanlar, T. (Eds.). (2010). *Knowledge-based development for cities and societies: integrated multi-level approaches*. IGI Global, London. DOI:10.4018/978-1-61520-721-3

Millar, C.C., & Ju-Choi, C. (2010). Development and knowledge resources. *Journal of Knowledge Management*, 14(5), 759–776. DOI:10.1108/13673271011074881

Morais, P., Miguéis, V.L., & Camanho, A.S. (2013) Quality of life experienced by human capital. *Social Indicators Research*, 110(1), 187–206. DOI:10.1007/s11205-011-9923-5

Murphy, E., Fox-Rogers, L., & Redmond, D. (2015). Location decision making of creative industries. *Growth and Change*, 46(1), 97–113. DOI:10.1111/grow.12086

Musterd, S., & Gritsai, O. (2013). The creative knowledge city in Europe. *European Urban and Regional Studies*, 20(3), 343–359. DOI:10.1177/0969776412439199

Norouzian-Maleki, S., Bell, S., Hosseini, S. B., & Faizi, M. (2015). Developing and testing a framework for the assessment of neighbourhood liveability in two contrasting countries. *Ecological Indicators*, 48, 263–271. DOI:10.1016/j.ecolind.2014.07.033

Nurick, R., & Johnson, V. (1998). Towards community-based indicators for monitoring quality of life and the impact of industry in south Durban. *Environment & Urbanization*, 10(1), 233–250. DOI: 10.1630/095624798101284400

Pancholi, S., Yigitcanlar, T., & Guaralda, M. (2014). Urban knowledge and innovation spaces. *Asia Pacific Journal of Innovation and Entrepreneurship*, 8(1), 15–38. DOI:10.1080/10630732.2016.1164443

Pancholi, S., Yigitcanlar, T., & Guaralda, M. (2015). Place making facilitators of knowledge and innovation spaces. *International Journal of Knowledge-Based Development*, 6(3), 215–240. DOI:10.1504/IJKBD.2015.072823

Pancholi, S., Yigitcanlar, T., & Guaralda, M. (2017). Governance that matters. *Journal of Place Management and Development*, 10(1), 73–87. DOI:10.1108/JPMD-08-2016-0053

Pancholi, S., Yigitcanlar, T., & Guaralda, M. (2019). Place making for innovation and knowledge-intensive activities. *Technological Forecasting & Social Change*, 146(Sep), 616–625. DOI: 10.1016/j.techfore.2017.09.014.

Parker, R. (2010). Evolution and change in industrial clusters. *European Planning Studies*, 17(3), 245–260. DOI:10.1177/0969776409358244

Pelkonen, A. (2005). State restructuring, urban competitiveness policies and technopole building in Finland. *European Planning Studies*, 13(5), 685–705.

Perveen, S., Kamruzzaman, M., & Yigitcanlar, T. (2017). Developing policy scenarios for sustainable urban growth management. *Sustainability*, 9, 1787. DOI:10.3390/su9101787

Porter, M.E. (1998). Location, clusters, and the new microeconomics of competition. *Business Economics*, 33(1), 7–13.

Petrov, A.N. (2008) Talent in the cold? *Arctic*, 61(2), 162–176. DOI:10.14430/arctic15

Ruppert, J., & Duncan, R.G. (2017). Defining and characterizing ecosystem services for education. *Journal of Research in Science Teaching*, 54(6), 737–763. DOI:10.1002/tea.21384

Rust, N.A. (2017). Can stakeholders agree on how to reduce human-carnivore conflict on Namibian livestock farms? *Oryx*, 51(2), 339–346. DOI:10.1017/S0030605315001179

Saaty, T.L. (1980). *The analytic hierarchy process*. McGraw-Hill, New York.

Santos, L.D., & Martins, I. (2007) Monitoring urban quality of life. *Social Indicators Research*, 80(2), 411–425. DOI:10.1007/s11205-006-0002-2

Sarimin, M., & Yigitcanlar, T. (2012). Towards a comprehensive and integrated knowledge-based urban development model. *International Journal of Knowledge-Based Development*, 3(2), 175–192. DOI:10.1504/IJKBD.2012.047035

Schuckmann, S.W., Gnatzy, T., Darkow, I.L., & Von der Gracht, H.A. (2012). Analysis of factors influencing the development of transport infrastructure until the year 2030. *Technological Forecasting and Social Change*, 79(8), 1373–1387. DOI:10.1016/j.techfore.2012.05.008

Scott, A.J. (2010). Jobs or amenities? *Papers in Regional Science*, 89(1), 43–63. DOI: 10.1111/j.1435-5957.2009.00263.x

Sepe, M. (2010). Liveability, quality and place identity in the contemporary city. *Journal of Place Management and Development*, 3(3), 221–246. DOI:10.1108/17538331011083952

Sepe, M., & Pitt, M. (2014). The characters of place in urban design. *Urban Design International*, 19(3), 215–227. DOI:10.1057/udi.2013.32

Singhal, S., McGreal, S., & Berry, J. (2013). An evaluative model for city competitiveness: Application to UK cities. *Land Use Policy*, 30(1), 214–222. DOI:10.1016/j.landusepol.2012.03.018

Smith, T., Nelischer, M., & Perkins, N. (1997). A framework for understanding the relationship between quality and physical form. *Landscape and Urban Planning*, 39, 229–241.

Sohn, D.W., & Kenney, M. (2007). Universities, clusters, and innovation systems. *World Development*, 35(6), 991–1004. DOI:10.1016/j.worlddev.2006.05.008

Soria-Lara, J.A., & Banister, D. (2017). Participatory visioning in transport backcasting studies. *Journal of Transport Geography*, 58, 113–126. DOI:10.1016/j.jtrangeo.2016.11.012

Sotarauta, M. & Linnamaa, R. (2001). Urban competitiveness and management of urban policy networks. *Technology, Society and Environment*, 2, 1–17.

Storper, M., & Manville, M. (2006). Behaviour, preferences and cities. *Urban studies*, 43(8), 1247–1274. DOI:10.1080/004209806007756

Storper, M., & Scott, A.J. (2009). Rethinking human capital, creativity and urban growth. *Journal of Economic Geography*, 9(2), 147–167.

Sutterluty, A., Hesser, F., Schwarzbauer, P., Schuster, K. C., Windsperger, A., & Stern, T. (2017). A Delphi approach to understanding varying expert viewpoints in sustainability communication. *Journal of Industrial Ecology*, 21(2), 412–422. RePEc:bla:inecol:v:21:y:2017:i:2:p:412-422

Taddeo, R., Simboli, A., Ioppolo, G., & Morgante, A. (2017). Industrial symbiosis, networking and innovation. *Sustainability*, 9(2), 169. DOI:10.3390/su9020169

Trip, J.J. (2007). Assessing quality of place. *Journal of Urban Affairs*, 29(5), 501–517. DOI: 10.1111/j.1467-9906.2007.00362.x

Van Winden, W., Van den Berg, L., & Pol, P. (2007). European cities in the knowledge economy. *Urban Studies*, 44(3), 525–549. DOI:10.1080/00420980601131886

Van Winden, W., De Carvalho, L., Van Tuijl, E., Van Haaren, J., & Van den Berg, L. (2013). *Creating knowledge locations in cities*. Routledge, London. DOI: 10.4324/9780203127162

Viitanen, J., & Kingston, R. (2014). Smart cities and green growth: outsourcing democratic and environmental resilience to the global technology sector. *Environment and Planning A*, 46(4), 803–819. DOI:10.1068/a46242

Von der Gracht, H.A. (2012). Consensus measurement in Delphi studies. *Technological Forecasting & Social Change*, 79(8), 1525–1536. DOI:10.1016/j.techfore.2012.04.013

Yang, Z., Shi, X., & Su, Q. (2016). Knowledge-based raster mapping approach to wetland assessment. *Wetlands*, 36(1), 143–158. DOI:10.1007/s13157-015-0725-4

Yigitcanlar, T., Baum, S., & Horton, S. (2007). Attracting and retaining knowledge workers in knowledge cities. *Journal of Knowledge Management*, 11(5), 6–17. DOI:10.1108/13673270710819762

Yigitcanlar, T., & Velibeyoglu, K. (2008). Knowledge-based urban development. *Local Economy*, 23(3), 195–207. DOI:10.1080/02690940802197358

Yigitcanlar, T., O'Connor, K., & Westerman, C. (2008a). The making of knowledge cities. *Cities*, 25(2), 63–72. DOI:10.1016/j.cities.2008.01.001

Yigitcanlar, T., Velibeyoglu, K., & Martinez-Fernandez, C. (2008b). Rising knowledge cities. *Journal of Knowledge Management*, 12(5), 8–20. DOI:10.1108/13673270810902902

Yigitcanlar, T. (2010). Making space and place for the knowledge economy. *European Planning Studies*, 18(11), 1769–1786. DOI:10.1080/09654313.2010.512163

Yigitcanlar, T., Metaxiotis, K., & Carrillo, F. J. (Eds.). (2012). *Building prosperous knowledge cities*. Edward Elgar, London. DOI:10.4337/9780857936042

Yigitcanlar, T., & Dur, F. (2013). Making space and place for knowledge communities. *Australasian Journal of Regional Studies*, 19(1), 36–63.

Yigitcanlar, T., & Bulu, M. (2015). Dubaization of Istanbul. *Environment and Planning A*, 47(1), 89–107. DOI: 10.1068/a130209p

Yigitcanlar, T., Edvardsson, I., Johannesson, H., Kamruzzaman, M., Ioppolo, G., & Pancholi, S. (2017). Knowledge-based development dynamics in less favoured regions. *European Planning Studies*, 25(12), 2272–2292. DOI:10.1080/09654313.2017.1358699

Yoon, K., & Kim, C. (2016). Components of expertise of public officials in emergency management in Korea. *International Review of Public Administration*, 21(1), 37–56. DOI:10.1080/12294659.2015.1111127

Youtie, J., & Shapira, P. (2008). Building an innovation hub. *Research Policy*, 37(8), 1188–1204. DOI:10.1016/j.respol.2008.04.012

Zeeman, H., Wright, C.J., & Hellyer, T. (2016). Developing design guidelines for inclusive housing. *Journal of Housing and the Built Environment*, 31(4), 761–772. DOI: 10.1007/s10901-016-9499-0

APPENDIX 6.A REGIONAL SCALE FRAMEWORK OF INNOVATION DISTRICT PLACE QUALITY

Scale	Theme	Category	Indicator	Description
Region	**Instrumental capital**	Tangible	Natural	Existing before the settlement: geographic location, longitude and latitude, climate, orography and hydrography, geology and seismography, accessibility, natural surroundings—How good are the city's location, climate and physical landscape? How advanced are the city's environmental policies?
			Artificial	Created or incorporated by settlers: land scaping, underground infrastructure, urban structure, zoning, neighbourhoods, districts, civil infrastructure, highways, streets, bridges, tunnels, traffic configuration, public transport networks—How good is the quality of urban physical infrastructure? How good is the city's transportation and telecommunications connectivity?
		Intangible/ Symbolic	Physical and symbolic systems	Organisation and production systems in non-electronic repositories: government systems and procedures, commercial institutions, NGO's, manual registration, processing, storing, retrieval and distribution systems, archaeological and historical sites and records, museums and collections—How robust are the governance and social organisation structure?
			Digital systems	Organisation and production systems in electronic repositories: How good is the coverage and quality of public information services, electronic registration, processing, storing, retrieval and distribution systems like

(*Continued*)

Scale	Theme	Category	Indicator	Description
				e-government and electronic media, electronic databases like government databases, public services and other databases belonging to civic organisations?
	Meta capital	Articulation	Financial	Economic sustainability, financial institutions and resources available to people: the region's articulation of monetary denomination of production value dimensions, which elicit economic sustainability within the capital system (macro indicators: investment, gross domestic product, tax system, un/employment)—How healthy are public financial accounts?
			Relational	Social integration and cohesion, political state and institutions: political demarcation (national and other encompassing authority, structural and functional limits, surrounding or closely related places), institutions (government, industry and trade)—How strong is the city's social cohesion? How good are the city's regional, national and international relations?
		Referential	Identity	Formal and informal elements that contribute to determine the region's identity, its clarity and differentiation (historic profile, city characterisation, belonging, physical)—How distinctive and well positioned is the city's identity? How valuable is the city's brand and reputation? How good is the city knowledge-based development strategy and the capability to enact it?
			Intelligence	Region's systems capacity, make sense of and respond to external agents and events, which are significant to the city's welfare:

(Continued)

Scale	Theme	Category	Indicator	Description
				city's strategic planning agencies, city-public/private future centres, prospective studies—How good are the city's strategic intelligence systems? Does it have a future centre or equivalent initiative?
	Human capital	Collectives	Cultural liveliness	Wealth of cultural inheritance and cultural fitness: languages, religions, customs, clothing, food, celebrations and rituals, age and history, cultural heritage—How diverse and creative is the city's cultural environment? How good is the city's cultural capacity to foster economic and political innovation?
			Evaluative capacity	Developing an evaluation culture of valuing evidence, valuing questioning, and valuing evaluative thinking: resilience and adaptability, ethnic diversity, tolerance, participation, linguistic diversity, demographic configuration, socio-economic configuration, labour force and immigration.
		Individuals	Health	A state of complete physical, mental, and social well-being, and not merely the absence of disease or infirmity: biological inheritance and physical development, health facilities available to public, health insurance, health care system, social assistance—How good are public health standards?
			Education	Training and holistic personal development: educational systems in schools and quality of universities—How good is the educational system in schools and universities? How 'youthful' is the city? How equal are social and economic opportunities for individuals' citizens to develop their full potential?

APPENDIX 6.B CITY SCALE FRAMEWORK OF INNOVATION DISTRICT PLACE QUALITY

Scale	Theme	Category	Indicator	Description
City	**Environment**	Natural environment	Physical environment	The part of the human environment that includes purely physical factors: green spaces, climate, air quality, water quality, noise.
			Spatial conditions	Unique natural conditions that make a city distinctive from others: waterfront, natural resources, locating in remote continents.
		Built environment	Physical urban development	Physical patterns which are necessary to promote the economic betterment, convenience and welfare: land use, urban design, architectural design, location and character of public buildings and structures, design of streets.
			City image and brand	The process employed by public administrations to create place brands, based on the visual, verbal, and behavioural expression of a place: smart city, innovative and creative city, culture-creative city.
		Infrastructure and facilities	infrastructure	The basic physical and organisational structures and facilities: transportation, roads, bridges, tunnels, loading docks, cold storage chambers, electrical capacity, fuel tanks, cranes, overhead clearances, components of water supplies, sewers, electrical grids, telecommunications.
			Public facilities	Urban objects those are designated to fulfil supportive functions related to the health and well-being of the citizens of a modern society or urban area: Sport, health, social and educational facilities, recreational Centres and Parks, Fire stations and safety services, shopping facilities.
	Economy	Affordability	Properties	Availability and affordability of properties in a city: housing affordability, home ownership, properties' prices, rental prices.

(Continued)

Scale	Theme	Category	Indicator	Description
			Services and amenities	Accessibility of services and amenities for all people: diversity of housing, accessibility to public facilities, accessibility to healthcare system.
		Economic development	Investment and production	The phenomenon of investment and market productivity and rise in GDP: growth rate of GDP, labour production efficiency, knowledge-based products.
			Employment and income	The percentage of the labour force that is employed: job properties, labour market, income and consumption, expenses, cost of living.
		Internationalisation	Connections	The capability of a city to engage in the regional and international relations: centre of national & international events, international transport connection, international reputation.
			Tourism	Refers generally to leisure vacations in cities, visiting them and for carrying out diverse activities: urban attractions, cultural, religious, business and sports tourism, tourist facilities (hotels, transport, airport).
	Society	Community structure	Population structure	The makeup or composition of a population: median age, live births, foreign residents, gender, age, occupations, diversity of population.
			Lifestyle opportunities	The diversity of lifestyle: gay index, bohemians, ethnic diversity.
		Culture	Cultural facilities	Facilities which are used for the programming, production, presentation, exhibition of any of the arts and cultural discipline: music, dance, theatre, creative writing, literature, painting, sculpture, folk arts, photography, crafts, media arts, visual arts, programs of museums.
			Cultural dynamics	The dynamic character of culture in a city: heritage, cultural diversity, cultural performances and events, usage of cultural facilities.
		Social life	Social services & activities	Asset, services and activities those aim to promote the welfare of others: access to social services for children, elderly,

(Continued)

Scale	Theme	Category	Indicator	Description
				handicapped and poor families, low rate of youth unemployment, and equal opportunity for men and women.
			Social participation	Refer to one's participation in the activities of a social group: political participation, voluntary associations, sense of community.
	Wellbeing	Health	Physical	It is the soundness of the body, freedom from disease or abnormality, and the condition of optimal well-being: premature death rate, access to facilities and activities that empower the physical health and wellbeing like hospitals, sports facilities, healthy food.
			Mental	A person's condition with regard to their psychological and emotional well-being: suicide and homicide rate, active use of rehabilitation centres for drug addicts, homeless population.
		Personal development	Education	The process of receiving or giving systematic instruction, especially at a school or university: students in higher education, students pursuing postgraduate studies, master's degree and doctorates, rate of early school leavers.
			Individual growth	Covers activities that improve awareness and identity, develop talents and potential, build human capital and facilitate employability, enhance the quality of life and contribute to the realisation of dreams and aspirations: learn to be along with others, improving body language, acquire conflict resolution skills.
		Safety and security	Personal safety	The condition of being protected from or unlikely to cause danger, risk, or injury: road accidents, safety of urban cycling routes, safety of urban transport at night.
			Personal security	Freedom from violence and from the fear of violence: crime rate, no-go areas.

APPENDIX 6.C CLUSTER SCALE FRAMEWORK OF INNOVATION DISTRICT PLACE QUALITY

Scale	Theme	Category	Indicator	Description
Cluster	**Form**	Location	Centrality	Closeness to important parts of the city: inner city, suburban, close to the airport, universities, research centres.
			Unique surroundings	Surrounding by unique natural or human-made environment: waterfront locations, national parks, historical sites, anchor universities.
		Urban Form and structure	Urban form	The physical characteristics that make up built-up areas: the shape, size, density and configuration of settlements.
			Urban spatial structure	The arrangement of urban public space: accessibility to public spaces, land use.
		Design	Urban design	The style of designing unbuilt spaces: landscape, parks, streetscape, landmarks, spatial diversity.
			Architectural design	The style of designing buildings and other physical structures: appearance, well-made details & quality, built heritage, the convergence of old & new.
		Amenities	Essential amenities	Accessibility to basic facilities: schools, hospitals and elderly/child care facilities, well-served stores.
			Advanced amenities	Accessibility to advanced facilities: outdoor sports, cycling, jogging, well-equipped sports grounds, café, restaurant, bars, Wi- Fi.
	Function	Services	Urban mobility	All modes and forms of transport: public transportation, traffic, connections.
			Management	Body corporate to organise open areas & events, linking businesses, research organisations and industry associations.
		Land Use	Mixed-use development	A combination of residential, commercial, cultural, institutional, & industrial uses.
			Property availability and value	Value and availability of residential & commercial properties: renting and purchasing rates, affordable housing.
		Work Condition	Thick labour market	Offering many and varied employment opportunities, access to a thick pool of the highly educated and professionally talented workforce.
			Professional networks	The flexibility of timetables, friendly colleagues, attractive work environment, job security, level of income.
		Company Profile	Technology adoption	The design and construction of communications technology, digital infrastructure and cutting-edge services: smart services.
			Knowledge/ Creative industries	Percentage of high-tech, biotech, IT, media and design industries, company productivity.

(Continued)

Scale	Theme	Category	Indicator	Description
	Ambiance	Cultural milieu	Public and cultural spaces	Social and cultural spaces that is generally open or accessible to people of all levels of society: meeting places, cinemas, libraries, theatres.
			Cultural events	Planned events which involve large groups of people: live performance venues, cultural festivals, local cultural venues.
		Social context	Work climate	Interactions and relationships of a business nature: Work-related peer networks, spill-over of knowledge.
			Social interaction	The relationships you have with the people around you: Social communities, social networking, the mating market.
		Diversity	Diverse communities	Racial, linguistic, income, religious & sexual diversity: percentage of foreign-born people.
			Tolerance	Acceptance of different lifestyles: percentage of bohemians, LGBIT.
		Creativity	Creative communities	Artistic, cultural, technological & economic creative atmosphere: percentage of artists & inventions.
			Openness	Openness to new idea, people, immigrants: art streets.
	Image	Lifestyle	Pace of life	The speed at which urban life happens: peaceful and quiet, boring or active, street life.
			Variety of lifestyle	Access to different cultural & social activities: restaurants, urban amenities, dynamic nightlife.
		Personal safety	Informal social control	Societal mechanisms that regulate individual and group behaviour, leading to conformity and compliance to the rules of a given society, or social group.
			Safety and security	How safe is to live in this place: the risk of going around late evening, street safeness for pedestrians & cycling, noise level.
		Sense of place	Place attachment	Passion for staying in the place: considering the intended location as an ideal place for you.
			Social cohesion	People feel connected through similar work, educational and religious training & lifestyle: having common things with the ongoing lifestyle in the place.
		Identity	Authenticity	The quality of being novel or unusual: monuments, historic buildings, established neighbourhoods, distinctive music scene.
			Brand	Creating a unique name & image for a cluster in the users' mind (both knowledge workers and industries): representing well-known companies & brands.

CHAPTER 7

Place Quality Attributes of Innovation Districts

7.1 INTRODUCTION

Creativity, digital disruption, and knowledge-based activities have increasingly characterised contemporary global markets (Lee *et al.*, 2014; Cooke, 2017). In a knowledge economy, productivity is mostly characterised by intangible and symbolic values (Boddy, 1999; Meijer & Thaens, 2018). Those values include the capacity of knowledge generation; institutional networks; knowledge flows; and quality of scientific, high-tech, and artistic outputs (Cooke, 2004; Carrillo *et al.*, 2014). Productivity is mainly related to innovative ideas generated by educated and talented workforces who are being considered as key assets of stimulating economic growth (Clifton, 2008; Pancholi *et al.*, 2019). Consequently, investments to foster, attract, and retain human assets have become a key priority in knowledge economy (Glaeser, 2005).

Urban development strategies play a prominent role on economic transition (Benneworth & Hospers, 2007; Yigitcanlar & Velibeyoglu, 2008). At first, it was assumed knowledge workers and industries could locate wherever they would like due to advancements in information and communication technologies (ICTs) (Hall, 1996). However, real-world experiences, such as Silicon Valley, Cambridge Science Park, and Sophia Antipolis along with other examples, revealed the opposite (Carvalho & Van Winden, 2017). Knowledge-based industries clustered with universities and research institutes around specific locations to share knowledge, workforce, and facilities (Sohn & Kenney, 2007; Youtie & Shapira, 2008).

DOI: 10.1201/9781032657431-10

Place and lifestyle were recognised as significant factors to foster, attract, and retain knowledge workers (Wolfe, 1999; Storper & Venables, 2004). Global cities, as well, embraced knowledge-based urban development (KBUD) strategies to make their local economies thrive (Metaxiotis *et al.*, 2010; Yigitcanlar *et al.*, 2012).

KBUD not only targets infrastructural, institutional, and financial requirements of industries, but also considers socio-spatial desires of workers (Sarimin & Yigitcanlar, 2012). This development paradigm encourages clustering of creative and knowledge-based activities (Benneworth & Ratinho, 2014). Clustering empowers firms to increase their innovative capacity by sharing ideas, products, services and work-forces (Reve *et al.*, 2015). Simultaneously, it gathers communities of creative and knowledge workers, who are the architects of economic growth (Martin *et al.*, 2015). These clusters represent neighbourhood-scale places, such as innovation districts, knowledge precincts, creative hubs, technology parks, and the like (Yigitcanlar & Bulu, 2015). Economic growth is accelerated by many urban elements at urban and regional scales. However, the cluster-scale is recognised as specifically important as it is the scale where the tangible place is designed and day-to-day activities take place (Trip, 2007; Durmaz, 2015).

The academic literature clearly underlines the importance of clustering for economic growth. However, our understanding on the role that place plays at the cluster-scale along with the essential place characteristics are still limited. This chapter aims to identify the essential place characteristics of innovation districts that can support a flourishing knowledge economy performance. The chapter attempts to address two research questions. Firstly, whether place quality matters for innovation districts; and if yes, secondly, the essential place characteristics of innovation districts. In order to tackle these issues, the research undertakes an empirical study, analysing three innovation districts from Brisbane, Australia—i.e., Kelvin Grove Urban Village (KGUV), Diamantina Knowledge Precinct (DKP), Brisbane Technology Park (BTP).

7.2 LITERATURE BACKGROUND

Innovation districts, also referred to as knowledge precincts, particularly in the context of Australia, are local industrial specialisations that are generally clustered around universities, research institutions, and knowledge-based industries with a high internal and external networking and knowledge sharing capabilities (Clark *et al.*, 2010; Millar & Ju-Choi,

2010). Innovation districts require knowledge workers, who are in theory no longer attracted to locations solely by employment opportunities (Yigitcanlar *et al.*, 2007, Van Winden *et al.*, 2013).

Many knowledge workers chose their locations, particularly considering the place characteristics and lifestyle options, which fulfil their sophisticated needs and creative identities—e.g., amenities, authenticity, quality of meeting places for business and leisure (McCann, 2004; Baum *et al.*, 2009; Kloosterman & Trip, 2011). These place characteristics influence where knowledge industries locate, as they seek places with the highest concentration of talent (Florida, 2005). Therefore, depicting specific place characteristics that help the knowledge economy grow is beneficial for gathering an agglomeration of talent, attracting investment, and forming an appealing socioeconomic platform (Lonnqvist *et al.*, 2014).

Studies emphasised that the most critical place characteristics are the intangible and quality-based conditions or soft factors—e.g., urban ambiance, vibrancy, socio-cultural diversity, tolerance (Bereitschaft & Cammack, 2015). However, the importance levels of these factors do not totally detract the relevance of hard factors—e.g., investment availability, job opportunity, cost of living, salary level (Storper & Scott, 2009; Scott, 2010; Alfken *et al.*, 2015). There is also a third approach, which recently gained popularity (Bontje & Musterd, 2009; Boren *et al.*, 2013; Durmaz, 2015); an attractive place should be capable of offering ideal conditions both in terms of hard and soft factors (Brown & Mczyski, 2009; Yigitcanlar & Dur, 2013).

In line with the third approach, Chapter 5 introduced a multi-dimensional conceptual framework for investigating place characteristics in innovation districts—including the most significant hard factors, i.e., form and function, and soft factors, i.e., ambiance and image—that shape a place (Figure 7.1). In light of the literature, these factors are discussed below by following the structure of the conceptual framework.

7.2.1 Context

Innovation districts should not be planned in isolation from their urban and regional contexts. Excellence of the 'capital system' and 'quality of life' in cities and their regions influence the appeal of their innovation districts (Trip, 2007; Durmaz, 2015; Pancholi *et al.*, 2015). The capital system focuses on the assets of cities and regions as a whole. These assets provide the macro-scale requirements of KBUD of cities and regions—e.g., natural, financial, relational, digital, intellectual, human assets (Yigitcanlar & Dur, 2013). Quality of life is the umbrella concept for the mezzo-scale factors that

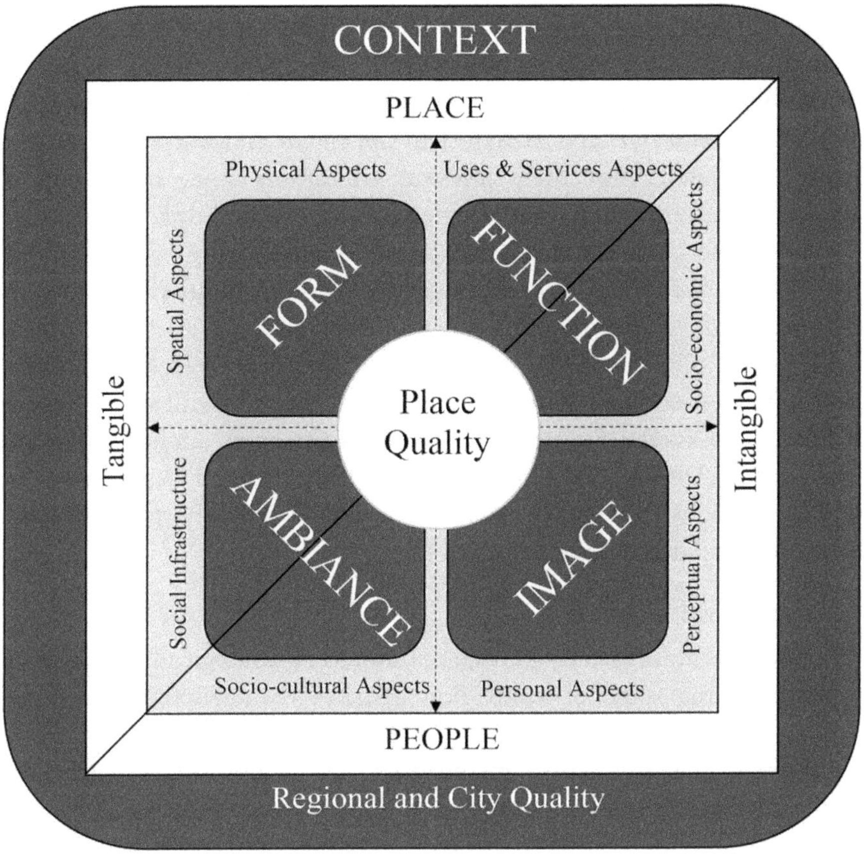

FIGURE 7.1 Conceptual framework for characterising place in innovation districts (see Chapter 5).

contribute to the general experience of urban living—e.g., quantity and quality of amenities (Santos & Martins, 2007; Morais *et al.*, 2013). Improved 'capital system' and 'quality of life' enhance the ability of cities and regions to attract a thick pool of talent and investment—which in return empowers the knowledge economy growth. One successful example of such context for nourishing and flourishing innovation districts, is the City of Barcelona. Barcelona has transformed itself into a knowledge city in the early-2000s—and a smart city in the late-2010s (Yigitcanlar, 2015). Urban regeneration in Barcelona not only helped meeting the infrastructural requirements for knowledge-based activities, but also augmented the cultural and social activities. Today, Barcelona and its world-class innovation districts, such as 22@Barcelona, are global magnets for talent and investment (Charnock *et al.*, 2014).

7.2.2 Form

Form, in terms of physical patterns, layouts, structures, and so on, should be carefully determined in the planning stage of innovation districts (Esmaeilpoorarabi *et al.*, 2016a). The appeal of innovation districts is highly correlated with their physical conditions—e.g., inner city or suburban location, urban form and design, architectural design, amenities. These tangible place characteristics create an attractive physical environment for the wellbeing of knowledge workers and industries (Brown & Mczyski, 2009; Kloosterman & Trip, 2011; Frenkel *et al.*, 2013). Innovation districts also offer advanced amenities and physical spaces for creating unique user experiences. Exemplar innovation districts with such offerings include One-North (Singapore) with its sustainable architectural and urban design, Dumbo (Brooklyn) with its luxury lofts and warehouses, and Sophia Antipolis (Nice) with its environmentally friendly spaces (Parker, 2010; Yigitcanlar & Dizdaroglu, 2014).

7.2.3 Function

Function of buildings and open spaces should be strategically determined at the planning stage of innovation districts (Esmaeilpoorarabi *et al.*, 2016b). Functions are assigned to buildings and open spaces by considering the following: mixed-used or single-use developments; type and variety of innovative activities; significance of economic opportunities and company profiles; quantity and quality of labour pool; capability of knowledge generation and knowledge spill-over; proficiency of professional networks (Grant & Buckwold, 2013; Pancholi *et al.*, 2017). Dynamic life offerings of multifunctional and mixed-use innovation districts, such as Arabianranta (Helsinki), Macquarie Park Innovation District, and Westmead Innovation District (Sydney), are an advantage in comparison to those traditional and single-purpose technology parks—e.g., Australian Technology Park (Sydney) and Dandenong National Employment and Innovation Cluster (Melbourne) (Pancholi *et al.*, 2014; Makkonen *et al.*, 2018).

7.2.4 Ambiance

Ambiance, in terms of urban, cultural, creative, digital, and so on, should be deliberated at the planning stage of innovation districts as a key attractor. Creative workers seek diversity of people, variety of lifestyles, multiculturality, and openness—briefly the ambience. Talented workers prefer variety of scenes enhancing social vitalities, encouraging social and business interactions—e.g., music events, street art, nightlife, third places. Subsequently, providing such an ambiance attracts and retains talent, which in turn supports the performance of an innovation district (Florida, 2005; Clifton & Cooke, 2009; Heebels & Van Aalst, 2010). One of the exemplar practices is Brooklyn Tech Triangle New York. The district combined the educational and innovative opportunities with a variety of public places, as well as cultural and artistic and recreational activities in the street level; e.g., Brooklyn Bridge Park, cultural district, free Wi-Fi.

7.2.5 Image

Image building should be considered at the planning stage of innovation districts for brand and reputation development (see Chapter 1). Creating an appealing image in the minds of knowledge workers is recognised as a strategic asset. Generally, image is an authentic identification, which distinguishes an innovation district from others through physical features or appearance (i.e., form), observable activities and purposes (i.e., function), and meaning or symbols (i.e., ambiance). These features together create the specific image of an innovation district (Florida, 2005; D'Mello & Sahay, 2007; Sepe, 2010). All successful innovation district best practices created a unique image through branding exercises; for instance, the image varies from the hectic New York–style living of Dumbo to green and serene suburban living of Australian innovation districts (Yigitcanlar *et al.*, 2017).

7.3 METHODOLOGY

7.3.1 Research Design

This research follows a case study approach to highlight the essential place characteristics of innovation districts that attract and retain talent and investment that subsequently accelerate their socioeconomic performance. The study relies on multiple sources of evidence, and follows Eisenhardt's (1989) approach that is an inductive approach with a pre-planned structure. The chapter proposes an analytical framework—derived from Chapter 5 and Chapter 6—in order to empirically generate inductive insights on whether place quality matters for innovation districts and if so, what their essential place characteristics are. The analytical framework is shown in Table 7.1. As all cases are from the same city-region, the influence of context is considered equal (weight = 0.000). The collected data sets were analysed to determine the importance of place characteristics and what the specific

TABLE 7.1 Analytical Framework to Evaluate Place Characteristics of Innovation Districts

Theme Weight	Category Weight	Indicator	Weight	Description
Context W = 0.000	Urban quality Weight = 0.000	Regional quality	0.000	Characterised by regional capital system: human capital, financial capital, relational capital …
		City quality	0.000	Characterised by quality of life: environment, infrastructure, economic development, social life, health, education …
Form W = 0.235	Location W = 0.059	Centrality	0.030	Closeness to important parts of the city: inner-city, suburban, close to the airport, universities, research centres.
		Unique surroundings	0.029	Surrounding by unique natural or human-made environment: waterfront locations, bushlands, national parks, historical sites.
	Urban form and structure W = 0.057	Urban structure	0.029	The arrangement of urban spaces. The degree of connectivity and accessibility: accessibility to public and private spaces, land uses.
		Urban form	0.028	The physical characteristics that form built-up areas: the shape, size, density and configuration of settlements.
	Design W = 0.059	Urban design	0.030	The style of designing unbuilt spaces: landscape, parks, streetscape, landmarks, spatial diversity.
		Architectural design	0.029	The style of designing buildings and other physical structures: appearance, well-made details, the convergence of old and new.
	Amenities W = 0.061	Essential amenities	0.031	Accessibility to basic facilities: schools, hospitals and elderly/child care, well-served stores.
		Advanced amenities	0.030	Accessibility to advanced facilities: outdoor sports, cycling, jogging, café, restaurant, bars, Wi-Fi, hotels.

Function W = 0.255	Services W = 0.064	Urban mobility	0.033	All modes and forms of transport: public transportation, traffic, connections, parking.
		Management	0.031	Body corporate/management system to organise open areas/events, linking people, businesses, research organisations and industry.
	Land Use W = 0.062	Mixed-use development	0.031	A combination of residential, commercial, cultural, institutional, and industrial uses.
		Property availability and value	0.031	Value and availability of residential and commercial properties: renting and purchasing rates.
	Company profile W = 0.065	Knowledge/creative industries	0.033	Active presence of high-tech, biotech, information technology, media, and design industries and so on and company productivity.
		Technology adoption/facilities	0.032	Design and construction of communications technology, digital infrastructure and cutting-edge services and facilities: smart services.
	Work condition W = 0.065	Thick labour markets	0.033	Offer many varied employment opportunities, access to highly educated and professionally talented workforce, job security, income.
		Professional networks	0.032	Business interactions, work-related peer networks, relations between universities, institutions and firms: spill-over of knowledge.
Ambiance W = 0.261	Public spaces and events W = 0.061	Public and cultural spaces	0.031	Public and cultural spaces that are open or accessible to people of all levels of society: meeting places, cinemas, libraries, theatres …
		Public and cultural events	0.030	Planned events: farmer markets, live performances, cultural festivals, children programs …

(*Continued*)

TABLE 7.1 *(Continued)* Analytical Framework to Evaluate Place Characteristics of Innovation Districts

Theme Weight	Category Weight	Indicator	Weight	Description
	Public engagement W = 0.067	Work climate	0.034	Attractive work environment: relaxed/stressful atmosphere, opportunity for interaction with other professionals, more than work place.
		Social interaction	0.033	Relationships with the people around you: social communities, social networking …
	Diversity W = 0.66	Diverse communities	0.032	Racial, linguistic, income, age, religious & sexual diversity: percentage of foreign-born people, or homogenous communities.
		Tolerance/openness	0.034	Acceptance of or opportunities for different lifestyles: presence of bohemians, LGBTQ, or different age groups.
	Creativity W = 0.066	Creative communities	0.033	Active presence of high educated and talented people: artists, scientists, inventors …
		Creative atmosphere	0.033	Artistic, cultural, technological and economic creative atmosphere: art streets, high-tech exhibitions, openness to new ideas.
Image W = 0.249	Buzz of place W = 0.062	Pace of life	0.030	The speed at which urban life happens: peaceful, quiet, hectic, boring or active.
		Variety of lifestyles	0.032	Vibrancy of street life, access to different cultural and social outdoor activities: restaurants, urban amenities, dynamic nightlife.
	Sense of safety W = 0.061	Subjective safety	0.028	Opportunity of feeling safe and being comfortable to be yourself.

		Objective safety and security	0.033	How safe is to live in this place: the risk of going around late evening, street safeness for pedestrians and cycling, noise.
	Sense of place W = 0.063	Place attachment	0.030	Passion for staying in the place: considering the intended location as an ideal place.
		Sense of community	0.033	Social cohesion, People feel connected together: having common things with the ongoing lifestyle.
	Place identity W = 0.063	Authenticity	0.032	The quality of being novel or unusual: monuments, historic buildings, distinctive music scene.
		Brand	0.031	Creating a unique name and image in users' mind: representing well-known companies and brands.
Sum = 1.000	Sum = 1.000		Sum = 1.000	

characteristics were. This study assumes all general similarities of the case innovation districts as fixed factors and focused its attention on tangible and place-related characteristics.

7.3.2 Data

Required data was collected through 27 semi-structured interviews and 32 online surveys conducted with a wide-range of key stakeholders and users. The interviewees were selected from five major interest groups of the case innovation districts, as shown in Table 7.2. In each group, interviewees were selected either as experts at key positions associated with the case innovation districts or as users. Prior to the interview, interviewees were requested to complete a Likert-scale online survey. In addition to 27 interviewees, five academics completed the survey to reflect theoretical viewpoints. However, as academics were not directly involved with the cases, they were not interviewed. Data collection was undertaken during the first half of 2018. Each interview lasted around 45–60 minutes, was digitally recorded, and then transcribed verbatim. Primary and secondary sources, i.e., policy documents, plans, observations, maps, photographs, were also used.

Collected data was analysed through qualitative content and quantitative statistical techniques. Indicators, shown in Table 7.1, were considered not only in designing interview and survey questions, but also employed for coding data. Survey questions were quantitative and same for all participants. However, depending on the background of each interviewee conversation tweaked to capture their specific expertise and experience with the case. For instance, government executives were asked about Brisbane's strategies for emerging or empowering innovation districts for fostering, attracting, and retaining talent—reflecting 'Context'. Conversely, interviews with knowledge workers were questioned about their perceptions and expectations of working, studying, living, and playing in the innovation districts—reflecting 'Ambiance' and 'Image'. This way, data from interviews with different groups complemented each other.

Qualitative data from interviews—verbatim transcripts—and quantitative data from surveys—Likert-scale scores—were manually coded following the indicators (Table 7.1) and then uploaded in NVivo and SPSS, respectively, for further analyses. Finally, results of the analyses were reported in a comparative way to indicate which characteristics of place were more important in fostering, attracting, and retaining talent and investment. The results of the comparative case study analysis were

TABLE 7.2 Interviewee Profiles

Group	No.	Expertise	Engagement
Group 1: Government Executives	Interviewee#1	Strategy Planner	Manager of Brisbane Marketing Precinct Development Division
	Interviewee#2	Economist	Facilitator of Queensland Department of Infrastructure and Planning
	Interviewee#3	Urban and Regional Planner	Manager of Queensland Land and Asset Management
	Interviewee#4	Strategy Planner	Director of Economic Development Queensland
Group 2: Developers, Body Corporates	Interviewee#5	External Relations	Manager of QUT's involvement in KGUV
	Interviewee#6	Systems Thinker	Chair of KGUV principal body corporate
	Interviewee#7	Science Infrastructure and Operations	Manager of Ecosciences Precinct in DKP
	Interviewee#8	Business and Technology Precincts	Manager of BTP services
Group 3: Architects, Urban Planners	Interviewee#9	Architect	Involved in/familiar with KGUV, DKP
	Interviewee#10	Architect and Urban Designer	Involved in/familiar with KGUV
	Interviewee#11	Architect and Urban Morphologist	Involved in/familiar with DKP, KGUV, BTP
	Interviewee#12	Urban Designer and Planner	Involved in/familiar with KGUV, BTP, DKP
	Interviewee#13	Place Specialist	Involved in/familiar with KGUV, DKP, BTP
Group 4: Knowledge Workers, Company Managers	Interviewee#14	International and Development Management	Works in KGUV
	Interviewee#15	Musician and Audio Engineer	Works in KGUV
	Interviewee#16	Optometrist	Works in KGUV

(Continued)

TABLE 7.2 *(Continued)* Interviewee Profiles

Group	No.	Expertise	Engagement
	Interviewee#17	Digital Mapping and Media Technologies	Works in KGUV
	Interviewee#18	Computational Biologist	Works in DKP
	Interviewee#19	Toxicology and Drug Analyst	Works in DKP
	Interviewee#20	Plant Biochemist	Works in DKP
	Interviewee#21	Ecologist	Works in DKP
	Interviewee#22	Performance Analyst	Works in BTP
	Interviewee#23	Software Developer	Works in BTP
	Interviewee#24	System Engineer	Works in BTP
	Interviewee#25	Solution Architect	Works in BTP
	Interviewee#26	Electronic Design Engineer	Works in BTP
	Interviewee#27	Infrastructure Operations Engineer	Works in BTP

utilised to establish insights for planning, designing, developing, and managing prosperous innovation districts.

7.3.3 Cases

Brisbane was selected as the case study city context and Kelvin Grove Urban Village (KGUV), Diamantina Knowledge Precinct (DKP) Brisbane Technology Park (BTP) as the case innovation districts (Figure 7.2). This selection is justified due to the following: (a) Brisbane is a new world city from Australia with high ambitions to achieve a thriving KBUD. (b) Over the last three decades, various policies were developed to support knowledge economy transformation, including 'Smart State Strategy', 'Advanced Queensland', 'Brisbane 2022 New World City Action Plan', and 'Brisbane's Global Precincts'. (c) Brisbane offers a high quality of life and place for (inter)national talent and investment. (d) Brisbane's innovation districts play a pivotal role in strengthening the city's reputation as a top-tier business and investment destination (see Chapter 1).

The chosen case innovation districts are Brisbane's leading knowledge and innovation hubs. They are planned and developed from scratch as opposed to spontaneous agglomerations and considered as medium-sized. On the other hand, they are all different in terms of specific place characteristics, such as development process, proximity to central business district (CBD), spatial design, major functions, and community aspects. Each case innovation district is introduced as follows.

KGUV was a joint initiative of Queensland Government Department of Housing (QGDH) and Queensland University of Technology (QUT)

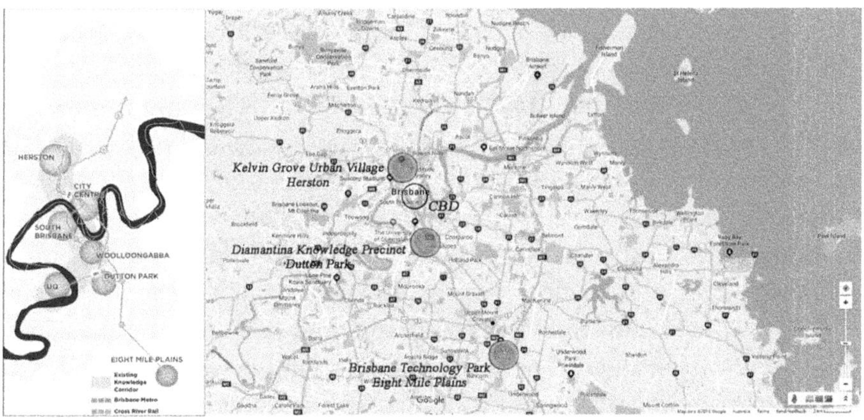

FIGURE 7.2 Brisbane's knowledge corridor and location of knowledge precincts (retrieved from https://www.google.com/maps/@-27.4707486,153.0677792,11.5z).

to create a sustainable urban community within 2 kilometres from CBD. The initial vision of QUT was to expand its campus through the 'new urbanism' model and integrating with surrounding communities. Meanwhile, QGDH was looking for showcasing effective urban renewal practices to provide diverse and affordable accommodation, liveable communities, and eco-friendly constructions. Hence, community-based aspects, such as integration of commercial, residential, and educational uses, were the priority. Various small and medium-scale private developers were involved in developing the village adhering to its overarching principles. The Smart State policies that intended to transform Brisbane's economy into a knowledge-based one also played a role in this development. Consequently, KGUV was established as a public-private-community-based innovation district.

DKP was built on land owned by the state government about 3.5 kilometres from CBD in close proximity to the University of Queensland (UQ) and a major hospital. Considering those proximity potentials, along with an existing public transport system, the state government decided to develop DKP in partnership with UQ and Commonwealth Scientific and Industrial Research Organisation (CSIRO). The development was governmental-driven and institutional-based. However, the government also intended to foster community-based values—e.g., developing on-site residential blocks. Formerly known as the Dutton Park Knowledge Precinct, DKP has been so far unsuccessful in establishing its own community and the district was mostly occupied by public research institutes. Consequently, due to the absence of dominant private investments and community-based aspects, DKP was established as a public institution-based innovation district.

BTP was conceived almost 30 years ago following the model of Silicon Valley, settled in suburbia 13 kilometres from CBD. The state government decided to engage the private sector in the establishment of BTP as the first example of a large-scale innovation cluster formation in Brisbane. BTP was assigned to one major development company—i.e., Graystone Group. This decision accelerated the district's growth and transformed it into the largest technology park in Brisbane. However, this decision also made BTP almost a single-purpose innovation district—dominantly technology and business workspaces. Accordingly, the market-driven nature of BTP forced it to concentrate on economic profitability with no recognition of community-based values. Consequently, BTP was established as a private firm-based innovation district—specialising in technology and business office supply.

7.4 RESULTS

7.4.1 Qualitative Analysis

In consensus, all interviewees responded to the question of whether place quality matters for the socioeconomic success of innovation districts. Most remarkably, Interviewee#3 stated, *"Quality of place is utmost importance while planning and developing knowledge precincts as without high standards it's not possible to compete with other overseas locations, particularly the emerging economy precincts'.* Likewise, Interviewee#4 noted, *"Making space and place for the knowledge economy activities require strategic investment and provision of high-quality environments to host knowledge precinct is one of the strategic priorities'.*

7.4.1.1 Context

Brisbane today houses a number of innovation districts. Early local policies for clustering knowledge activities were mainly government-driven and incentivised rather than organic or market-driven. The recent vision concentrates on a more natural, flexible, and infrastructure-driven approach. The purpose is to improve connectivity and liveability through investing in transportation, the city's digital network, and the city's main assets rather than directly funding innovation districts. For instance, as Interviewee#1 elaborated, *"Knowledge corridor, metro and river aren't just transport infrastructure projects; they're catalysts for economic exchange in innovation districts those need to be clustered'* (Figure 7.2). In this new context, governmental support, such as tax incentives and grants for start-ups, still plays a crucial role in sustaining KBUD.

In terms of lifestyle, Brisbane benefits from its subtropical climate, proximity to beaches, inner-city bushlands, hilly landscapes, and river. Brisbane River is the centre of recreational amenities. Environment is strategic, as argued by Interviewee#2, *"Brisbane's attraction rather than innovation districts. Knowledge workers move and live here for the environment'.* Furthermore, while Brisbane has global economic potentials in numerous sectors—i.e., higher education, health, digital economy—living costs and failure risk for start-ups are lower than other major Australian cities. Interviewee#13 stated, *"It's actually a nice place to live; it's cheaper than Melbourne and Sydney. It's a great lifestyle and education opportunities and it's affordable for start-ups; if you fail here, the consequences aren't as bad as the consequences when you've got higher rent'.* Basically, lifestyle options, entertainment and environment assets, specialised medical research, and tech-based employments make the city attractive for knowledge workers and industries. Investigated cases took advantage of these opportunities in their unique ways.

7.4.1.2 Form

Location: Interviewees suggested that the close proximity to CBD and hospitals, as well as co-locating with the university and its affiliated research institutes is a major strength of KGUV (Figure 7.3, top map). Proximity supports physical connections and facilitated professional and institutional networks. KGUV benefits from world-class tourism and recreational facilities in CBD with inner-city vibes. DKP is also located close to CBD, hospitals, research institutes, and universities. However, interviewees complained that DKP is impacted by several physical barriers, railways, roads, and river, both internally and externally. These significantly restricted social, professional, and institutional connections inside DKP, and separated DKP from the CBD (Figure 7.3, middle map). BTP is located in a suburb, and not co-located with universities or research institutes (Figure 7.3, bottom map). However, BTP is surrounded by business parks and benefits from their infrastructure and business atmosphere. In terms of natural environments, both KGUV and DKP's hilly topographies provide stunning views of CBD and all three innovation districts are co-located with great natural green spaces. Nonetheless, the fact that KGUV was an old army barrack and DKP hosts a historic prison provided an opportunity to restore heritage assets and create a unique ambiance. These natural and historical potentials contribute to the attractiveness of the place.

Urban form and structure: KGUV is connected to the existing city fabric, conceived as an extension of the city and surrounding neighbourhoods. In terms of blurring boundaries, Interviewee#17 described, "*You can't easily notice the boundary between the university and surrounding district*'. KGUV's urban form is medium to high density with a good connection to the streetscape; this characteristic prevents the sprawling and assures the presence of a right amount of people in KGUV (Figure 7.4). In contrast, DKP contains five extremely large buildings, extensive parking, and massive vacant land. The size of buildings and huge gaps between them are critical in integrating DKP with surrounding neighbourhoods. These problems vastly weakened walkability, liveability, and street life, and damaged the community formation efforts (Figure 7.5). BTP has medium to low density to stay in harmony with nearby low-density suburban neighbourhoods. In reality, BTP is purely

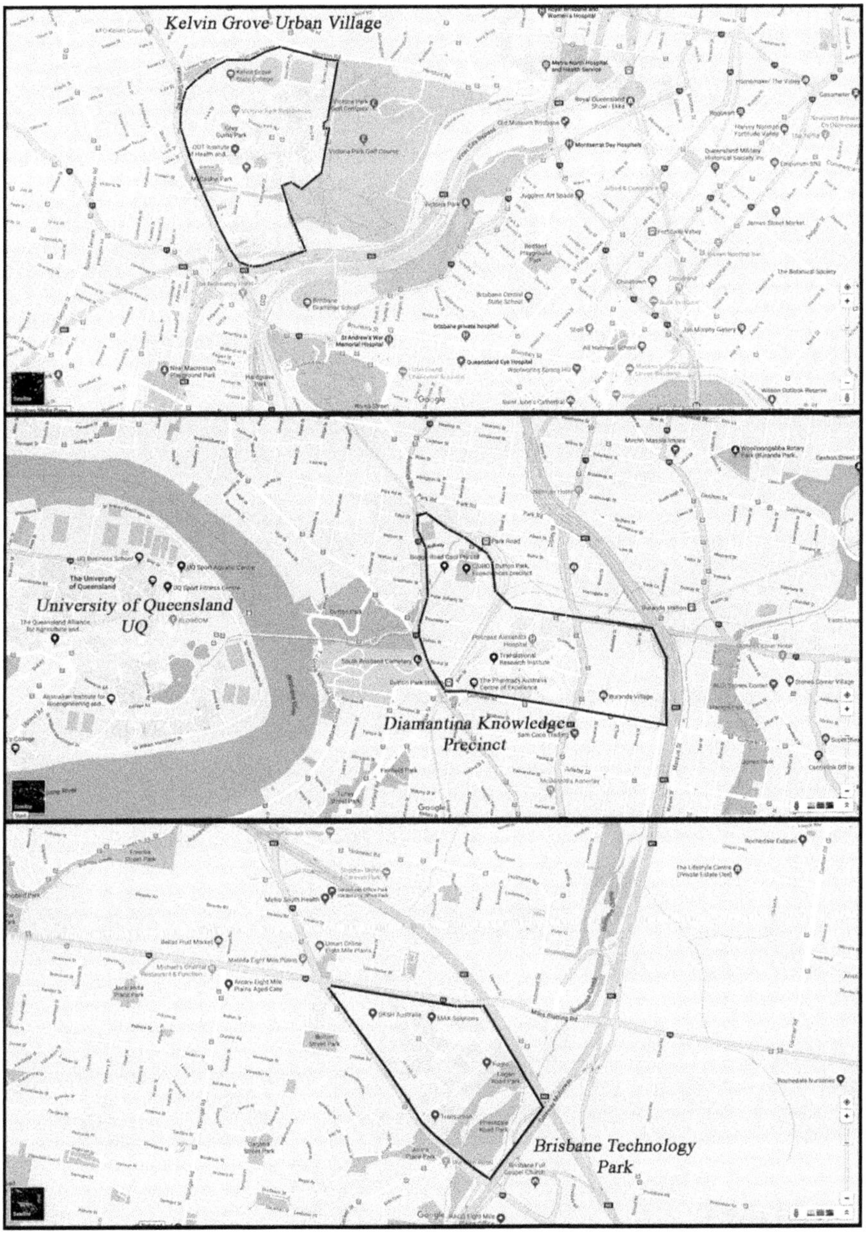

FIGURE 7.3 Layout and surrounding of KGUV, DKP, BTP (retrieved from https://www.google.com/maps/@-27.4548772,153.0238924,14.75z, https://www.google.com/maps/@-27.4960437,153.0279553,15z, https://www.google.com/maps/@-27.5769193,153.0998147,15.25z).

FIGURE 7.4 Dynamic street life of KGUV (Photograph by the authors).

FIGURE 7.5 Segmented urban structure and gloomy street life of DKP (Photograph by the authors).

FIGURE 7.6 Car-oriented and weak streetscape of BTP (Photograph by the authors).

occupied with many segmented workplaces, which are not properly connected to streetscapes and adjacent neighbourhoods. BTP being suburban and car-oriented worsens this situation and extremely weakens walkability (Figure 7.6).

Design: KGUV is developed around a main street with two QUT faculties as anchors at the two end-points. Residential, recreational, commercial and institutional uses are located around the main street. This structure supported a variety of movement between spaces and activates the street life. In contrast, DKP and BTP offer nothing at the street level to shape the district life. As Interviewee#12 noted, '*One of the biggest challenges for urban planners is to actually design places that creative urbanism*'. KGUV offers a variety of public green and open spaces. However, there are no purposefully designed public places in DKP and BTP. Unlike DKP and BTP, community-based considerations are strongly reflected in the urban and architectural design of KGUV and guaranteed a dynamic district life. In KGUV, the commercial nature of the majority of the residential developments provided generic spaces, which, although liveable, do not reflect the creative and cutting-edge identity of KGUV. In DKP, architectural design is high-tech and futuristic, reflecting the innovative sectors. Architectural design in all three districts generally lacks the visibility of innovative activities, which support a creative atmosphere—i.e., visibility of fashion studios or fishery labs at the street level.

Amenities: KGUV has successfully combined QUT's facilities with the community infrastructures. These include schools, health facilities, gym, swimming pool, sports courts, free public Wi-Fi, convenience stores, variety of restaurants, open and green spaces, as well as co-working, meeting, and research spaces. KGUV also benefits from CBD's social and recreational facilities within walking distance. Contrarily, DKP and BTP are not planned to provide such facilities and urban amenities. In this regard, Interviewee#8 claimed, *"It's very hard to create a social environment and bring people in, if you don't have all that infrastructure planned in advance'*. Subsequently, DKP and BTP are struggling with attracting and retaining knowledge workers.

7.4.1.3 Function

Services: As a pedestrian-oriented development, KGUV is adequately connected to CBD by sidewalks, bike paths, and public transport. While a lack of walkability in DKP is noticeable, it still benefits from a well-served public transport system. KGUV and DKP are placed in Brisbane's Knowledge Corridor, which is a transport route that connects Brisbane's innovation districts (Figure 7.2). However, BTP is located far from that corridor and lacks both walkability and public transport. Mobility in BTP entirely relies on the cars and the nearby highway. In terms of management, KGUV has a central management scheme, which links individual buildings' body corporates and ensures the liveability and community linkage. In BTP, the developer manages the entire district and links individual body corporates. However, the role of the manager is mainly restricted to maintenance and leasing control rather than community development and management. In contrast, each building of DKP is individually managed by a separate body corporate. Therefore, in comparison to KGUV, DKP and BTP lack facilitators for fostering their community.

Land use: KGUV is a mixed-use development that merges life, work, study, and entertainment. Interviewee#10 indicated, *"You can live, work, shop, play and your children can go to school here. Because of this it's a really attractive place'*. The presence of university and residential blocks in KGUV formed a dense population. This density becomes a key element for supporting retail, entertainment, and other commercial activities. The dynamic street life of KGUV is a result of this compactness. On the other hand, BTP and DKP are established as single-purpose innovation districts. DKP concentrated on research institutes and BTP on providing workplaces. Thus, DKP and BTP are struggling with a rising population to support other activities and activate the district life. However, while mixed-used development is ideal, the most critical challenge for these districts is getting the right balance between residential blocks and workspaces. In KGUV, except some co-working areas inside the university, there are no typical workplaces for knowledge companies. Interviewee#8 stated, '*Government forced BTP to follow firm-based development. This guarantees the availability of workplaces; however, such mono-functional development weakened the community aspects of BTP*'. Therefore, a balanced mixed-use development is required to please both work and community functions. In regard to rent and property prices, all interviewees commented negatively that the inner-city style of living around KGUV and DKP is too expensive for an average knowledge worker or a start-up. In contrast, BTP as a suburban innovation district benefits from lower rent and property prices of residential and work properties. This is a significant advantage for start-ups, small companies, and less-established knowledge workers.

Company profile: KGUV, DKP, and BTP specialise in, respectively, health and creative industries, health and eco-science, and ICT. The state government targeted health and digital media sectors as major actors of Brisbane's knowledge economy. Following this, QUT's faculties of health and creative industries were located on KGUV. Creative industries in KGUV were empowered by the inner-city buzz and, in return, their creative atmosphere flowed into the city. However, the lack of workplaces prevents KGUV from hosting new health and creative industry companies. Similarly, DKP's public institution intensity limits private sector workspace availability. KGUV and DKP lack the presence of adequate levels of private companies. The economy of KGUV and DKP generally relies on the university and governmental assets rather than private-sector investments. In contrast, in BTP, the demand of Brisbane for the ICT sector naturally gathers companies together. Availability of cost-effective workspaces moved the private sector into BTP; it lacks the presence of educational and research institutes.

Work condition: BTP offers a variety of job opportunities for knowledge workers through on-site companies. In contrast, employment in KGUV and DKP is mostly limited to research positions provided by the university or government. Due to the lack of diversity in professional organisations, professional interactions in KGUV and DKP are weak and homogenous. On the other hand, while BTP benefits from diversity of professions, it lacks the plan and place to make professional networks happen. BTP not only suffers from the absence of a community manager to plan for formal networks—e.g., conferences, symposium, exhibitions, public events—but also offers limited public places for informal networks—e.g., cafés, restaurants, parks, sports facilities. As Interviewee#14 mentioned, *"Sometimes informal networking is the best form of networking for professional businesses'*, but DKP and BTP neglected this opportunity. KGUV has plans and places; however, due to weak diversity in the professional population, the community benefits from them more than professionals.

7.4.1.4 Ambiance

Public and cultural spaces and events: KGUV contains variety of open and public spaces—e.g., parks, walking trails, playgrounds—that encourage outdoor activities and facilitate social interactions. As Interviewee#16 illustrated, *"Open spaces are fantastic community spots where people obviously enjoy meeting and socialising and there's always activity there'*. Theatre and heritage buildings are great additions to KGUV, contributing to the creation of an artistic ambiance. KGUV and QUT provided various public events to bring students, employees, and residents together and develop a casual and welcoming ambiance—e.g., Annual Brisbane Festival, Saturday markets, movie nights, fashion shows. However, DPKD and BTP do not provide such community-based amenities or events.

Public engagement: Interviewees reported three major advantages for KGUV that distinguishes it from DKP and BTP. Firstly, there is a right mass of people to occasionally, accidentally, or co-incidentally run into each other, socialise, and organise activities. However, in DKP and BTP, as Interviewee#7 experienced, *"There's nothing to really bring enough people to the district. They do need to turn things in the area to bring people'*. Secondly, since KGUV hosts commercial sectors as well as a variety of events all the time, there are various activities in the streets that people can attend freely. However, DKP and BTP purely focused on work and lack social activities. Interviewee#19 explained, *"It's quite uncommon for my colleagues to actually leave our building. You can go for weeks without seeing people if you wanted to'*. Lastly, KGUV offers interior and exterior incidental spaces where people meet each other—a circulation realm. KGUV shaped an energetic vibe through offering the right amount of people, activities, and places, where DKP and BTP did not.

Diversity: Brisbane embraces people from diverse cultures and ethnic backgrounds. In respect to the diversity of communities, it offers variety of multicultural amenities—e.g., international cuisine, festivals, community events—to bring communities together. Consequently, all three case studies nourish from such vibrant context and benefit from the culture of tolerance. Interviewee#17 emphasised, *"There's relatively a higher level of not just tolerance but a celebration of diversity'*. However, undeniably, KGUV is the most successful one in terms of reflecting the ambiance of diversity and tolerance. KGUV is a heterogenous community. In contrast, the community of DKP and BTP are excessively homogenous and, as Interviewee23# mentioned, *"The place is good for catching up with like-minded people. Probably not as good for meeting with new people'*. For example, a scientist in DKP has never had the opportunity to interact with an artist—this is happening in KGUV.

Creativity: Interviewees frequently emphasised that a major part of their creativity is connected to people who are based in their districts. For example, Interviewee#19 said, *"It's creative because I'm around other very, very smart people. Everyone that I work with is very talented. I find my colleagues inspiring people that help me to develop new ideas in my job'*. Due to the lack of places for social and professional interactions, organisations and companies in DKP and BTP are isolated. They are struggling to connect their creative communities and not successful in shaping their creative atmosphere. Interviewees highlighted that their creativity is also related to specific attributes of place, such as inspiring architecture, artistic features and performances, visibility of creative activities, and even natural environments for taking a walk and clearing their minds.

7.4.1.5 Image

Buzz of place: According to interviewees, the most advantageous character of KGUV for activating the place's buzz, again, is its dense population—both DKP and BTP are deprived of this. KGUV empowers the artistic buzz of creative industry activities. For instance, Interviewee#12 pointed out, *"In KGUV, you've got people performing and dancing and playing music and so it's very attractive'*. However, it is still much livelier and interactive during the day—fits with university hours—and lacks nightlife; besides, dining options are limited to the students' level. Subsequently, KGUV's buzz is mostly influenced by QUT as the anchor. KGUV requires the presence of other major actors to improve this vibe. DKP and BTP perform poorly in terms of offering various experiences—this justifies their boring and gloomy environment.

Sense of safety: At the city level, Brisbane is considerably safe; this is an advantage for all three case studies to shape a safe image in users' minds. At the cluster level, KGUV represents a better example of safety, since it embraces a dense population, an active street life, and a pedestrian-friendly design. The presence of residential blocks is another advantage for KGUV, which keep it alive and safe during the night. In contrast, safety in DKP and BTP is threatened by sprawling suburbia, lack of walkability, and low presence of people and activities at the street level; they are also entirely vacant at night.

Sense of place: In KGUV, there is no boundary between the district and the city fabric and even between internal usages—i.e., live, work, study, play. People feel free to come to KGUV, use the place in different ways, and develop a sense that they own the place. Consequently, KGUV shaped a coherent community who feel attached to the place. Such place attachment encouraged people to stay in KGUV for the long term. In contrast, since community aspects of DKP and BTP are insignificant, people do not feel belonging there and have been unable to retain people.

Place identity: Interviewees considered that all three cases poorly perform in terms of building up their local, national, and international reputation. The lack of reputation in these districts is due to their physical and institutional aspects. For example, from the perspective of physical aspects, KGUV is formed to support the community values. Thus, KGUV's architectural and urban design is generic, with no unique or monumental bold statement. On the other hand, interviewees argued that the village feel and the combination of live, work, study, and play activities in KGUV are highly unique; the historic feel in DKP is also attractive. In terms of institutional aspects, unfortunately, all cases neglected the importance of attracting and hosting distinguished companies or institutes.

7.4.2 Quantitative Analysis

In the survey, 32 participants were asked to evaluate the importance of place characteristics in the case studies through a 5-point Likert-scale (1–extremely low to 5–extremely high). Collected data—43 questions—were categorised under the 32 indicators of the assessment framework (Table 7.1). The mean value, as the most common analytical statistic, was calculated for each indicator. As shown in Table 7.3, mean values specify the central tendency of responses in each innovation

TABLE 7.3 Quantitative Results of the Survey

Theme Weight	Total Score			Total Score (Weight*Mean)				Indicator	Weight	Mean (N=32)		
	KGUV	DKP	BTPL		KGUV	DKP	BTPL			KGUV	DKP	BTPL
Form W = 0.235	0.948	0.741	0.579	Location W = 0.059	0.225	0.204	0.123	Centrality	0.030	4.0	3.6	2.3
								Unique surroundings	0.029	3.5	3.2	1.8
				Urban form and structure W = 0.057	0.240	0.171	0.150	Urban structure	0.029	4.0	2.8	2.3
								Urban form	0.028	4.0	2.9	2.7
				Design W = 0.059	0.231	0.195	0.159	Urban design	0.030	4.0	2.9	2.2
								Architectural design	0.029	3.7	3.6	3.1
				Amenities W = 0.061	0.252	0.171	0.147	Essential amenities	0.031	4.3	3.2	2.6
								Advanced amenities	0.030	4.1	2.5	2.3
Function W = 0.255	0.867	0.741	0.699	Services W = 0.064	0.219	0.198	0.162	Urban mobility	0.033	3.9	4.1	2.8
						Manag-ement		0.031	3.4	2.5	2.6	
				Land Use W = 0.062	0.219	0.147	0.144	Mixed-use development	0.031	4.0	2.4	1.9
								Property availability and value	0.031	3.3	2.5	2.9
				Company profile W = 0.0.65	0.204	0.207	0.204	Knowledge/ creative industries	0.033	3.2	3.3	3.7
								Technology adoption/ facilities	0.032	3.6	3.6	3.1
				Work condition W = 0.065	0.225	0.189	0.189	Thick labour markets	0.033	3.7	3.2	3.5
								Professional networks	0.032	3.8	3.1	2.8

(Continued)

TABLE 7.3 *(Continued)* Quantitative Results of the Survey

Theme Weight	Total Score			Total Score (Weight*Mean)				Indicator	Weight	Mean (N=32)		
	KGUV	DKP	BTPL		KGUV	DKP	BTPL			KGUV	DKP	BTPL
Ambiance W = 0.261	0.939	0.738	0.669	Public spaces and events W = 0.061	0.216	0.120	0.099	Public and cultural spaces	0.031	3.7	2.1	1.6
								Public and cultural events	0.030	3.5	1.9	1.7
				Public Engagement W = 0.067	0.228	0.186	0.174	Work climate	0.034	3.8	3.5	3.1
								Social interaction	0.033	3.8	2.7	2.7
				Diversity W = 0.066	0.252	0.228	0.204	Diverse communities	0.032	4.4	3.8	3.5
								Tolerance/ openness	0.034	4.0	3.8	3.3
				Creativity W = 0.066	0.243	0.204	0.192	Creative communities	0.033	4.1	3.4	3.0
								Creative atmosphere	0.033	4.0	3.4	3.4
Image W = 0.249	0.864	0.705	0.603	Buzz of place W = 0.062	0.210	0.150	0.126	Pace of life	0.030	3.7	2.8	2.6
								Variety of lifestyles	0.032	3.3	2.2	1.6
				Sense of safety W = 0.061	0.225	0.201	0.177	Subjective safety	0.028	3.6	3.5	3.1
								Objective safety and security	0.033	3.9	3.2	2.8
				Sense of place W = 0.063	0.231	0.174	0.153	Place attachment	0.030	3.9	2.6	2.2
								Sense of community	0.033	3.8	3.2	2.9
				Place identity W = 0.063	0.198	0.180	0.147	Authenticity	0.032	3.3	3.2	1.8
								Brand	0.031	3.3	2.8	3.1
Total	3.618	2.925	2.550		3.618	2.925	2.550		1.000	3.76	3.05	2.65

district—highlighted cells indicate the high and low scores received. Then, the weighting system was applied by multiplying weights and mean values. The weighting system is the reflection of the fact that some themes—i.e., 'Function' and 'Ambiance'—and some indicators—i.e., 'Urban mobility' and 'Work climate'—have more influence on shaping the general characteristic of a place (see Chapter 6). In general, the quantitative results demonstrated the weakness and strength points of each case and identified the innovation district that presents the most important characteristics in each category.

7.4.2.1 Context

As all three cases are located in the same city, the impact of context is considered equal and eliminated from Table 7.3. The overall score in each innovation district demonstrated that KGUV is the most successful one—KGUV = 3.618, DKP = 2.925, BTP = 2.550. These results confirmed that involving community-based values in planning, designing, developing, and managing of KGUV improved the place characteristics.

7.4.2.2 Form

The study confirmed that KGUV is the most successful one in terms of location, urban form, design, and amenities. They highlighted that KGUV's spatial characteristics are the most effective ones in empowering and activating the place—i.e., proximity to CBD, collocating with the university, unique natural environment, integration with city fabric, medium- to high-density urban form, connection with street life, walkability, variety of facilities and amenities. On the other hand, the results revealed that the form-related characteristics in BTP are not capable of fulfilling the requirements of innovation districts—i.e., lack of accessibility and connectivity, low-density development, segmented urban form, generic architectural design, car-oriented system, lack of urban amenities.

7.4.2.3 Function

The quantitative findings, as shown in Table 7.3, revealed that the management system in KGUV—allied community management—was the most efficient and DKP—individual body corporates—was the weakest model. The results confirmed that mixed-used development in KGUV was a better model for shaping innovation districts. The findings clarified that BTP was the most successful one in attracting knowledge companies; such a characteristic is caused by the availability and more affordable workspaces as well as the flexibility of BTP to adapt with the changing economic trends. However, due to the lack of an active street life and a community manager, BTP has the weakest professional networks.

7.4.2.4 Ambiance

The findings emphasised the presence of a variety of public places and events in KGUV that remarkably facilitated social and professional interaction. This encourages people to outdoor activities, guarantees a dynamic street life, and boosts the socio-dynamic ambiance. Due to the multicultural nature of Brisbane, the results represented an acceptable level of diversity and tolerance in all case studies. However, the findings still confirmed the superiority of KGUV; collocating with QUT and its creative industries faculty significantly led to higher levels of diversity, openness, and artistic atmosphere.

7.4.2.5 Image

The low scores of DKP and BTP indicated the lack of street life has formed a less desirable environment. Among the case studies, KGUV presented the most successful performance in creating an attractive buzz through its dynamic and artistic ambiance. However, in KGUV, lifestyles are still limited to the student level, which justifies the medium level of satisfaction in participants. The results demonstrate that innovation districts with higher levels of mixed-use development, dynamic street life, and walkability are more successful in creating a sense of safety and a sense of place. The findings clarified that higher levels of these senses lead to greater place attachments and retain people in innovation districts. In terms of uniqueness and brand, the results signified approximately low-level equal scores in all cases.

7.5 DISCUSSION AND CONCLUSION

This study confirms, in line with the literature, that knowledge economy growth is facilitated by a combination of both hard and soft factors (Darchen & Tremblay, 2010; Scott, 2010). It highlights the predominance of hard factors for attracting workers and industries—e.g., job opportunities, infrastructure, commuting time, cost of living and running a business. On the other hand, soft factors also strongly contribute to the sense of place and attachment, which help in retaining workers and industries. Accordingly, the study pinpoints important contributions of hard and soft factors in a flourishing knowledge economy performance.

In addressing the first research question, the qualitative analysis revealed that place quality matters for the competitiveness and socio-economic success of innovation districts. This finding is also supported by the literature (Sepe, 2010; Durmaz, 2015; Pancholi *et al.*, 2015). In terms of the second question, the qualitative and quantitative analyses generated insights into the specific place characteristics of innovation districts. The analyses confirmed that place characteristics of innovation districts contain various and complex spatial and socioeconomic elements. The chapter categorised these essential place characteristics under comprehensible themes (Table 7.1).

The analytic framework, shown in Table 7.1, was tested and validated as an efficient tool for investigating place characteristics, comparing different innovation districts, and revealing their weaknesses and strength points. The findings from the comparative analysis of investigated cases disclose the key place characteristics that could be generalised and implemented by other innovation districts (Table 7.4). While this research suggested various place characteristics supporting innovation districts, one characteristic is overarching. This study emphasises that innovation districts are no longer restricted to clusters of industries or even talented workers.

In fact, innovation districts need to be considered an extension of the city fabric and involve the general public. Innovation districts properly integrated in the city and respecting requirements of the public are more likely to boost a dynamic street life. Accordingly, they are more likely to shape a vibrant ambiance for attracting and retaining talent and investment. They are also highly likely to encourage the spontaneous agglomeration of creative and knowledge-based activities around their blurred boundaries. Consequently, integration of innovation districts with the city fabric and general public is a remarkable mechanism for rising and empowering the knowledge economy in cities. Over time, this mechanism could transform the entire city into a platform for creative and knowledge-based activities rather than limit them to specific zones.

The findings also resonate with the human capital theory. Based on this theory, economic growth in innovation districts increasingly relies on their capacity for fostering, attracting, and retaining talent (Glaeser, 2005; Gabe *et al.*, 2013). The outcomes of the analyses indicate that innovation districts seeking to invest on the essential place characteristic, which are appealing for talented workforce, would be more successful.

TABLE 7.4 Essential Place Characteristics of Innovation Districts

Theme	Category	Characteristics
Context	Urban quality	• Cities with global reputation, warm climate, lower costs of living or running a business are more likely to attract talent and investment. • Cities with high-quality hard and soft infrastructures are more capable of attracting talent and investment. • Innovation districts to be developed and operated by the private sector in the light of the strategies developed by the public sector.
Form	Location	• While inner-city districts benefit from variety of facilities and city vibes, suburban districts provide affordability and tranquil environments. • Creative districts are more likely to survive and flourish in inner-city locations. • Tech- and research-based innovation districts can be located either in the inner city or suburbs, as long as suburban locations are well connected to the inner-city amenities. • As a fully isolated district, it is too hard to survive; accessibility and connectivity are priorities. • Innovation districts are those collocated with universities, research institutes, and other knowledge-hubs benefit from collective facilities as well as knowledge spill-over.
	Urban form and structure	• Successful innovation districts are mostly planned to integrate into city fabrics by blurring boundaries between where people work, study, and live. • Successful innovation districts can naturally flow into the city and share facilities with surrounding areas and grow freely. • Medium-/high-density development, accessibility to green and public spaces, and high level of walkability encourage the general public to step into innovation districts.
	Design	• Innovation districts need to connect their workplaces to a dynamic and attractive district life by offering various activities at the street level. • Innovation districts need to not only provide a variety of well-equipped places for outdoor activities, but also motivate large amounts of movement in the streetscape. • Innovation districts require attractive and unique architectural design to reflect their creative, high-tech, and cutting-edge identity that supports the visibility of activities.

(*Continued*)

TABLE 7.4 *(Continued)* Essential Place Characteristics of Innovation Districts

Theme	Category	Characteristics
	Amenities	• Innovation districts have the potential to become the centre of essential and advanced facilities and urban amenities in their neighbourhoods. • Offering public amenities prevents innovation districts from being isolated and invites the general public to spontaneously use the place and activate the district life.
Function	Services	• Considering the importance of commuting time in locational choices of workers, effective mobility services that minimise the travelling time will enhance attractiveness. • Presence of a community manager, along with allied body corporates, is a beneficial service—accelerates community development and professional/social networks.
	Land use	• Mixed-use districts that embrace different functions—educational, institutional, residential, commercial—are likely to link community-based and economic values. • Mixed-use districts are more capable of rising a dense and diverse community for supporting commercial and social activities. • Mixed-use development accelerates sustainable socioeconomic growth in innovation districts—but a right balance between functions is needed.
	Company profile	• The presence of research institutes along with various knowledge companies, who consume the produced knowledge would be helpful for innovation districts' growth. • Collaboration between research institutes and industry needs to be flexible to address (inter) national economic demands. • Innovation districts that provide place and opportunity for different scales of companies and institutes potentially facilitate a sustainable economic growth.
	Work condition	• Presence of different companies, as well as various educational and research institutes, provides a thick pool of job opportunities for knowledge workers. • Variety of professional activities shapes a diverse professional community that makes knowledge spill-over and networks more meaningful. • Business networks accelerate by the presence of formal spaces and events as well as places that encourage informal and incidental networks.

(Continued)

TABLE 7.4 *(Continued)* Essential Place Characteristics of Innovation Districts

Theme	Category	Characteristics
Ambiance	Public spaces and events	• Public places and events are natural nodes of incidental interactions that help develop informal professional networks and social interactions with the broader community. • Availability of such places and events for the general public encourages innovation districts' community as well as surrounding neighbourhoods to freely attend activities.
	Public engagement	• As knowledge workers prefer casual and dynamic ambiances that offer more than official work environments, innovation districts need to involve a variety of activities. • Innovation districts that combine adequate amount of people, activities, and interactions are more likely to shape an attractive ambiance.
	Diversity	• Innovation districts need to establish a heterogenous community from different ages, backgrounds, and fields of expertise. • Mixed-used development and prevention from locating too many homogenous innovative sectors in a district contribute to the diversity of people and professionals. • Innovation districts need to be capable of providing an adventurous opportunity for interacting with non-like-minded people and experiencing different lifestyles.
	Creativity	• Innovation districts naturally gather a variety of highly educated and talented workers, who inspire each other with creative minds and nourish a creative ambiance. • Visibility of innovative actives, artistic atmosphere, and presence of the state-of-the-art ideas and amenities inspire talents and flourish a creative atmosphere.
Image	Buzz of place	• Combining environmental, economic, and social characteristics empowers innovation districts to fulfil desires of knowledge workers for experiencing different vibes. • Gloomy atmosphere of mono-functional innovation districts that isolate talents in their workplaces do not fit the requirements of the world-class innovation districts.

(Continued)

TABLE 7.4 *(Continued)* Essential Place Characteristics of Innovation Districts

Theme	Category	Characteristics
	Sense of safety	• Sense of safety is mostly enhanced by specific architectural and urban design characteristics that contribute to walkability and street life. • High presence of people and activities on the ground and collocating with special functions such as residential blocks are critical factors in shaping a safety image in mind. • Activating the nightlife and preventing vacancy during nighttime are effective mechanisms for improving the sense of safety.
	Sense of place	• Innovation districts that respect community-based values, provide public amenities, and develop dense and coherent communities are likely to establish place attachment. • While economic aspects are major factors in attracting talent and investment, community aspects are significantly effective in shaping the sense of place and retaining them.
	Place identity	• Branded innovation districts potentially attract more talented workers and investments, which also reduce the risk of failure in businesses. • Reputation is supported by the presence of renowned companies and institutes as well as authentic architectures and unique vibes—i.e., futuristic, heritage, bohemian.

Consequently, innovation districts are no longer seen as a pure agglomeration of firms and institutes, such as in DKP and BTP. The study exposed that community-based innovation districts, such as KGUV, fulfil the desires of talents better.

The insights generated are useful for urban policymakers, urban and economic development planners, and urban designers to better understand critical place characteristics of innovation districts. This can lead to an informed policy and plan making to boost the knowledge economy performance of their cities through prosperous innovation districts. Today, successful innovation districts offer activities beyond solely focusing on economic accomplishments. At this point, further research is needed to map out the reciprocal relation between innovation districts and local communities and their contribution to city life.

This study has the following limitations and solutions: (a) The results highlighted are based on a limited number of case studies; some characteristics of innovation districts are highly place specific and hard to generalise. In address this limitation, the study undertook a comparative analysis to minimise the location specific factors. Besides, global best practices are linked where appropriate to support and generalising the findings. This helped develop the general characteristics shown in Table 7.4. (b) This study focused on innovation districts located in the same city to eliminate the impact of the context. Thus, the findings are mostly relevant at the cluster or neighbourhood scale rather than regional and city scales. For the regional and city scales, the use of an alternative framework proposed in Chapter 6 will be useful. (c) Generated insights are based on innovation districts that are developed from scratch—not spontaneously emerged ones. However, proposed characteristics are useful not only for the planning, design and development stages of planned districts, but also for the management and redevelopment stages of the existing innovation districts. (d) The framework could be applied to real-world cases by using different methods, although the mixed methods—quantitative and qualitative—applied in this research are invaluable. (e) Even though a software is used for coding and content analysis, there still might be some bias in interpretations. In order to reduce bias, four members of the research team analysed the collected data (Esmaeilpoorarabi *et al.*, 2018).

ACKNOWLEDGEMENTS

This chapter, with permission from the copyright holder, is a reproduced version of the following journal article: Esmaeilpoorarabi, N., Yigitcanlar, T., Guaralda, M., & Kamruzzaman, M., (2018). Does place quality matter for innovation districts? Determining the essential place characteristics from Brisbane's knowledge precincts. Land Use Policy, 79(1), 734–747.

REFERENCES

Alfken, C., Broekel, T., & Sternberg, R. (2015). Factors explaining the spatial agglomeration of the creative class. *European Planning Studies*, 23(12), 2438–2463. DOI:10.1080/09654313.2014.979767

Baum, S., O'Connor, K., & Yigitcanlar, T. (2009). The implications of creative industries for regional outcomes. *International Journal of Foresight and Innovation Policy*, 5, 44–64. DOI:10.1504/IJFIP.2009.022098

Benneworth, P., & Hospers, G.J. (2007). Urban competitiveness in the knowledge economy. *Progress in Planning*, 67(2), 105–197. DOI:10.1016/j.progress.2007.02.003

Benneworth, P., & Ratinho, T. (2014). Reframing the role of knowledge parks and science cities in knowledge-based urban development. *Environment and Planning*, 32(5), 784–808. DOI:10.1068/c1266r

Bereitschaft, B., & Cammack, R. (2015). Neighborhood diversity and the creative class in Chicago. *Applied Geography*, 63, 166–183. DOI:10.1016/j.apgeog.2015.06.020

Boddy, M. (1999). Geographical economics and urban competitiveness. *Urban Studies*, 36(5-6), 811–842. DOI: 10.1080/0042098993

Bontje, M., & Musterd, S. (2009). Creative industries, creative class and competitiveness. *Geoforum*, 40(5), 843–852. DOI:10.1016/j.geoforum.2009.07.001

Boren, T., Young, C., Kulturgeografiska, I., Stockholms, U., & Samhallsvetenskapliga, F. (2013). Getting creative with the creative city? *International Journal of Urban and Regional Research*, 37, 1799–1815. DOI:10.1111/j.1468-2427.2012.01132.x

Brown, J., & Mczyski, M. (2009). Locational choices of creative knowledge workers. *Built Environment*, 35(2), 238–252. DOI:10.2148/benv.35.2.238

Carrillo, J., Yigitcanlar, T., Garcia, B., & Lonnqvist, A. (2014). *Knowledge and the city*. New York: Routledge. DOI:10.4324/9781315856650

Carvalho, L., & Van Winden, W. (2017). Planned knowledge locations in cities. *International Journal of Knowledge-Based Development*, 8(1), 47–67. DOI:10.1504/IJKBD.2017.082429

Charnock, G., Purcell, T., & Ribera-Fumaz, R. (2014). City of rents. *International Journal of Urban and Regional Research*, 38(1), 198–217. DOI:10.1111/1468-2427.12103

Clark, J., Huang, H.I., & Walsh, J.P. (2010). A typology of innovation districts. *Cambridge Journal of Regions, Economy and Society*, 3(1), 121–137. DOI:10.1093/cjres/rsp034

Clifton, N. (2008). The creative class in the UK. *Geografiska Annaler*, 90(1), 63–82. DOI:10.1111/j.1468-0467.2008.00276.x

Clifton, N., & Cooke, P. (2009). Creative knowledge workers and location in Europe and North America. *Creative Industries Journal*, 2(1), 73–89. DOI:10.1386/cij.2.1.73/1

Cooke, P. (2004). The role of research in regional innovation systems. *International Journal of Technology Management*, 28(3-6), 507–533. DOI:10.1504/IJTM.2004.005307

Cooke, P. (2017). Complex spaces. *Journal of Open Innovation*, 3(2), 9–32. DOI: 10.1186/s40852-017-0060-5

D'Mello, M., & Sahay, S. (2007). I am kind of a nomad where I have to go places and places. *Information and Organization*, 17(3), 162–192. DOI:10.1016/j.infoandorg.2007.04.001

Darchen, S., & Tremblay, D. (2010). What attracts and retains knowledge workers/students. *Cities*, 27(4), 225–233. DOI:10.1016/j.cities.2009.12.009

Durmaz, S. (2015). Analyzing the quality of place. *Journal of Urban Design*, 20(1), 93–124. DOI:10.1080/13574809.2014.972348

Eisenhardt, K.M. (1989). Building theories from case study research. *Academy of Management Review*, 14(4), 532–550. DOI: 10.2307/258557

Esmaeilpoorarabi, N., Yigitcanlar, T., & Guaralda, M. (2016a). Place quality and urban competitiveness symbiosis? *International Journal of Knowledge-Based Development*, 7(1), 4–21. 21. DOI:10.1504/IJKBD.2016.075444

Esmaeilpoorarabi, N., Yigitcanlar, T., & Guaralda, M. (2016b). Towards an urban quality framework. *International Journal of Knowledge-Based Development*, 7(3), 290–312. DOI:10.1504/IJKBD.2016.078556

Esmaeilpoorarabi, N., Yigitcanlar, T., Guaralda, M., & Kamruzzaman, M., (2018). Does place quality matter for innovation districts? Determining the essential place characteristics from Brisbane's knowledge precincts. *Land Use Policy*, 79(1), 734–747. DOI: 10.1016/j.landusepol.2018.09.016

Florida, R. (2005). *Cities and the creative class*. New York: Routledge. DOI:10.4324/9780203997673

Frenkel, A., Bendit, E., & Kaplan, S. (2013). Residential location choice of knowledge-workers. *Cities*, 35(1), 33–41. DOI:10.1016/j.cities.2013.06.005

Gabe, T., Florida, R., & Mellander, C. (2013). The creative class and the crisis. *Cambridge Journal of Regions, Economy and Society*, 6(1), 37–53. DOI:10.1093/cjres/rss012

Glaeser, E. (2005). Review of Richard Florida's the rise of the creative class. *Regional Science and Urban Economics*, 35(5), 593–596. DOI:10.1016/j.regsciurbeco.2005.01.005

Grant, J., & Buckwold, B. (2013). Precarious creativity. *Cambridge Journal of Regions, Economy and Society*, 6(1), 1–17. DOI:10.1093/cjres/rss008

Hall, P. (1996). Revisiting the non-place urban realm. *International Planning Studies*, 1(1), 7–15. DOI:10.1080/13563479608721640

Heebels, B., & Van-Aalst, I. (2010). Creative clusters in Berlin. *Geografiska Annaler*, 92(4), 347–363. DOI:10.1111/j.1468-0467.2010.00357.x

Kloosterman, R., & Trip, J. (2011). Planning for quality? *Journal of Urban Design*, 16(4), 455–470. DOI:10.1080/13574809.2011.585863

Lee, J., Hancock, M., & Hu, M. (2014). Towards an effective framework for building smart cities. *Technological Forecasting and Social Change*, 89(1), 80–99. DOI:10.1016/j.techfore.2013.08.033

Lonnqvist, A., Kapyla, J., Salonius, H., & Yigitcanlar, T. (2014). Knowledge that matters. *European Planning Studies*, 22(10), 2011–2029. DOI:10.1068/a1302

Makkonen, T., Merisalo, M., & Inkinen, T. (2018). Containers, facilitators, innovators? *European Urban and Regional Studies*, 25(1), 106–118. DOI:10.1177/0969776417691565

Martin, R., Florida, R., Pogue, M., Mellander, C., Gugler, P., & Ketels, C. (2015). Creativity, clusters and the competitive advantage of cities. *Competitiveness Review*, 25(5), 482–496. DOI:10.1108/CR-07-2015-0069

McCann, E. (2004). Best places. *Urban Studies*, 41(10), 1909–1929. DOI:10.1080/0042098042000256314

Meijer, A., & Thaens, M. (2018). Urban technological innovation. *Urban Affairs Review*, 54(2), 363–387. DOI: 10.1177/1078087416670

Metaxiotis, K., Carrillo, F.J., & Yigitcanlar, T. (2010). Knowledge-based development for cities and societies. *IGI Global*. DOI:10.4018/978-1-61520-721-3

Millar, C., & Ju-Choi, C. (2010). Development and knowledge resources. *Journal of Knowledge Management*, 14(5), 759–776. DOI:10.1108/13673271011074881

Morais, P., Miguéis, V., & Camanho, A., 2013. Quality of life experienced by human capital. *Soc. Indic. Res.* 110(1), 187–206. DOI:10.1007/s11205-011-9923-5

Pancholi, S., Yigitcanlar, T., & Guaralda, M. (2014). Urban knowledge and innovation spaces. *Asia Pacific Journal of Innovation and Entrepreneurship*, 8(1), 15–38. DOI:10.1080/10630732.2016.1164443

Pancholi, S., Yigitcanlar, T., & Guaralda, M. (2015). Place making facilitators of knowledge and innovation spaces. *International Journal of Knowledge-Based Development*, 6(3), 215–240. DOI:10.1504/IJKBD.2015.072823

Pancholi, S., Yigitcanlar, T., & Guaralda, M. (2017). Governance that matters. *Journal of Place Management and Development*, 10(1), 73–87. DOI:10.1108/JPMD-08-2016-0053

Pancholi, S., Yigitcanlar, T., & Guaralda, M. (2019). Place making for innovation and knowledge-intensive activities. *Technological Forecasting & Social Change*, 146(Sep), 616–625. DOI: 10.1016/j.techfore.2017.09.014

Parker, R. (2010). Evolution and change in industrial clusters. *European Urban and Regional Studies*, 17(3), 245–260. DOI:10.1177/0969776409358244

Reve, T., Sasson, A., Gugler, P., & Ketels, C. (2015). Theoretical and methodological advances in cluster research. *Competitiveness Review*, 25(5), 524–539.

Santos, L.D., & Martins, I. (2007). Monitoring urban quality of life. *Social Indicators Research*, 80(2), 411–425. DOI:10.1007/s11205-006-0002-2

Sarimin, M., & Yigitcanlar, T. (2012). Towards a comprehensive and integrated knowledge-based urban development model. *International Journal of Knowledge-Based Development*, 3(2), 175–192. DOI:10.1504/IJKBD.2012.047035

Scott, A. (2010). Jobs or amenities? *Regional Science*, 89(1), 43–63. DOI: 10.1111/j.1435-5957.2009.00263.x

Sepe, M. (2010). Liveability, quality and place identity in the contemporary city. *Journal of Place Management and Development*, 3(3), 221–246. DOI:10.1108/17538331011083952

Sohn, D., & Kenney, M. (2007). Universities, clusters, and innovation systems. *World Development*, 35(6), 991–1004. DOI:10.1016/j.worlddev.2006.05.008

Storper, M., & Venables, A. (2004). Buzz: face-to-face contact and the urban economy. *Journal of Economic Geography*, 4(4), 351–370. DOI:10.1093/jnlecg/lbh027

Storper, M., Scott, A. (2009). Rethinking human capital, creativity and urban growth. *Journal of Economic Geography*, 9(2), 147–167.

Trip, J. (2007). Assessing quality of place. *Journal of Urban Affairs*, 29(5), 501–517. DOI: 10.1111/j.1467-9906.2007.00362.x

Van Winden, W., De Carvalho, L., Van Tuijl, E., Van Haaren, J., & Van den Berg, L. (2013). *Creating knowledge locations in cities*. London: Routledge. DOI: 10.4324/9780203127162

Wolfe, M. (1999). The wired loft. *Urban Affairs Review*, 34(5), 707–728. DOI:10.1177/10780879922184158

Yigitcanlar, T., Baum, S., & Horton, S. (2007). Attracting and retaining knowledge workers in knowledge cities. *Journal of Knowledge Management*, 11(5), 6–17. DOI:10.1108/13673270710819762

Yigitcanlar, T., & Velibeyoglu, K. (2008). Knowledge-based urban development. *Local Economy*, 23(3), 195–207. DOI:10.1080/02690940802197358

Yigitcanlar, T., Metaxiotis, K., & Carrillo, F.J. (2012). *Building prosperous knowledge cities*. Edward Elgar, London. DOI:10.4337/9780857936042

Yigitcanlar, T., & Dur, F. (2013). Making space and place for knowledge communities. *Australasian Journal of Regional Studies*, 19, 36–63.

Yigitcanlar, T., & Dizdaroglu, D. (2014). Ecological approaches in planning for sustainable cities. *Global Journal of Environmental Science and Management*, 1(2), 159–188. DOI:10.7508/gjesm.2015.02.008

Yigitcanlar, T. (2015). Smart cities. *Australian Planner*, 52(1), 27–34. DOI:10.1080/07293682.2015.1019752

Yigitcanlar, T., & Bulu, M. (2015). Dubaization of Istanbul. *Environment and Planning A*, 47(1), 89–107. DOI: 10.1068/a130209p

Yigitcanlar, T., Edvardsson, I., Johannesson, H., Kamruzzaman, M., Ioppolo, G., & Pancholi, S. (2017). Knowledge-based development dynamics in less favoured regions. *European Planning Studies*, 25(12), 2272–2292. DOI:10.1080/09654313.2017.1358699

Youtie, J., & Shapira, P. (2008). Building an innovation hub. *Research Policy*, 37(8), 1188–1204. DOI:10.1016/j.respol.2008.04.012

CHAPTER 8

Community Engagement with Innovation Districts

8.1 INTRODUCTION

Over the last two decades, knowledge-based and technology-driven sectors have driven the economy globally (Lonnqvist *et al.*, 2014). Many cities around the world have transitioned from industrial economic activities to knowledge-based ones, characterised by digital disruption and creativity (Bontje & Musterd, 2009). This transition has forced cities to embrace new strategies and structural changes for sustained economic progress (Metaxiotis *et al.*, 2010). These strategies are mainly interwoven under the knowledge-based urban development (KBUD) paradigm (Carrillo *et al.*, 2014). KBUD aims to pursue a more economically, socially, and environmentally viable and balanced urban future, utilising knowledge-intensive potentials (Sarimin & Yigitcanlar, 2012; Yigitcanlar *et al.*, 2012). It implements strategies to upgrade infrastructures, improve the quality of life, and create an attractive environment for knowledge workers (Yigitcanlar & Dur, 2013).

In this context, a new type of land use, the innovation district, has emerged and become the nucleus of knowledge-based activities in cities. Silicon Valley, Cambridge Science Park, and Sophia Antipolis are among the pioneering early examples of innovation districts (Zou & Zhao, 2014). In the 1980s and 1990s, the focus of innovation districts—previously called technology/science/innovation parks—was only on occupational and industrial development (Porter, 1990); at the time, innovation districts were secluded and isolated clusters of knowledge-intensive or creative

DOI: 10.1201/9781032657431-11

activities with limited professional and social networks or urban configuration.

Recently, the importance of other components, such as quality of place, mixed-used development, walkability, heterogeneity, diversity, and authenticity, were also recognised (Florida, 2005; Glaeser, 2005; Hawken & Han, 2017). The domination of these urban values has shaped a distinctive type of innovation district in comparison to traditional technology/science/innovation parks (Benneworth & Ratinho, 2014; Blakely & Hu, 2019). These urban innovation districts 'cluster and connect leading-edge anchor institutions and cutting-edge innovative firms with supporting and spin-off companies, business incubators, mixed-use housing, office and retail, and twenty-first-century amenities and transport' (Katz & Bradley, 2013, p. 114).

This shift was caused by the following: (a) The peculiar profile of knowledge workers—who seek authentic scenes, amenities, and quality of living, not just a lucrative occupation. They prefer vibrant urban locations, where they can socialise with like-minded people as well as diverse communities (Florida, 2005; Baum *et al.*, 2009; Lawton *et al.*, 2013; Reve *et al.*, 2015; Esmaeilpoorarabi *et al.*, 2016). (b) The certain need for informal networks—as informal networks is the easiest type of networking opportunity for businesses. Urban innovation districts facilitate firm linkage, knowledge exchange, and synergy among people, firms, and place through offering a casual environment (Esmaeilpoorarabi *et al.*, 2016; Yun *et al.*, 2018). (c) The change in the nature of creative/smart/innovative works—where work is no longer a separate event of daily activities. Knowledge workers tend to prefer flexible working hours and work environment. Knowledge workers require new configurations of place with blurred boundaries between work and daily activities (Blakely & Hu, 2019).

In this new approach, innovation districts have become an extension of the urban fabric arising as dense and mixed-use districts. The vibrant ambiance in these districts relies on the active presence of local communities, who bring energy and enthusiasm to the public life (Katz & Bradley, 2013; Katz & Wagner, 2014). Therefore, in contrast to the models of the 1980s and 1990s exurban technology/science/innovation parks, these innovation districts are less likely to be located outside metropolitan areas. During the last two decades, these districts started to emerge in the vibrant ambiance of downtowns (Baum *et al.*, 2009; Edvardsson *et al.*, 2016; Engel *et al.*, 2018); also in consideration of their prime amenities (Rutland, 2015)—e.g., diversity of lifestyles, variety of cafés and restaurants, dynamic nightlife, active street life, improved aesthetic, vibrant cultural opportunities (Asheim & Hansen, 2009).

This situation, however, has risen certain socio-cultural conflicts. Firstly, the concentration of amenities and jobs in downtowns has increasingly left even inner-city neighbourhoods lacking knowledge-intensive employment (Kelly *et al.*, 2014). This situation not only increased pressures on downtowns, but also dramatically threatened the jobs-housing balance in cities (Martinez-Fernandez & Potts, 2009). Secondly, these locations have been occupied by privileged knowledge workers, which marginalise other groups. The rise in land prices, along with the high cost of living, undesirably pushed the lower and middle classes away from the urban core. It led to a negative impact on the city's polyvalent character by making it less democratic and diverse (Yigitcanlar & Bulu, 2015; Stehlin, 2016; Pancholi *et al.*, 2017).

In response to this unbalanced growth of knowledge-based activities, innovation districts have been considered a fundamental element for shaping polycentricity in cities—multiple business districts surrounded with residential areas (Kinch, 2019). Innovation districts are compact and amenity-rich urban configurations, which can bring facilities, knowledge sectors, and jobs at least to inner-city neighbourhoods/suburbs (Van Winden *et al.*, 2013; Katz & Wagner, 2014). In return, innovation districts can benefit from the pre-existing population and urban values of these neighbourhoods/suburbs for shaping the required vibrant ambiance. These urban values contain mixed-use, transit-oriented, and walkable areas as well as locations around anchor institutions and post-industrial or historic precincts (Katz & Bradley, 2013; Hawken & Han, 2017).

In sum, urban innovation districts, located in midtown and close-in suburban districts, have been planned to not only fulfil the requirements of the knowledge sectors, but also to provide a jobs-housing balance and social equality—also taking advantage of the more affordable land prices. These districts are expected to propound open-door policies and benefit diverse groups of society. However, current innovation districts have been unsuccessful in integrating with surrounding neighbourhoods, and convincing the local communities to be involved in ongoing activities in these districts. It appears to be that open-door policies and multi-functional facilities of innovation districts are not sufficient to assure community engagement (Lawton *et al.*, 2013). Furthermore, there is lack of evidence in the literature on how to promote community engagement with innovation districts; while community engagement is a vital essence for innovation districts, those aim to: (a) shape a dynamic and heterogeneous district life for attracting knowledge workers and industries; (b) balance work-housing urban setting to benefit both knowledge workers and the local community.

In order to bridge this knowledge gap, this chapter aims to identify the likely features of urban innovation districts—including downtown, midtown and close-in suburban districts—that can enhance community engagement. With this aim in mind, the chapter attempts to address the research question of: How can an enhanced community engagement with innovation districts be established? The research examines public opinions from three major Australian cities—i.e., Sydney, Melbourne, Brisbane—that house a number of globally renowned prosperous innovation districts. Public views are captured to identify the characteristics of innovation districts that either encourage or discourage community engagement.

8.2 LITERATURE BACKGROUND

Community engagement is a widely used concept, applied in various disciplines including sociology, economics, psychology, and urban studies. Literature focuses on community socioeconomic engagement in several scales. The first scale is the macro-scale—or city level—and is characterised by 'social capital' (Pooley *et al.*, 2004). Social capital concerns the communities' socioeconomic and democratic wellbeing (Putnam, 2000). The social capital theory emphasises that economic transactions—e.g., job search, employment—are more efficient when people are part of social networks (Granovetter, 2011). Social capital, therefore, is an explicit attempt for creating and enhancing social networks; those accelerate the accessibility of the public to socioeconomic benefits (Hero, 2003; Lynn, 2006; Farrell, 2007).

From the KBUD perspective, cities should put technology truly at the service of the public to motivate socioeconomic interactions, and derive the city into its knowledge-based future—e.g., e-governance, equal access, and knowledge of ICT usage. In this regard, created networks aim to empower the society to get involved with, and benefit from, the knowledge economy (Millar & Ju-Choi, 2010). Socioeconomic networks either directly or indirectly engage the public in knowledge-intensive activities, and in return foster public trust, since people who are familiar and involved with the knowledge economy are more likely to approve the transition toward a knowledge-based future. Trust improves community engagement with knowledge-based activities (Longhi, 1999; Wal, 2013; Yigitcanlar, 2015).

The second scale is the micro-scale—or firm level—and characterised by 'social licence'. Social licence is an informal consent within the local community and stakeholders on ongoing activities of a firm/project. It includes three different approval levels—i.e., legitimacy, credibility, trust (Dare *et al.*, 2014). Similar to the macro-scale, the aim of firms/projects is to gain the highest level of social trust through sharing socioeconomic benefits with the local community in particular and with the public at large (Head, 2007). Public trust is stimulated through the following five levels of engagement models.

(a) *Inform:* To provide the public with objective information to assist them in understanding problems, alternatives, and opportunities. This is often facilitated through information booths, media releases, newsletters, brochures, mail out programs, websites, and hotlines. The use of these techniques is often perceived as a way to present basic information to a wide range of stakeholders. (b) *Consult:* To obtain public feedback. This one-way interaction involves public meetings, discussion groups, polls, surveys, and focus groups. (c) *Involve:* To work directly with the public to ensure their concerns and aspirations are considered. This stage establishes a two-way interaction. (d) *Collaborate:* To partner with the public in developing alternatives and identifying solutions. This step represents more active and stakeholder-driven interaction. Activities include workshops, discussion groups, learning circles, interviews, reference groups, and community consultative committees. (e) *Empower:* To place decision making in the hands of the public. This level of engagement includes participation in planning and decision making for operation, management, and community-development process (Bryson, 2004).

Knowledge-based companies are accustomed to these strategies as they engage with the public to gain their trust. For instance, the Arabianranta innovation district in Helsinki hosts Europe's first living lab driven by Nokia and Alto University—that engages local communities with the district's activities (Carvalho & Van Winden, 2017). A living lab examines technical systems and their socioeconomic influences; hence, it focuses on the social needs of people. This process simultaneously promotes sustainable growth of companies, and at the same time, generates social benefits to the public (Bergvall & Stahlbrost, 2009; Liedtke *et al.*, 2012).

There is also a third scale of socioeconomic engagement that is the mezzo-scale—or neighbourhood level. While social engagement at this scale has been widely investigated (Jarvis *et al.*, 2012), socioeconomic engagement of the public with knowledge-based activities is an under-studied area. From the knowledge-based activity perspective, the mezzo-scale is mainly represented by innovation districts (Pancholi *et al.*, 2015).

This study combines two models of socioeconomic engagement in the macro- (Hawkins & Maurer, 2009) and micro-scales (Bass, 1990) to establish an innovation district community engagement model at the mezzo-scale (Figure 8.1). In this model, innovation districts are enquired through three types of socioeconomic relationships—i.e., bonding, bridging, linking—that are supported by a range of community engagement strategies—i.e., transactional, transitional, transformational.

Bonding refers to relationships amongst members of a network who are homogenous (Putnam, 2000)—e.g., bonds between co-workers in innovation districts. *Bridging* refers to relationships amongst people who are dissimilar in a demonstrable fashion—e.g., bridges knowledge workers, companies, and local communities located around innovation districts. *Linking* is the extent to which individuals build relationships with institutions and individuals, who have relative power over them

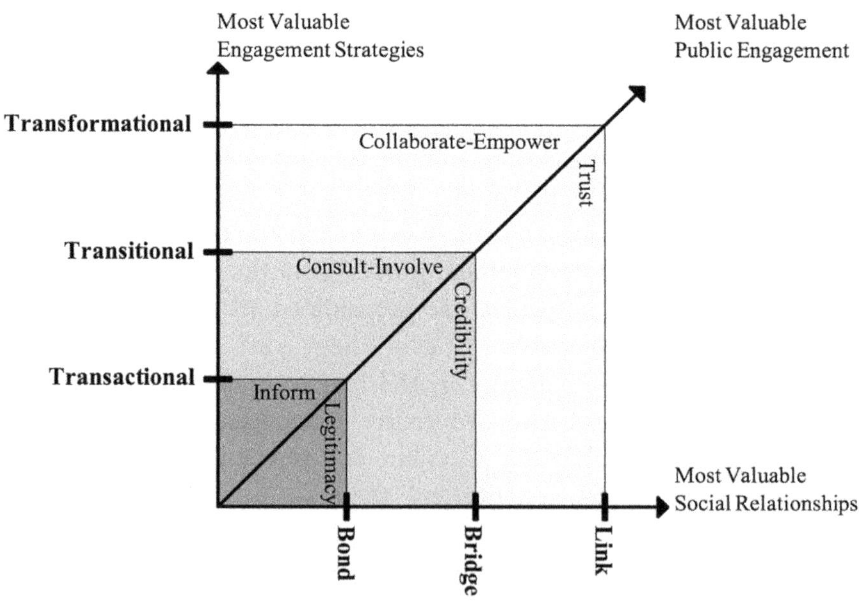

FIGURE 8.1 Innovation district community engagement model (by the authors).

(Woolcock, 2001)—e.g., link between the public and authorities, who plan, design, develop, and manage innovation districts.

Linking is the weakest relationship, but it is the most valuable outcome, as linkage provides public access and connection to power structures and institutions (Hawkins & Maurer, 2009). Unlike bonding, which basically relies on existing similarities, bridging and linking are characterised by the development of new ideas, values, and perspectives that are particularly invaluable for innovation districts (Szreter & Woolcock, 2004). Innovation districts need to foster and retain not only bonding, but also bridging and linking relationships among the dissimilar community of users. Consequently, for sustainable socio-economic growth, companies, knowledge workers, and local communities need to be encouraged for effective interactions. In order to encourage such interactions, engagement strategies should be clearly planned and thoroughly followed.

Community engagement, however, is not a one-time action; continuous engagement strategies are required to nurture public trust and involvement. Transactional engagement strategies are related to one-way communication (Figure 8.1). At this level, innovation districts may engage with the public by providing information, employee volunteer time, free training, and philanthropy donations (Van den Berg *et al.*, 2004). Transitional engagement is characterised by two-way communication, consultation, and collaboration. However, control of resources and benefits in these interactions remains with innovation districts, rather than being fully shared with the local community (Bowen *et al.*, 2010).

On the other hand, transformational engagement is represented by joint learning and sense-making (Hart & Sharma, 2004), joint management of resources and socioeconomic profits (Newman *et al.*, 2004), and community leadership in decision making (Rasche & Esser, 2006). The transformational engagement is the most proactive level of engagement strategies; hence, it should be considered a goal for innovation districts. These three levels of engagement strategies align well with the five-step micro-scale engagement model—i.e., inform, consult, involve, collaborate, and empower (Figure 8.1).

A balanced development of social relationships under effective engagement strategies will lead to gaining public trust and enhancing the community engagement. The proposed model (Figure 8.1) suggests the strongest involvement would be achieved under the direct involvement of the public in driving socioeconomic profits of innovation districts.

This chapter emphasises that community engagement with innovation districts can be improved by promoting social interactions (i.e., bridging, linking) and engagement strategies (i.e., transitional, transformational). At present, the integration of innovation districts with the public has been only limited to some basic physical and functional factors—e.g., blurring boundaries with surrounding neighbourhoods, sharing facilities, mixed-use development with diversified functions (Charnock & Ribera-Fumaz, 2011). While these strategies are fundamental, it is necessary to understand to what extent such physical and functional approaches are effective to attract communities to innovation districts. Furthermore, it is important to recognise social and relational factors that improve the effectiveness of physical and functional approaches.

With these matters in mind, this research is a starting point to consult with the public and collect their opinions and feelings about getting involved with innovation districts. As stated in the proposed model (Figure 8.1), the foundation of an advanced community engagement is the consultation. To achieve the highest level of community engagement and public trust, innovation districts initially need to commence the consultation process. Following this, necessary strategies can be identified and promoted by directly involving the public in the district life.

This chapter firstly investigates attributes of innovation districts that gained the public trust and attributes that threaten this trust, since public trust is the foundation of successful community engagement. The chapter examines whether the general public believes in the effectiveness of innovation districts in improving their quality of life. The idea here is that if people believe in the positive impact of innovation districts on their life, they would show a higher level of trust toward them. Conversely, negative aspects of innovation districts on liveability would damage public trust. For instance, this chapter investigates the effectiveness of innovation districts in providing job opportunities, and at the same in protecting the natural environment. The chapter collects public opinions to highlight these positive and negative aspects, and explores whether they are physical, functional, or relational. This study, then, proposes approaches to improve public trust, as improving the public trust would enhance community engagement.

Secondly, the study focuses on attributes of innovation districts that either discourage the public to be present in innovation districts or encourage them to be involved in the activities offered. This study attempts to discover physical, functional, and emotional factors that influence community engagement. The idea here is to consult the public, collect their opinions, and reflect their expectations in planning, designing, and developing innovation districts. To do so, this research gathers public views around some common attributes, such as the type of activities that they expect, and their feelings about visiting or living in or near innovation districts. Besides, the study attempts to collect hidden factors that influence the public decision for involving the district life. These attributes can be useful for policymakers, urban planners, and community developers in enhancing community engagement with innovation districts.

8.3 METHODOLOGY

8.3.1 Case Study

This research utilises a case study approach (Yin, 2003) to investigate which factors or attributes of innovation districts fulfil public expectations, and which factors or attributes harm the public trust. Australia is selected as the case study context with three case cities—i.e., Sydney, Melbourne, Brisbane—that are the most populated and globalised cities of the country. The selection is justified as: (a) Australia is a leading country in the knowledge and innovation activities in the world (Yigitcanlar *et al.*, 2019). The economy has grown at a healthy pace for the past three decades on the back of its global network readiness, ease of starting a business, and enhancing the number of firms and employments in the knowledge industry (Schwab *et al.*, 2006; Dutta & Mia, 2007). (b) State and local governments planned a number of KBUD projects in the case of cities to boost knowledge economy performance (Yigitcanlar *et al.*, 2017). (c) Decentralisation strategies were implemented in the case cities—due to a high concentration of innovation activities in the downtowns—resulting in developing innovation districts in midtown and close-in suburban districts. (d) The case cities host a number of renowned and prosperous innovation districts (Figure 8.2) specialising in knowledge-intensive and creative activities (Pancholi *et al.*, 2019).

In all three cases, early attempts for clustering knowledge activities were mainly in the form of exurban technology parks—e.g., Macquarie Park, Sydney in the 1970s; Brisbane Technology Park in the 1980s. However, like other global cities, over the years, the concentration of highly productive activities in city centres has escalated issues against fulfilling economic potentials of these cities—for example, too many workers live too far away (Baum, 1997; Hu, 2012; Baum *et al.*, 2004; Kelly

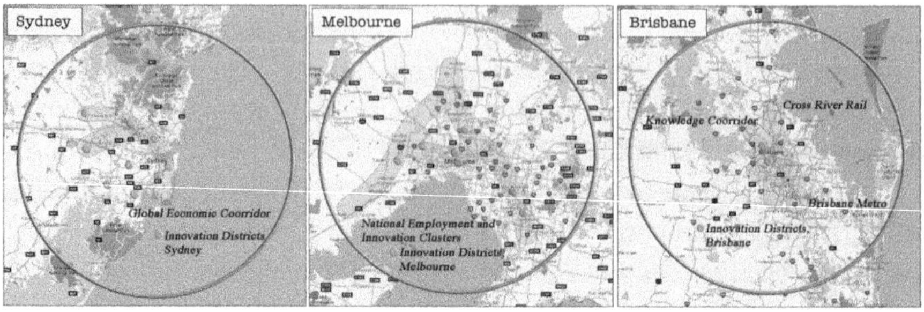

FIGURE 8.2 Innovation district locations (retrieved from https://www.google.com/maps/@-33.6389603,151.2393985,9z, https://www.google.com/maps/@-37.9498383,145.1736595,9.5z, https://www.google.com/maps/@-27.4500785,153.1933654,9z).

et al., 2014). Currently, all three case studies endorse strategies to plan innovation districts across their inner-city neighbourhoods and shape polycentricity—for instance, health precincts in Brisbane, art and music precincts in Melbourne, and ICT precincts in Sydney (Sharpe, 2007; Martinez-Fernandez & Potts, 2009; Kinch, 2019).

Despite formal and functional differences of innovation districts in the case study cities, all the districts have been properly connected through transportation systems, networking systems, and above all collaboration plans. The Global Economic Corridor of Sydney, National Employment and Innovation Clusters (NEIC) of Melbourne, and Knowledge Corridor of Brisbane are some examples of these plans. Most of these innovation districts have blurred boundaries and are considered an extension of surrounding neighbourhoods (Statement of Principles for Australian Innovation Precincts, 2018). They are, in general, mixed-use centres of knowledge economy located in inner-city and close-in suburbia. The purpose of innovation districts is to revitalise inner-city neighbourhoods' economy, reduce commuting to work through a heterogeneity of functions, and provide a more balanced land-use distribution (Hawken & Han, 2017).

Aforementioned similarities of innovation districts in Sydney, Melbourne, and Brisbane made them appropriate cases for this research. Innovation districts in these three cities have also some differences in the cases of major function, economic performance, size, spatial configuration, and location. However, this study has only focused on the generic concept of innovation districts as mixed-use knowledge centres with blurred boundaries located in inner cities and close-in suburbia, regardless of their differences in functioning.

8.3.2 Data Collection

An online survey was conducted to collect data from the general public across Sydney, Melbourne, and Brisbane. The Key Survey tool—an enterprise survey platform—was employed to distribute the questionnaire. The target group was selected based on residential location—i.e., within 30 km from the respective CBDs. The 30 km boundary was determined because it includes downtown, midtown, and close-in suburbs. Exurban innovation districts were not included, since they do not fit Australia as a highly urban nation with pre-existing growth centres (Blakely & Hu, 2019). In particular, sprawling development is a major concern in Australian cities (Flew, 2006). Recent development plans in Australia mostly encourage compacted cities, but with several business districts to balance jobs-housing (Kinch, 2019). In this regard, a 30 km radius from CBD represents urban districts with access to pre-existing infrastructure. The 30 km was also recognised as the radius containing the highest density of innovation districts in most Australian cities (Figure 8.2).

The number of survey participants was determined by Krejcie & Morgan's (1970) sample size method. Based on this method, the required sample size for populations above 100,000 is 384 based on 95% confidence level and 5% margin of error. People under 18 years of age were excluded from data collection; people who were 18 years old and above were targeted as participants. Services of a professional survey company were acquired to identify participants who live in targeted postcodes—30 km from CBDs. The survey link was sent to individuals' email addresses in May 2018. Participants were provided with a brief description of the project and definition of innovation districts to reduce the risk of biased interpretations and misunderstandings; 396, 385, and 384 participants completed the survey from Sydney, Melbourne, and Brisbane, respectively (in total 1,165). The participation rate was 82%. This high response rate was achieved due to the participants being registered members of the survey company. The responses were checked for completeness; cases with missing data were excluded from further analysis. This resulted in 1,040 valid responses, which reduced the response rate to 73%—non-response rate of 27%. In order to ensure participants' suitability, the representativeness of participants was checked against the most recent census data, 2016 Census (Table 8.1). The comparison confirmed that a heterogeneous sample of the public was assembled—ensuring a fair range of opinions from all groups of the society.

TABLE 8.1 Salient Characteristics of Survey Participants

Attribute	Category	Frequency	Survey %	Census %
City	Sydney	351	33.8	–
	Melbourne	346	33.2	–
	Brisbane	343	33.0	–
Age	18–29	159	15.3	15
	30–49	376	36.2	29.1
	49–65	295	28.4	17.1
	65+	210	20.2	15.7
Gender	Female	534	51.3	50.7
	Male	506	48.7	49.3
Born	Australia	744	71.5	71.8
	Other	296	28.5	28.2
Marital status	Married, living with a partner	652	62.7	58.1
	Single, widowed, divorced	388	37.3	41.9
Have children (under 18 years old)	Yes	325	31.3	37
	No	715	68.8	63
Highest education	Up to year 12	278	26.7	53.4
	Diploma/vocational	270	26.0	24.6
	Graduate	492	47.3	22.0
Income	AU$ 0–599 per week	469	45.1	50.7
	AU$ 600–999 per week	310	29.8	36.6
	AU$ 1,000–1,499 per week	170	16.3	10.9
	AU$ 1,500+ per week	91	8.8	1.8

8.3.3 Data

Attributes that improve community engagement with innovation districts were investigated through: (a) public trust factor, (b) restrictive factors, and (c) welcoming factors. Based on these classifications, the participants were asked to indicate the extent to which they agree with the following claims on innovation districts:

Public trust factor:
Contribution of innovation districts to the liveability of your neighbourhood and city.
Restrictive factors:
Attributes of innovation districts that discourage you from getting involved with them.

Attributes of innovation districts that discourage you from living near/around them.

Welcoming factors:

Attributes of innovation districts that encourage you getting involved with them.

Attributes of innovation districts that encourage you living near/around them.

Each of the above-mentioned five questions contained a set of attributes (items)—13, 10, 8, 16, and 14 items, respectively (in total 61). Participants were asked to rate their level of agreement with these 61 items on a 5-point Likert scale (from 1–strongly disagree to 5–strongly agree). The 5-point scale was chosen based on the literature (Lozano *et al.*, 2008). Moreover, all five survey questions were accompanied with an open-ended one, giving the opportunity of adding other factors, attributes, or comments. The open-ended question was shown as the 'other' option. These questions were designed to collect the required data for qualitative analysis.

A control question was included to identify people who were not familiar with innovation districts. Participants were provided with three options: 'I have visited an innovation district', 'I have not visited any innovation district, but I am familiar with them', 'I have never heard about innovation districts'; 229, 617, and 194 participants selected these items, respectively, and 194 participants who had no familiarity with innovation districts were removed from data analysis, since including opinions of people with no knowledge about the subject would reduce the validity. This resulted in 846 suitable responses for data analysis. Additionally, the control question confirmed the initial claim of the research on the low-level engagement of the public with innovation districts, as only 229 (22%) of the participants had visited an innovation district.

8.3.4 Empirical Analysis

The methodological approach includes mixed methods of qualitative and quantitative analyses. For the purpose of quantitative analysis, a factor analysis was conducted in addition to descriptive methods. While the descriptive method aims to indicate the priority and importance of investigated attributes, the factor analysis categorises these attributes under fundamental dimensions; it facilitates the interpretation of the data. SPSS was utilised for the analysis.

The descriptive analysis was based on calculating the level of agreement on investigated attributes. To calculate the level of agreement, 'agree' and 'strongly agree' categories were combined to obtain the percentage of agreement, and 'disagree' and 'strongly disagree' categories were combined to obtain the percentage of disagreement (Perveen *et al.*, 2019). The percentage of the midpoint—number 3—was directly reported as neutral. Then, the investigated attributes were ranked based on the percentage of agreement or disagreement. The attributes that reached the highest percentage—either agreement or disagreement—were marked as important.

Factor analyses were conducted for each set of attributes collected by the five main survey questions (see Section 8.3.3). Factor analysis is a widely utilised and broadly applied statistical technique in social sciences. The aim is to reveal any latent variables that cause the manifest variables to co-vary (Costello & Osborne, 2005). However, there are several factor analysis extraction methods to choose from. This research conducted factor analyses based on polychoric correlations matrix, in order to consider the ordinal nature of the variables. The number of factors were extracted by examining respective scree plot (Velicer & Jackson, 1990). Only the cleanest factor structures were retained: (a) item loadings above 0.3—suppress small coefficients under 0.3; (b) no or few item cross-loadings; (c) no factors with fewer than three items (Tabachnick & Fidell, 2001). After retrieving the factors, Cronbach's alpha was calculated among items of each factor; it confirms the internal reliability of the extracted factors. Values above 0.7 mean that the classification is strongly reliable.

Prior to factor analysis, measures of sampling adequacy (MSA) and Kaiser-Meyer-Olkin (KMO) were also calculated. While MSA reveals whether an individual variable is acceptable for a factor analysis, KMO explains whether all the included variables together are adequate to a factor analysis. It is advisable to remove any item with MSA—communality score—less than 0.2 (Child, 2006). The value less than 0.50 for KMO shows that the results of factor analysis are not accurate. Furthermore, the Bartlett test of sphericity was conducted to explore a sufficient degree of correlation between variables. Small values—less than 0.05—indicate that factor analysis is useful.

The qualitative component, on the other hand, aims to gain a deep understanding of public expectation towards innovation districts through open-ended questions. These questions retrieved further effective attributes that those neglected in the survey. As answering these questions was optional, only few participants completed them. Provided responses have been directly sorted and reported as additional factors that needed to be considered.

This study encountered a number of limitations and offered some solutions to tackle them: (a) The findings were obtained by unifying data from three different case studies. To ensure no impact of local context on the results, the data from three cases were carefully examined by chi-square test of association. A 'Crosstab' procedure was conducted in SPSS to calculate the chi-square. Note that the categorical nature of the data (5-point Likert scale and three independent cities) justifies the selection of the chi-square test. The aim was to determine the factors that were dependent on the context of each case study city. All the factors (61 in total) that were considered important to enhance community engagement were used as dependent variables and tested their variability among the case study cities. If the scores of these factors differ significantly (at the 0.05 level) between the cases, this means that the said factors are dependent on context and are not considered for further investigation. This selection enables the study to inform strategic-level policy (rather than local level) for community engagement with innovation district. (b) The survey questionnaire covered limited economic, educational, social, physical, functional, and relational attributes. While open-ended questions collected additional factors, other attributes may also need to be taken into consideration. However, collected factors are still helpful in improving the community engagement with innovation districts. (c) The findings could not be crosschecked with any other empirical research, since this is a pioneering study. This study was a direct consultation with the general public to collect their opinions and feelings about innovation districts, while previous studies had mostly focused on the preferences of knowledge workers and companies.

8.4 RESULTS

8.4.1 Public Trust

Factor analysis was conducted without the requirements of dropping any variables for question 1, because all MSAs (13 items) were found above 0.6. KMO is also found high—0.946. The Bartlett test of sphericity (chi-square = 7808.132 with df = 78; p = 0.00) illustrated a sufficient degree of correlation between variables. Three factors were selected using the Scree test and all factors consisted of items scoring above 0.3; only one cross-loadings item was observed; and all factors embraced at least three items. These three factors are labelled as supporting factors for 'Knowledge economy', 'Wellbeing/quality of life', and 'Advanced social desires' of the public (Table 8.2). The retrieved factors were utilised to interpret the results.

Descriptive analysis revealed that the public trust towards innovation districts is risked by two key items (Table 8.2). The public believes that innovation districts have negative impacts on 'Preserving the natural environment', and 'Providing reasonably priced properties' in their neighbourhoods—respectively ranked 1, 2 (disagree). The positive impact of innovation districts on providing 'Infrastructure and public facilities', 'Mental and physical health', and 'Safety and security' in cities are also insignificant—ranked 8, 10, 11 (agree). Based on the result of factor analysis, these five items are all classified under the 'Wellbeing'. This indicates that the public does not believe in the positive impact of innovation districts on their quality of life.

On the other hand, innovation districts have been extremely successful in attracting public trust as 'Centres of high-quality facilities for education and individual growth'—ranked 1 (agree). Moreover, the public believes in positive impacts of innovation districts in terms of: 'Helping the city to become smart'—ranked 2; and 'Improve national and international reputation and connections'—ranked 3. Regarding factor analysis, these three items are categorised as fundamental elements for supporting 'Knowledge economy'. Two more attributes, 'Facilitate economic development' (rank 7), and 'Provide high-quality building and spaces' (rank 6), are also classified under this category. This means that the public strongly believes that innovation districts facilitate knowledge-based economic growth. Conversely, the impact of innovation districts on fulfilling 'Advanced social desires of the public' was found moderate.

Besides the above-mentioned factors, participants repeatedly stated their concern about the ongoing inequality in innovation districts, which disappoint them and lower their trust. They specified that innovation districts are places for the privileged few and advanced resources for certain businesses. One of the participants has explained this inequality as: *"Innovation district seems to be cultivating a new elite group. It does not increase the value of the existing community. It just gentrifies the*

TABLE 8.2 Innovation District Contributions to Liveability

Attributes	Descriptive Analysis					Factor Analysis (Pattern Matrix)		
	DA%	Neutral%	A%	R-DA	R-A	Support Knowledge Economy	Support Wellbeing	Support Social Desires
Improves national/international reputation and connections	13.2	45.9	40.9		3	.859		
Facilitates economic development	14.2	48.1	37.7		6	.780		
Helps the city to become smart	14.1	44.8	41.1		2	.743		
Provides high-quality facilities for education and individual growth	10.2	43.0	46.0		1	.666		
Provides new types of high-quality buildings and spaces	13.2	50.4	36.4		7	.462		
Provides reasonably priced housing and offices	35.5	44.1	20.4	1			.797	
Preserves the natural environment and local assets	26.1	50.8	23.1	2			.791	
Improves safety and security of the city	24.9	47.0	28.0		11		.685	
Improves physical and mental health of society	22.0	48.1	29.9		10		.559	
Enhances infrastructure and public facilities	13.8	50.0	36.2		8	.328	.392	
Supports cultural dynamics	16.2	48.7	35.1		9			−.954
Supports diversity of lifestyle opportunities	17.7	43.9	38.4		5			−.644

(Continued)

TABLE 8.2 *(Continued)* Innovation District Contributions to Liveability

Attributes	**Descriptive Analysis**					**Factor Analysis (Pattern Matrix)**		
	DA%	**Neutral%**	**A%**	**R-DA**	**R-A**	**Support Knowledge Economy**	**Support Wellbeing**	**Support Social Desires**
Facilitates social interactions	17.1	44.3	38.5		4			–.491
Sum of squared loadings (rotated)						6.493	5.988	6.042
% of variance explained						57.482	5.051	2.866
Reliability (Standard Cronbach's alpha)						.884	.884	.887
Kaiser-Meyer-Olkin (KMO)								.946
N								846
Extraction method (polychoric correlation matrix)							Principal Axis Factoring	
Rotation method							Oblimin Quartimin	

* DA = disagree, A = agree, R = rank

neighbourhood'. Another participant described it as: *"In theory innovation district sounds good for the locals, but in practice no opportunities are provided for local community members to participate in work and education activities there'*. Moreover, participants hesitated about the existence of strategies and plans for driving innovation districts in their cities. In line with this, they frequently compare their cities to Singapore, Berlin, London, and other global cities. One of the respondents commented sorrowfully: *"Simply calling someplace an innovation district is not enough. Need real planning as in Singapore'*; or another participant defined innovation districts as: *"Expensive White Elephants, which will not generate useful jobs for the locals ... they are strictly home for foreign scientists and managers'*.

8.4.2 Restrictive Factors

Factor analysis of the second question did not require dropping any variables—i.e., all 10 items were included. MSAs were found above 0.4. However, the communality score for one of the items in question 3, was found under 0.2. This item was removed and MSAs was recalculated for the remaining items—7 items. This time the MSAs were found above 0.4. KMO is also found high for both questions—0.894 and 0.823. The Bartlett test of sphericity (chi-square = 3838.857 with df = 45; p = 0.00; and chi-square = 1020.515 with df = 21; p = 0.00) demonstrated an appropriate correlation between variables of both questions. For both questions, two factors were selected as fundamental restrictive elements, using the Scree test. The first factor represents 'Functions and facilities', weakening community engagement with innovation districts. The second factor signifies 'Senses and feelings' that repulse the public (Tables 8.3 and 8.4).

Descriptive results (Table 8.3) indicated that 'Senses and feelings' that limit public involvement—question 2—mostly related to lack of the following attributes: (a) Participants feel boundaries between innovation districts and their neighbourhoods (rank 2). These boundaries do not relate to physical attributes; since participants believe to the appropriate physical connection of innovation districts to surrounding neighbourhood. (b) Respondents stated that they do not feel banned to cross the physical boundaries. Therefore, it seems lack of social connections is shaping mental boundaries; also, participants claimed that they do not have anything in common with the present in-house community of

TABLE 8.3 Innovation District Attributes Discouraging Community Engagement

Attributes	Descriptive Analysis					Factor Analysis (Pattern Matrix)	
	DA%	Neutral%	A%	R-DA	R-A	Functions and Facilities	Senses and Feelings
Attracts a large population to the neighbourhood, which makes it overcrowded	26.0	54.8	19.2	5		.881	
Hosts too many young people, which make it unattractive for other age groups	28.9	50.9	20.2	4		.785	
Hosts too many migrants from different cultures, which I do not like	38.1	48.7	13.1	1		.761	
Does not provide additional facilities to the neighbourhood	20.5	57.4	22.0		5	.522	
Increases the cost of living, rent/property prices	11.9	55.8	32.2		1	.407	
Does not have good connectivity with the neighbourhood	32.6	46.8	20.6	3			.794
Obvious boundary, which separates surrounding neighbourhood	20.1	49.3	30.6		2		.679
Feel banned from being there	34.0	46.9	19.1	2			.533
Nothing special to attract/encourage me to spend time there	23.7	49.6	26.7		4		.525
Its people have nothing in common with the neighbourhood's community	25.4	47.1	27.5		3		.515
Sum of squared loadings (rotated)						4.092	3.797
% of variance explained						45.673	5.770
Reliability (Standard Cronbach's alpha)						.840	.821
Kaiser-Meyer-Olkin (KMO)							.894
N						846	
Extraction method (based on polychoric correlation matrix)						Principal Axis Factoring	
Rotation method						Oblimin Quartimin	

* DA = disagree, A = agree, R = rank

TABLE 8.4 Innovation District Attributes Discouraging Public to Live around Them

Attributes	**Descriptive Analysis**					**Factor Analysis (Pattern Matrix)**	
	DA%	**Neutral%**	**A%**	**R-DA**	**R-A**	**Function and Facilities**	**Senses and Feelings**
Not good places for raising children	28.6	42.5	28.9			−.352	.429
Size, shape and density of buildings are not suitable for residing	15.6	47.5	36.9		2	−.959	
Inner-city ones are overcrowded, others are overloaded by cars/parking	14.3	41.9	43.8		1	−.607	
Hard to be accepted and become part of the community	26.5	50.4	23.1	3			.851
Street life is too dull or hectic	28.9	48.5	22.6	2			.690
Hard to separate work and life	26.3	43.2	30.5		3		.634
Hard to use new technologies available in them	32.6	44.8	22.6	1			.630
Sum of squared loadings (rotated)						2.443	2.933
% of variance explained						8.981	46.234
Reliability (Standard Cronbach's alpha)						.769	.799
Kaiser-Meyer-Olkin (KMO)							.823
N						846	
Extraction method (based on polychoric correlation matrix)						Principal Axis Factoring	
Rotation method						Oblimin Quartimin	

* DA = disagree, A = agree, R = rank

innovation districts (rank 3). (c) Innovation districts are not as unique and attractive as they should be (rank 4)—i.e., lack of offering cutting-edge scenes.

On the other hand, 'Functions and facilities' that weaken community engagement are caused by the following: (a) Innovation districts are too expensive for the public and increase the prices in surrounding neighbourhoods (rank 1). (b) Amenities provided at innovation districts are restricted, and are not in balance with resources that they consume from neighbourhoods (rank 5). Oppositely, high population density, cultural diversity, and high presence of youth are not restrictive factors from the viewpoint of the public.

Three attributes were signified important for discouraging the public to live in or around innovation districts—question 3 (Table 8.4). Two of them are related to 'Functions and facilities', and another represented by 'Senses and feelings'. In regards to 'Functions and facilities', firstly, participants believe that inner-city innovation districts are overcrowded and suburban ones are overloaded by cars and parking. Secondly, respondents scored that size, shape, and density of innovation district buildings are not suitable for residential areas. Regarding 'Senses and feelings', the results revealed that it is challenging to separate life and work in mixed-use innovation districts.

Participants also suggested that housing in innovation districts are limited to flats in high-rise buildings; while they expect diversity of living options. Furthermore, they were unsatisfied with the rent and property prices. For instance, one of the participants commented: *"Don't try to sell innovation districts to the normal folk, who actually work for a living and struggle to survive on minimalist wages, while the elites live in luxury'.*

8.4.3 Welcoming Factors

Factor analyses for questions 4 and 5 were conducted without dropping any variables—respectively 16 and 14 items; since all MSAs were found above 0.5. KMOs were also found high for both questions—0.938 and 0.940. The Bartlett test of sphericity (chi-square = 7081.833 with df =120; p = 0.00; and chi-square = 4064.031 with df = 91; p = 0.00) also showed a satisfactory degree of correlation between variables of both questions. Two factors were extracted through the Scree test for both questions. The first factor, similar to restrictive factors, relates to 'Functions and facilities' provided at innovation districts; it mainly contains essential facilities and functions that public expected from an innovation district. However, the second factor mostly represents 'Diverse and authentic scenes' that the public experiences in innovation districts; it contains advanced and high-quality recreational amenities and expectations.

Descriptive analysis demonstrated that all 16 investigated items in question 4 are effective in attracting the public (Table 8.5). However, based on the level of agreement, all 6 items categorised under 'Diverse and authentic scenes' were found among priorities. It includes variety of cafes, restaurants, and bars (rank 1); cultural facilities (rank 2); cultural events (rank 3); shopping centres and retail stores (rank 4); free Wi-Fi (rank 5); and farmer markets (rank 6). On the other hand, 'Facilities and functions' that gained the highest scores were conferences, cutting-edge technology exhibitions, and other scientific events (rank 7); healthcare services (rank 8); and school, college, and university (rank 9). While the first six items selected by the public mainly characterise advanced amenities, the rest focus on the essential needs of the society. Consequently, the expectation of the public from innovation districts undoubtedly is beyond basic and essential facilities.

Besides, participants declared that 'Free educational/training opportunities' and 'Free parking' are among encouraging factors for the public visit innovation districts. However, they believed that existing innovation districts do not have such facilities.

Descriptive results confirmed that all 14 investigated items in question 5 have positive impacts on encouraging the public for living in or around innovation districts (Table 8.5). In addition to advanced amenities that attract the public into the districts' boundaries, essential 'Functions and facilities' are still required to persuade them for living in or around these districts—safe and secure to walk or cycle around (rank 1), positive and recognisable reputation (rank 3) accessibility to other parts of the city through a wide range of roads and public transport (rank 4), accessibility to both essential and advanced amenities (rank 5), and accessibility to a variety of professional networks and educational facilities (rank 7). However, 'Diverse and authentic scenes' are also found effective—accessibility to a variety of restaurants, cafés, bars, and nightlife (rank 2), and openness to new ideas and people (rank 6). In contrast, living in a racially, linguistically, religiously, and sexually diverse community (rank 13) and living and working in the same place (rank 12) have less priority for the public. (Table 8.6)

TABLE 8.5 Innovation District Facilities/Activities Encouraging Community Engagement

Attributes	**Descriptive Analysis**					**Factor Analysis (Pattern Matrix)**	
	DA%	**Neutral%**	**A%**	**R-DA**	**R-A**	**Functions and Facilities**	**Diverse and Authentic Scenes**
Childcare facilities and playgrounds	23.6	39.7	36.7		15	.860	
Schools, colleges, and universities	14.3	38.9	46.8		9	.820	
Science competitions	18.9	43.2	37.9		14	.768	
Social and cohesive communities	13.1	42.7	44.2		10	.582	
Outdoor sports	17.1	39.2	43.7		12	.568	
Smart technology use	15.4	43.9	40.7		13	.546	
Dynamic nightlife	20.9	43.5	35.6		16	.518	
Street art and street performances	17.4	38.8	43.8		11	.491	
Conferences, exhibitions, and scientific events	12.0	36.4	51.6		7	.470	
Healthcare services	10.8	39.0	49.3		8	.409	
Cafes, restaurants, bars	7.1	28.6	64.3		1		.760
Cultural facilities	6.8	29.9	63.3		2		.744
Cultural events	8.1	31.2	60.7		3		.721
Shopping centres and retail stores	12.6	31.8	55.6		4		.644
Farmer markets	12.6	34.3	53.1		5		.532
Free Wi-Fi	12.7	35.5	51.8		6		.413
Sum of squared loadings (rotated)						6.496	5.974
% of variance explained						45.164	4.932
Reliability (Standard Cronbach's alpha)						.907	.841
Kaiser-Meyer-Olkin (KMO)							.938
N						846	
Extraction method (based on polychoric correlation matrix)						Principal Axis Factoring	
Rotation method						Oblimin Quartimin	

* DA = disagree, A = agree, R = rank

TABLE 8.6 Innovation District Attributes Encouraging Public to Live around Them

Attributes	Descriptive Analysis					Factor Analysis (Pattern Matrix)	
	DA%	Neutral%	A%	R-DA	R-A	Functions and Facilities	Diverse and Authentic Scenes
Accessibility to high-quality buildings and open spaces	6.9	39.0	54.1		9	.828	
Accessibility to essential and advanced amenities	4.7	37.3	58.0		5	.816	
Safe and secure to walk or cycle	6.2	26.4	67.4		1	.813	
Accessibility to other parts of city through private/ public transport	7.9	33.9	58.2		4	.768	
A place with a positive and recognisable reputation	6.4	33.5	60.1		3	.727	
Close proximity to employment provided by innovation district	8.5	36.7	54.8		8	.598	
No boundaries between work, accommodation, recreation	11.9	38.6	49.4		12	.523	
Accessibility to professional networks and educational facilities	9.3	35.5	55.2		7	.507	
Accessibility to smart technologies and facilities	10.9	36.9	52.2		10	.489	
Accessibility to restaurants, cafes, bars, and nightlife	7.3	31.2	61.5		2	.417	.326
A culturally vibrant community	16.0	39.7	44.3		13		.881
Accessibility to diverse socio-cultural activities	10.0	35.2	54.8		8		.668
An artistically or technologically creative atmosphere	8.8	39.4	51.8		11		.642
Openness to new ideas and people	6.8	36.5	56.7		6		.591

(Continued)

TABLE 8.6 *(Continued)* Innovation District Attributes Encouraging Public to Live around Them

Attributes	**Descriptive Analysis**					**Factor Analysis (Pattern Matrix)**	
	DA%	**Neutral%**	**A%**	**R-DA**	**R-A**	**Functions and Facilities**	**Diverse and Authentic Scenes**
Sum of squared loadings (rotated)						6.997	6.061
% of variance explained						52.787	3.980
Reliability (Standard Cronbach's alpha)						.913	.877
Kaiser-Meyer-Olkin (KMO)							.940
N						846	
Extraction method (based on polychoric correlation matrix)						Principal Axis Factoring	
Rotation method						Oblimin Quartimin	

* DA = disagree, A = agree, R = rank

8.5 DISCUSSION AND CONCLUSION

The chapter investigated the key attributes of innovation districts that can help attract the public to this new land-use type, gain their trust, and enhance community engagement. The results revealed that public trust is still limited to educational and professional aspects of innovation districts. The public believes that innovation districts threaten the natural environment, increase property prices, and escalate inequality in neighbourhoods. The public does not see innovation districts as an effective function for increasing their wellbeing—i.e., health, safety, cultural dynamism, public facilities. It seems that in reality innovation districts solely consider the needs of their internal communities rather than cater to the broader and external community. In order to achieve desired community engagement outcomes, innovation districts need to ensure that the expectations of the broader community are delivered. Offered activities do not need to be entirely scientific or knowledge-related; conversely, these activities need to be social in nature for the public to enjoy. These social activities can be gradually combined with scientific and artistic events.

As mentioned in the proposed engagement model (Figure 8.1), the first step in gaining the public trust and encouraging their engagement is consulting with them. As participants stated, lack of effective engagement strategies in innovation districts dramatically weakens their abilities in consulting with local communities and delivering public expectations. These strategies need to be clearly identified in prospective studies.

The study at hand is a pioneer one that surveys public views on factors that encourage or discourage their engagement decisions with innovation districts. The results disclosed that the most significant restrictive factor for involvement is related to mental barriers—'Senses and feelings'. In order to eliminate this barrier, innovation districts require to do more than just removing their physical boundaries. These districts need to change the public perception and address the bias that separates local communities from their knowledge workers. Basically, the public and knowledge workers are perceived as two heterogeneous groups with rare natural interactions.

Social interaction is unlikely to happen, unless people feel that they have something in common with each other. At the moment, it seems that the public considers knowledge workers as privileged members of an elite community. On the other hand, knowledge workers implicitly consider the presence of the public unnecessary in innovation districts. However, this study revealed that effective collaboration between innovation districts and the public is beneficial for both. Innovation districts should play a catalyser role to foster a sense of community (bridge, link). Continuous engagement strategies are required to bridge the gap between knowledge workers and the public. It is necessary to build trust as the foundation of future interactions.

The findings disclosed that the welcoming factors considered in the literature have no substantial relation to knowledge workers. The public similarly engages with high-quality spaces, authentic scenes, and dynamic street life (Alfken *et al.*, 2015). However, the public also expects affordability. In line with these expectations, reasonably priced housing and diversity of living options could be effective to foster a social mix. The findings indicated that innovation districts need to maintain a creative flavour to differentiate themselves from other parts of the city. This characteristic attracts both knowledge workers and the public that expect authenticity and uniqueness.

The public is satisfied with neither hectic inner-city lifestyles of innovation districts, nor gloomy suburban environments of them. They expect a balanced division of population and facilities. An effective mechanism for achieving this expectation is to promote a polycentric urban model based on different work-living nodes and hubs in cities (Frenkel *et al.*, 2013; Kinch, 2019). However, while locals expect innovation districts to bring jobs and facilities into their neighbourhoods—rather than concentration of jobs in downtowns—the results revealed that people do not tend to work and live in the same locations. These results are in contrast to the initial goal of innovation districts for improving jobs-housing balance across inner-city neighbourhoods. This research suggests that innovation districts need to adapt a new urban-setting model and creative hierarchy of spaces in neighbourhoods to overcome this conflict.

The current problem is that innovation districts are still modelled on the traditional structure of science/technology/innovation parks as a centralised cluster of workplaces. This old model has stayed unchanged and literally moved into inner-city neighbourhoods. These districts have accommodated a few residential blocks, retails, and open public spaces inside the old model. In this situation, the best scenario is that people are living inside a luxury commercial or educational village/park/hub; it increases the unpleasant sense of living and working in the same place. This structure still can be distinguished from the rest of the neighbourhood and weakens the integration with the broader community. In such an urban setting, social and class segregation is still the norm (Peck, 2005, 2010; Florida, 2017).

This study suggests that innovation districts need to move away from this traditional structure, and be seen as distributed clusters of innovative activities across the neighbourhood. A successful example of such an urban setting is the 'One-North' innovation district in Singapore, where seven clusters of innovative activities are located around the main spine of the district. The spine acts as a networking path between the clusters, including strong pedestrian linkages and servicing the entire neighbourhood (Carvalho & Van Winden, 2017).

The flexible zoning will assist innovation districts to truly blur themselves inside the neighbourhoods. This setting is far beyond merely removing the walls and gates of traditional districts, and opening recreational activities for the public. People feel attached and engaged to flexible locations where they can shape place identity (Sepe & Pitt, 2014). Consequently, innovation districts need to be considered a hierarchy of places, from the most planned and designed to the more informal and casual. In a flexible urban setting, the general public will spontaneously find their role and shape their desirable place. Additionally, it hopefully will reduce the sense of being trapped inside a living-working hub.

In sum, the findings explain currently experienced low-level community engagement with innovation districts caused by the mentioned physical, functional, and relational limitations in innovation districts. Innovation districts need to be physically integrated with the neighbourhoods, consider the public expectations, recognise and cater to their expectations and skills of the public, link wider communities with knowledge workers, and develop a sense of attachment in locals. Innovation districts have never been pictured as the facilitators of socioeconomic activities for the whole society; or as accelerators to remove socioeconomic inequality; or as incubators for providing a hierarchy of job opportunities from the most advanced to the simplest. This study suggests that if cities truly wish to improve community engagement with innovation districts, they need to revisit their strategies. At this point, further research is needed to map out how innovation districts can be transformed into inclusive and beneficial places for the whole society (Esmaeilpoorarabi *et al.*, 2020).

ACKNOWLEDGEMENT

This chapter, with permission from the copyright holder, is a reproduced version of the following journal article: Esmaeilpoorarabi, N., Yigitcanlar, T., Kamruzzaman, M., & Guaralda, M., (2020). How can an enhanced community engagement with innovation districts be established? Evidence from Sydney, Melbourne and Brisbane. Cities, 96(1), 102430.

REFERENCES

Alfken, C., Broekel, T., & Sternberg, R. (2015). Factors explaining the spatial agglomeration of the creative class. *European Planning Studies*, 23(12), 2438–2463. DOI:10.1080/09654313.2014.979767

Asheim, B., & Hansen, H.K. (2009). Knowledge bases, talents, and contexts. *Economic Geography*, 85(4), 425–442. DOI:10.1111/j.1944-8287.2009.01051.x

Bass, B. (1990). From transactional to transformational leadership. *Organizational Dynamics*, 18(3), 19–32. DOI:10.1016/0090-2616(90)90061-S

Baum, S. (1997). Sydney, Australia: a global city? *Urban Studies*, 34(11), 1881–1901.

Baum, S., Van Gellecum, Y., & Yigitcanlar, T. (2004). Wired communities in the city: Sydney, Australia. *Australian Geographical Studies*, 42(2), 175–192. DOI:10.1111/j.1467-8470.2004. 00274.x

Baum, S., O'Connor, K., & Yigitcanlar, T. (2009). The implications of creative industries for regional outcomes. *International Journal of Foresight and Innovation Policy*, 5, 44–64. DOI:10.1504/IJFIP.2009.022098

Benneworth, P., & Ratinho, T. (2014). Reframing the role of knowledge parks and science cities in knowledge-based urban development. *Environment and Planning C*, 32(5), 784–808. DOI:10.1068/c1266r

Bergvall, K.B., & Stahlbrost, A. (2009). Living lab: an open and citizen-centric approach for innovation. *International Journal of Innovation and Regional Development*, 1(4), 356–370. DOI:10.1504/IJIRD.2009.022727

Blakely. E., & Hu. R. (2019). *Crafting innovative places for Australia's knowledge economy*. Singapore: Springer. DOI:10.1007/978-981-13-3618-8

Bontje, M., & Musterd, S. (2009). Creative industries, creative class and competitiveness. *Geoforum*, 40(5), 843–852. DOI:10.1016/j.geoforum.2009.07.001

Bowen, F., Newenham, A., & Herremans, I. (2010). When suits meet roots. *Journal of Business Ethics*, 95(2), 297–318. DOI:10.1007/s10551-009-0360-1

Bryson, J. (2004). What to do when stakeholders matter. *Public Management Review*, 6(1), 21–53. DOI:10.1080/14719030410001675722

Carrillo, J., Yigitcanlar, T., Garcia, B., & Lonnqvist, A. (2014). *Knowledge and the city*. Routledge, New York. DOI:10.4324/9781315856650

Carvalho, L., & Van Winden, W. (2017). Planned knowledge locations in cities. *International Journal of Knowledge-Based Development*, 8(1), 47–67. DOI:10.1504/IJKBD.2017.082429

Charnock, G., & Ribera-Fumaz, R. (2011). A new space for knowledge and people? *Environment and Planning D*, 29(4), 613–633. DOI:10.1068/d17009

Child, D. (2006). *The essentials of factor analysis*. New York: Continuum.

Costello A.B., & Osborne J.W. (2005). Best practices in exploratory factor analysis. *Practical Assessment, Research & Evaluation*, 10(7), 1–9.

Dare, M., Schirmer, J., & Vanclay, F. (2014). Community engagement and social licence to operate. *Impact Assessment and Project Appraisal*, 32(3), 188–197. DOI:10.1080/14615517.2014.927108

Dutta, S., & Mia, I. (2007). *The global information technology report 2006–2007*. Davos: World Economic Forum.

Edvardsson, I., Yigitcanlar, T., & Pancholi, S. (2016). Knowledge cities research and practice under the microscope. *Knowledge Management Research and Practice*, 14(4), 537–564. DOI:10.1057/s41275-016-0003-0

Engel, J.S., Berbegal-Mirabent, J., & Pique, J.M. (2018). The renaissance of the city as a cluster of innovation. *Cogent Business & Management*, 5(1), 1–20. DOI:10.1080/23311975.2018.1532777

Esmaeilpoorarabi, N., Yigitcanlar, T., & Guaralda, M. (2016). Towards an urban quality framework. *International Journal of Knowledge-Based Development*, 7(3), 290–312. DOI:10.1504/IJKBD.2016.078556

Esmaeilpoorarabi, N., Yigitcanlar, T., Kamruzzaman, M., & Guaralda, M., (2020). How can an enhanced community engagement with innovation districts be established? Evidence from Sydney, Melbourne and Brisbane. *Cities*, 96(1), 102430. DOI:10.1016/j.cities.2019.102430

Farrell, C. (2007). Thinking critically about social capital. *Irish Journal of Sociology*, 16(2), 27–49. DOI:10.1177/079160350701600203

Flew, T. (2006). The new middle class meets the creative class. *International Journal of Cultural Studies*, 9(3), 419–429. DOI:10.1177/1367877906066887

Florida, R. (2005). *Cities and the creative class*. New York: Routledge. DOI:10.4324/9780203997673

Florida, R. (2017). *The new urban crisis*. New York: Basic Book.

Frenkel, A., Bendit, E., & Kaplan, S. (2013). Residential location choice of knowledge-workers. *Cities*, 35(1), 33–41. DOI:10.1016/j.cities.2013.06.005

Glaeser, E. (2005). Review of Richard Florida's the rise of the creative class. *Regional Science and Urban Economics*, 35(5), 593–596. DOI:10.1016/j.regsciurbeco.2005.01.005

Granovetter, M. (2011). *The sociology of economic life*. New York: Routledge.

Hart, S.L., & Sharma, S. (2004). Engaging fringe stakeholders for competitive imagination. *Academy of Management Executive*, 18(1), 7–18. DOI:10.1109/EMR.2004.25105

Hawken, S., & Han, J. (2017). Innovation districts and urban heterogeneity: 3D mapping of industry mix in downtown Sydney. *Journal of Urban Design*, 22(5), 568–590. DOI:10.1080/13574809.2017.1301203

Hawkins, R.L., Maurer, K. (2009). Bonding, Bridging and Linking. *British Journal of Social Work*, 40(6), 1777–1793. DOI:10.1093/bjsw/bcp087

Head, B. (2007). Community engagement. *Australian Journal of Political Science*, 42(3), 441–454. DOI:10.1080/10361140701513570

Hero, R. (2003). Social capital and racial inequality in America. *Perspectives on Politics*, 1(1), 113–135. DOI:10.1017/S1537592703000094

Hu, R. (2012). Clustering: concentration of the knowledge-based economy in Sydney. In *Building prosperous knowledge cities*, edited by Yigitcanlar, T., Metaxiotis, K., & Carrillo, J. pp. 195–212. UK: Edward Elgar.

Jarvis, D., Berkeley, N., & Broughton, K. (2012). Evidencing the impact of community engagement in neighbourhood regeneration. *Community Development Journal*, 47(2), 232–247. DOI:10.1093/cdj/bsq063

Katz, B., & Wagner, J. (2014). *The rise of innovation districts*. Washington, DC: Brooklyn Institution.

Katz, B., & Bradley, J. (2013). *The metropolitan revolution*. Washington, DC: Brooklyn Institution.

Kelly, J.F., Donegan, P., Oberklaid, M., & Chisholm, C. (2014). *Mapping Australia's economy*. Melbourne: Grattan Institute.

Kinch, J. (2019). Microeconomic reform to create activation and improve liveability. In: *Contemporary urban design thinking* (pp. 119–132). Springer, Cham.

Krejcie, R.V., & Morgan, D.W. (1970). Determining sample size for research activities. *Educational and Psychological Measurement*, 30(3), 607–610. DOI:10.1177/001316447003000308

Lawton, P., Murphy, E., & Redmond, D. (2013). Residential preferences of the creative class. *Cities*, 31(1), 47–56. DOI:10.1016/j.cities.2012.04.002

Liedtke, C., Jolanta, W.M., Rohn, H., & Nordmann, J. (2012). Living lab: user-driven innovation for sustainability. *International Journal of Sustainability in Higher Education*, 13(2), 106–118. DOI:10.1108/14676371211211809

Lonnqvist, A., Kapyla, J., Salonius, H., & Yigitcanlar, T. (2014). Knowledge that matters. *European Planning Studies*, 22(10), 2011–2029. DOI:10.1068/a1302

Longhi, C. (1999). Networks, collective learning and technology development in innovative high technology regions. *Regional Studies*, 33(4), 333–342. DOI:10.1080/713693559

Lozano, L., García-Cueto, E., & Muñiz, J. (2008). Effect of the number of response categories on the reliability and validity of rating scales. *Methodology*, 4(2), 73–79. DOI:10.1027/1614-2241.4.2.73

Lynn, M. (2006). Discourses of community: Challenges for social work. *International Journal of Social Welfare*, 15(2), 110–130. DOI: 10.1111/j.1468-2397.2006.00431.x

Martinez-Fernandez, C., & Potts, T. (2009). Quality of life through innovation indicators: the case of peripheral suburbs of Sydney. In: *Community quality-of-life indicators: best cases IV* (pp. 191–208). Springer, Dordrecht.

Metaxiotis, K., Carrillo, J., & Yigitcanlar, T. (2010). *Knowledge-based development for cities and societies*. Hersey, PA: IGI Global. DOI:10.4018/978-1-61520-721-3

Millar, C.C., Ju-Choi, C. (2010). Development and knowledge resources: a conceptual analysis. *Journal of Knowledge Management*, 14(5), 759–776. DOI:10.1108/13673271011074881

Newman, J., Barnes, H., Sullivan, H., & Knops, A. (2004). Public participation and collaborative governance. *Journal of Social Policy*, 33(2), 203–223. DOI:10.1017/S0047279403007499

Pancholi, S., Yigitcanlar, T., & Guaralda, M. (2015). Placemaking facilitators of knowledge and innovation spaces. *International Journal of Knowledge-Based Development*, 6(3), 215–240. DOI:10.1504/IJKBD.2015.072823

Pancholi, S., Yigitcanlar, T., & Guaralda, M. (2017). Governance that matters. *Journal of Place Management and Development*, 10(1), 73–87. DOI:10.1108/JPMD-08-2016-0053

Pancholi, S., Yigitcanlar, T., & Guaralda, M. (2019). Placemaking for innovation and knowledge-intensive activities. *Technological Forecasting & Social Change*, 146(Sep), 616–625. DOI:10.1016/j.techfore.2017.09.014.

Peck, J. (2005). Struggling with the creative class. *International Journal of Urban and Regional Research*, 29(4), 740–770. DOI:10.1111/j.1468-2427.2005.00620.x

Peck, J. (2010). *Constructions of neoliberal reason*. Oxford: Oxford University Press.

Perveen, S., Kamruzzaman, M., Yigitcanlar, T. (2019). What to assess to model the transport impacts of urban growth? *International Journal of Sustainable Transportation*, 13(8), 597–613. DOI:10.1080/15568318.2018.1491077

Pooley, J.A., Cohen, L., & Pike, L. (2004). Can sense of community inform social capital? *Social Science Journal*, 42(1), 71–79. DOI:10.1016/j.soscij.2004.11.006

Porter, M.E. (1990). The competitive advantage of nations. *Harvard Business Review*, 68(2), 73–93.

Putnam, R. (2000). *The collapse and revival of American community*. New York: Simon and Schuster. DOI:10.1145/358916.361990

Rasche, A., & Esser, D. (2006). From stakeholder management to stakeholder accountability. *Journal of Business Ethics*, 65, 251–267. DOI:10.1007/s10551-005-5355-y

Reve, T., Sasson, A., Gugler, P., & Ketels, C. (2015). Theoretical and methodological advances in cluster research. *Competitiveness Review*, 25(5), 524–539.

Rutland, T. (2015). Enjoyable life: planning, amenity and the contested terrain of urban biopolitics. *Environment and Planning D*, 33(5), 850–868. DOI:10.1177/0263775815599312

Sarimin, M., & Yigitcanlar, T. (2012). Towards a comprehensive and integrated knowledge-based urban development model. *International Journal of Knowledge-Based Development*, 3(2), 175–192. DOI:10.1504/IJKBD.2012.047035

Schwab, K., Porter, M., & Lopez-Claros, A. (2006). *The global competitiveness report 2006-2007*. London: Palgrave Macmillan.

Sepe, M., & Pitt, M. (2014). The characters of place in urban design. *Urban Design International*, 19(3), 215–227. DOI:10.1057/udi.2013.32

Sharpe, S. (2007). *Regional dimensions of innovative activity in outer Western Sydney*. Sydney: University of Western Sydney.

Stehlin, J. (2016). The post-industrial shop floor. *Antipode*, 48(2), 474–493. DOI:10.1111/anti.12199

Szreter, S., & Woolcock, M. (2004). Health by association? *International Journal of Epidemiology*, 33(4), 650–667. DOI:10.1093/ije/dyh013

Tabachnick, B.G., & Fidell, L.S. (2001). *Using multivariate statistics.* Boston: Allyn and Bacon.

Van den Berg, L., Braun, E., & Otgaar, A.H. (2004). Corporate community involvement in European and US cities. *Environment & Planning C*, 22(4), 475–494. DOI:10.1068/c0246

Van Winden, W., De Carvalho, L., Van Tuijl, E., Van Haaren, J., & Van den Berg, L. (2013). *Creating knowledge locations in cities.* London: Routledge. DOI:10.4324/9780203127162

Velicer, W.F., & Jackson, D.N. (1990). Component analysis versus common factor-analysis. *Multivariate Behavioural Research*, 25(1), 97–114. DOI:10.1207/s15327906mbr2501_12

Wal, A.L. (2013). Cluster emergence and network evolution. *Regional Studies*, 47(5), 651–668. DOI:10.1080/00343401003614258

Woolcock, M. (2001). The place of social capital in understanding social and economic outcomes. *Canadian Journal of Policy Research*, 2, 11–17.

Yigitcanlar, T., Metaxiotis, K., & Carrillo, J. (2012). *Building prosperous knowledge cities.* London: Edward Elgar. DOI:10.4337/9780857936042

Yigitcanlar, T., & Dur, F. (2013). Making space and place for knowledge communities. *Australasian Journal of Regional Studies*, 19, 36–63.

Yigitcanlar, T. (2015). Smart cities: an effective urban development and management model? *Australian Planner*, 52(1), 27–34. DOI:10.1080/07293682.2015.1019752

Yigitcanlar, T., & Bulu, M. (2015). Dubaization of Istanbul. *Environment and Planning A*, 47(1), 89–107. DOI:10.1068/a130209p

Yigitcanlar, T., Edvardsson, I.R., Johannesson, H., Kamruzzaman, M., & Ioppolo, G., Pancholi, S. (2017). Knowledge-based development dynamics in less-favoured regions. *European Planning Studies*, 25(12), 2272–2292. DOI:10.1080/09654313.2017.1358699

Yigitcanlar, T., Sabatini-Marques, J., Costa, E., Kamruzzaman, M., & Ioppolo, G. (2019). Stimulating technological innovation through incentives. *Technological Forecasting and Social Change*, 146(c), 403–412. DOI: 10.1016/j.techfore.2017.05.039.

Yin, R.K. (2003). *Case study research, design and methods.* Newbury Park: Sage.

Yun, J., Zhao, X., Yigitcanlar, T., Lee, D., & Ahn, H. (2018). Architectural design and open innovation symbiosis: insights from research campuses, manufacturing systems, and innovation Districts. *Sustainability*, 10(4495), 1–23. DOI:10.3390/su10124495

Zou, Y., & Zhao, W. (2014). Anatomy of Tsinghua University Science Park in China. *The Journal of Technology Transfer*, 39(5), 663–674. DOI:10.1007/s10961-013-9314-y

4

District Classification

This part of the book reveals that innovation clusters have evolved over time in terms of their functions, features, and spatial characteristics from 'science and technology park' model to 'innovation district' model. The latter model's key functional and spatial characteristics vary due to the local contextual factors. Furthermore, this part argues that investing in innovation districts is a risky business, as there is no one-size-fit-all innovation district type due to their varying characteristics. However, the part confirms that innovation districts can be holistically classified through a multidimensional framework consisting of numerous attributes—i.e., dimensions, indicators, and measures. The part reports on the process of framework development and validation by a panel of international experts through a Delphi survey. The result, a multidimensional innovation district classification framework, is the first of its kind. This part also presents the results of the real practical application of the validated framework on three eminent Australian innovation districts—i.e., Macquarie Park Innovation District (Sydney), Monash Technology Precinct (Melbourne), and Kelvin Grove Urban Village (Brisbane). The results confirmed that innovation district performances can be measured, and typologies can be developed through a novel approach.

DOI: 10.1201/9781032657431-12

CHAPTER 9

Innovation District Taxonomy

9.1 INTRODUCTION

During the last two decades, land-use policies and plans—targeting the formation of innovation spaces—have become central for gaining competitiveness in the innovation economy, and achieving sustained socioeconomic growth (Benneworth & Ratinho, 2014; Aldieri & Vinci, 2017; Esmaeilpoorarabi *et al.*, 2018b; Aldieri *et al.*, 2019). Issues around the need for boosting the economy, fostering new opportunities arising because of technological advances, searching for solutions to address climate emergency, and tackling unemployment and socioeconomic inequity problems have compelled many cities to make transition to innovation economy (Pareja-Eastaway & Piqué, 2011; Rabelo & Bernus, 2015). These cities embrace new strategies and structural changes under the development paradigm of knowledge-based urban development (KBUD) (Yigitcanlar, 2014). Through KBUD, cities implement strategies to upgrade their infrastructures, improve their quality of life, and create an attractive environment for talent and investment (Metaxiotis *et al.*, 2010).

Innovation districts, being the spatial nexus of KBUD, are a new land-use type that have become a global phenomenon for many cities. Silicon Valley in the United States and Sophia-Antipolis in France are traditional examples of innovation districts (Pancholi *et al.*, 2015a; Esmaeilpoorarabi et al., 2020a), where more recent contemporary examples include Singapore's One-North, and Spain's 22@ Barcelona Innovation District. Innovation districts were traditionally developed as closed innovation systems for a single-purpose use that includes a controlled environment within enclosed district walls. Nonetheless, contemporary innovation districts have become a significant urban land-use type as they adopted

DOI: 10.1201/9781032657431-13

open innovation systems with mixed-use and boundary blurred environments (Van Winden & Carvalho, 2016; Jones, 2017). These new generation innovation districts thrive as the growth nodes of metropolitan cities—not only stimulating the economic development of cities, but also boosting technological, sociocultural, and environmental developments (Yigitcanlar *et al.*, 2017; Pancholi, 2018a). They provide work-live-learn-play-cyber environments for knowledge workers and their families within the district (Yigitcanlar *et al.*, 2015; Pancholi *et al.*, 2019). Such an environment encourages networking and collaboration amongst knowledge workers, and hence contributes to the success of innovation districts (Kovacs & Petruska, 2014).

Despite being the nexus of innovation economy, the key functional and spatial characteristics of innovation districts vary due to different local contextual factors. Consequently, a variety of innovation districts were developed, differing from each other in many ways (Chapple *et al.*, 2010; Forsyth, 2014; Hsieh *et al.*, 2014). In general, their distinctions were identified mainly by only limited features, such as district size, sectoral specialisation, or locational factors (Markusen, 1996; Katz & Wagner, 2014). Existing literature confirms the fact that classification of innovation districts is an understudied area of research.

Against this backdrop, the study aims to expand our understanding on the classification of innovation districts based on their key characteristics. The methodological approach of the study employs a systematic review of the literature on innovation districts, and places global case studies reported in the literature under the microscope—to address the research question of: How can contemporary innovation districts be classified? Through the examination of global innovation district key characteristics reported in the literature, this chapter determines the key factors and indicators of innovation districts and develops a conceptual typology framework. This classification and the conceptual framework are important inputs for the development of a holistic typology framework for contemporary innovation districts—that will inform policymaking, planning, and development processes.

9.2 LITERATURE BACKGROUND

9.2.1 Evolution of Innovation Districts

During the late 20th century, cities primary focused on hosting the occupational and industrial developments—mostly in the form of science and technology parks—to boost economic activities and employment levels (Jolly & Zhu, 2012). The space configurations of these developments turned them into isolated, introverted, and single purpose controlled environments (Esmaeilpoorarabi *et al.*, 2020a). Nevertheless, in the 21st century, knowledge generation and innovation have taken prevalence in most global cities. This resulted in the adoption of a more systemic and holistic way to address interrelated issues of economic, societal, and spatial/environmental aspects of the economic hubs of cities (Rabelo & Bernus, 2015). Subsequently, this resulted in the formation of new innovation spaces that were branded as innovation districts (Lee, 2001; Engel & Palacio, 2011). This new land-use type is extravert and mixed-use in nature (Yigitcanlar *et al.*, 2016).

Due to the socioeconomic heterogeneity of cities and vast distinctions in regional economies, various types of innovation districts were developed to support a range of innovation economy activities across cities globally (Hawken & Han, 2017). In this study, the term 'innovation districts' is used as an umbrella term inclusive of 'knowledge and innovation spaces' (Pancholi *et al.*, 2019), 'innovation clusters' (Huggins, 2008), 'innovation milieu' (Evans & Hutton, 2009), 'knowledge (community) precincts' (Yigitcanlar *et al.*, 2008), 'innovation precincts' (Esmaeilpoorarabi *et al.*, 2018a), and the like—that are mostly (sub) urban mixed-function land uses in nature. Definitions of innovation district-related terms are listed in Table 9.1.

9.2.2 Characteristics of Innovation Districts

Despite most innovation districts having some common characteristics—in terms of general economic, spatial, and networking assets (Katz & Wagner, 2014; Yigitcanlar & Inkinen, 2019)—they are distinctive in possessing specific functions, features, and spatial qualities.

Firstly, innovation districts have different industry and business sector compositions and business functions. They mainly operate either as a: (a) high-technology-intensive innovation district (e.g., Silicon Valley, USA) or (b) creativity-intensive innovation district (e.g., Ontario Media Innovation District, Canada). They can also contain both industrial functions at the same time (e.g., Strijp-S, the Netherlands). There is also a third function, namely knowledge-intensive service activities (KISAs). They are commonly found in both aforementioned innovation district types (Martinez-Fernandez, 2010). Nevertheless, KISAs are support functions that do not operate as

TABLE 9.1 Definitions of Innovation District–Related Terms

Term	Definition	Reference
Innovation cluster	Geographic concentration of interconnected companies in a field that encompasses an array of linked industries and entities important to competition, including suppliers of specialised inputs.	Lee (2001)
Innovative milieu	Location that concerns with an incubation place of new innovation, and characterised by a set of collective and dynamic processes incorporating actors that lead to networks of synergy producing interrelationship.	Simmie (2005)
High-technology district	High-technology industry cluster that consists of a series of buildings set amidst impeccable landscaping in a campus-like atmosphere.	Forsyth (2014)
Science and technology park	Location that improves local innovation outcomes by promoting knowledge development and transmission among the co-located firms.	Diez-Vial & Olmos (2015)
Knowledge (community) precinct	Mixed-use urban settings that include a critical mass of knowledge enterprises and advanced network infrastructures, and developed with the aim of collecting and benefits of blurring the boundaries of living, shopping, recreation, and the working facilities of a knowledge community.	Yigitcanlar *et al.* (2016)
Innovation and cultural district	District that showcases innovation, research, training, and entrepreneurship as the hub of innovation, cultural creation, and entertainment.	Jones (2017)
Innovation district	Nexus of knowledge-based development in cities, where public and private actors work towards fostering, attracting, and retaining investment and talent with an aim of revitalising urban areas, and boosting knowledge and innovation economy activities.	Esmaeilpoorarabi *et al.* (2020c)

independent districts (Millar *et al.*, 2005; Yang & Wang, 2008); hence, they alone do not form a separate innovation district type.

Secondly, innovation districts display variances in their features. They show dissimilarities in terms of: (a) innovation and R&D companies and talented workforce of the district (Boja, 2011); (b) spatial features and territorial jurisdiction of the district and host city (Forsyth & Crewe, 2010); (c) proximity and connectivity of actors or stakeholders within the district (Crevoisier, 2014); and (d) social features of the district, such as demographic composition and social capital (Gabaldon-Estevan & Ybarra, 2017).

Lastly, most innovation districts are distinctive in their space-use. They differ in terms of: (a) spatial design and space configurations (Loures *et al.*, 2007); (b) use of open and/or close innovation systems (Chesbrough, 2006); (c) development size and scale (Pancholi *et al.*, 2019); (d) land-use composition, such as single-use or mixed-use activities (Forsyth, 2014); and (e) type of governance models such as double, triple, and quadruple helix partnership models (Cheng *et al.*, 2019).

9.2.3 Types of Innovation Districts

Research on defining the common types of innovation districts starts with Roeland et al.'s (1996) study that identified industrial clusters based on their specialisation patterns and innovation characteristics. They classified industrial clusters into four categories: (a) function, (b) characteristics of knowledge activities, (c) formation process, and (d) behaviour—i.e., competition and collaboration. Furthermore, Markusen (1996) identified four industrial districts based on the firm configurations, internal versus external orientation, and their governance structures: (a) Marshallian industrial district, (b) hub and spoke industrial district, (c) satellite platform, and (d) state-anchored districts. Capitalising on this industrial district typology, Clark *et al.* (2010) re-classified Markusen's (1996) districts by utilising patent data, regional resilience, and replaced 'industrial' with 'innovation' districts: (a) Type 1: Marshallian districts; (b) Type 2a: hub and spoke district; and (c) Type 2b: Satellite Platform District.

Lately, Forsyth (2014) developed her typology purely on the physical features of innovation districts: (a) location, (b) physical scale of development, and (c) level of physical planning and urban design. Using this typology, she identified six types of innovation districts: (a) corridor, (b) clumps, (c) cores, (d) comprehensive campus, (e) technology sub-divisions, and (f) scattered technology sites. Katz & Wagner (2014) developed a generic typology based on general observations of the existing innovation districts: (a) anchor plus, (b) re-imagined urban areas, and (c) urbanised science park models.

In sum, while there have been some attempts to determine and categorise innovation district types, the literature confirms the fact that there is no holistic or widely accepted innovation district typology framework. Such typology development requires comprehensive classification of innovation districts. There is, hence, a need for further investigations on innovation district typologies—as (sub)urban land-use types—to advance the knowledge in the field. This inconsequence will inform innovation district policymaking, planning, and development processes.

9.3 METHODOLOGY

This research utilises a systematic literature review to address the research question of: How can contemporary innovation districts be classified? The study adopted the methodological approach recommended by Bask & Rajahonka (2017), which contains a three-stage process. Stage 1 is the planning stage and involves objectives and review protocol for a systematic review, defining sources and procedures for literature searches. Stage 2 is the review stage and consists of descriptive and structural analysis. Stage 3 is the reporting and dissemination stage, and contains analysis and synthesis of the results according to the established objectives.

In Stage 1, the research aim, question, and set of inclusion and exclusion criteria were developed. The research aim was framed to identify the key characteristics of innovation districts in order to develop a framework. The keywords used were 'innovation districts', 'innovation clusters', 'innovation precincts', 'knowledge precincts', and 'industrial districts'. The inclusion criteria were the peer-reviewed academic journal articles that were available online in full text and published in English and relevant to the research aim. The exclusion criteria were those publications other than those mentioned in the inclusion criteria (Figure 9.1). The following databases were accessed to obtain the relevant articles: Scopus, Science Direct, Taylor & Francis Online, Web of Science, Wiley Online Library, and Directory of Open Access Journal.

In Stage 2, the search task of the relevant articles took place in October 2019. The starting date for publications was left open as our investigation focuses on establishing the evolution of innovation districts from as early

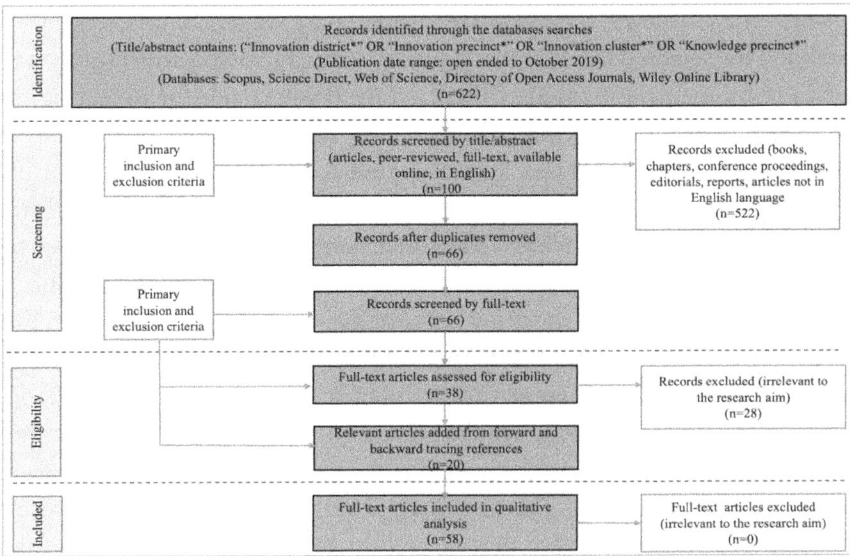

FIGURE 9.1 Literature selection procedure.

as possible to the date the search was conducted. The search results revealed that the earliest publication was in 1996; hence, covering a date range of well over two decades. The above-mentioned keywords were directed to the publication title of the articles. The search query used for the database search was 'Innovation district*' OR 'Innovation precinct*' OR 'Innovation cluster*' OR 'Knowledge precinct*'. Articles obtained from the search were reviewed as follows: Firstly, the abstract was read, and if found relevant, then the full text was read, and a decision was made on whether to include the article in the review pool. Initially, the search returned in total 622 articles. After filtering by inclusion and exclusion criteria, the articles were reduced to 100 potentially relevant articles. After screening against the research aim and removal of duplications, the number of articles was reduced to 66. The full-text articles were read against the research aim and further reduced to 38 articles. Since this figure was lower than the minimum target of 50 articles for the review pool—as suggested by Ruhlandt (2018)—a forward and backward tracing of references was conducted using the citation index of the 38 articles. This helped us find other seminal works that did not appear in our search and resulted in a total of 58 articles. All articles, then, were re-read, reviewed, categorised, and analysed. See Figure 9.1, which illustrates the literature selection process.

This study utilised a descriptive rather than statistical analysis of the results. The selected 58 literature pieces were reviewed thoroughly and critically by using a manual content analysis approach for detecting the indicators relating to typologies of innovation districts. For such an analysis, we adopted Yin's (2015) qualitative techniques of pattern matching and explanation building to identify and categorise selected articles into common themes. Pattern matching refers to scanning for commonalities and disparities in which an eye-balling technique is sufficiently convincing to draw a conclusion or categorisation (Yin, 1994). We adopted a four-step process for categorising the reviewed literature into specific themes (Yigitcanlar *et al.*, 2019a). The criteria for the formation of categories are presented in Table 9.2. In the first step, key critics and relevance to the aim of this research raised in the reviewed literature were tabulated. The most important themes that best represent the reviewed literature were determined in the second step. In the third step, these themes were cross-checked with other review studies and identified critics and relevance to the investigated topic to verify the common themes. In the final step, the categories were narrowed down to three common themes of innovation district 'classification by functions', 'classification by features', and 'classification by space-use'. The lists of these articles are presented in Tables 9.3 through 9.5, respectively.

In Stage 3, we concentrated on the write-up and reporting the findings from the 58 analysed articles. We also incorporated other highly relevant publications on the investigated topic as additional supporting literature

TABLE 9.2 Selection Criteria for the Formulation of Categories

Selection Criteria
1. Determine the key critics and relevance to innovation districts using the eye-balling technique in the literature
2. Detect the indicators relating to typologies of innovation districts
3. Identify the indicators relating to typologies of innovation districts
4. Group the identified indicators with similarities to form broader potential categories
5. Shortlist the categories and cross-check the reliability of these categories with other published literature
6. Reconsider the shortlisted categories by going through the selected and reviewed literature one more time
7. Confirm the selection and classification of the categories and finalise the formulation of categories
8. Place the reviewed literature pieces under the determined categories

evidence for a better analysis of the topic and elaboration of the literature findings. In the end, the total number of reviewed, cited, and quoted references was increased to 102 literature pieces.

9.4 RESULTS

9.4.1 General Observation

The search results revealed the increasing popularity of innovation districts during the last decade as almost three-quarters (74.1%) of the reviewed papers (n = 43) were published 2011 onwards. In terms of author affiliations, the leading countries in innovation districts studies were from the North American, European, and Oceanian regions. Nevertheless, there were also emerging interests in the innovation district research from the Southeast and Central Asia regions.

Two articles were published in each of the following journals: *Cities, Environment and Planning C: Government and Policy, Innovation: Management Policy and Practice, Journal of Technology Transfer, Journal of Urban Design, Journal of Urban Technology, Land Use Policy, Regional Studies,* and *Urban Studies.* The remaining 40 articles were published in 40 different journals; 23 papers were published in urban studies–focused journals; 11 in science and technology; 11 in business, management, and law; 8 in economic and social; and 5 in regional development. This indicates the broad interdisciplinary research perspectives on innovation districts.

After a stringent review, 58 papers were further categorised under three groups: (a) classification by function (n = 19 papers)—highlights the key functions of innovation districts; (b) classification by feature (n = 21)—points out the common features of innovation districts; and (c) classification by space-use (n = 18)—focuses on the plans, design, and development of innovation districts.

9.4.2 Classification by Function

About one-third (32.8%, n = 19) of the reviewed articles discussed innovation districts by their functions (see Table 9.3). The term 'function' has been identified by Esmaeilpoorarabi *et al.* (2018c) as one of the factors that defines the character of a place. According to the researchers, the following indicators compose the main functions of buildings and open spaces: 'Mixed-use or single-use developments; Type and variety of innovative activities; Significance of economic opportunities and company

TABLE 9.3 Classification of Innovation Districts by Functions

No.	Literature	Journal	Title	Relevance
1	Markusen (1996)	*Economic Geography*	Sticky places in slippery space: a typology of industrial district	Focuses on industrial districts and identification of typologies.
2	Lee (2001)	*Science Technology and Society*	From fragmentation to integration: development process of innovation clusters in Korea	Provides insights into innovation cluster's definition, typologies and characteristics in Korea.
3	Park (2003)	*Papers in Regional Science*	Economic spaces in the Pacific Rim: a paradigm shift and new dynamics	Focuses on a broader context, and the discussion on the evolution of types of industrial districts.
4	Millar *et al.* (2005)	*Technology Analysis and Strategic Management*	The state in science, technology and innovation districts: conceptual models for China	Identifies three models for innovation districts as well as their common characteristics.
5	Leon (2008)	*Innovation: Management Policy and Practice*	Attract and Connect: the 22@ Barcelona innovation district and the internalisation of Barcelona business	Provides relevant background information on the development process, characteristics and typology of 22@ Barcelona.
6	Yigitcanlar *et al.* (2008)	*Journal of Knowledge Management*	Rising knowledge cities: the role of urban knowledge precincts	Generates insights on the evolution of typologies of innovation districts.
7	Boix & Galleto (2009)	*Regional Studies*	Innovation and industrial districts: a first approach to the measurement and determinants of the I-district effect	Focuses on industrial districts and their effect on innovation and identifies seven types of local production systems for comparisons.
8	Davis *et al.* (2009)	*Innovation: Management Policy and Practice*	Applying innovation cluster frame to a creative industry: the case of screen-based media in Ontario	Identifies notable characteristics of clusters and significant differences between two case studies.

9	Clark *et al.* (2010)	*Cambridge Journal of Regions*	A typology of innovation districts: what it means for regional resilience	Provides insights into the typologies and characteristics of innovation districts.
10	Forsyth (2014)	*Environment and Planning C: Government and Policy*	Alternative forms of high-technology district: corridors, clumps, cores, campuses, subdivisions, and sites	Identifies typologies of high-technology innovation districts.
11	Fan (2015)	*Clinical Law Review*	Coming of age: innovation district and the role of law schools	Presents definitions of innovation districts, characteristics and types of models.
12	O'Dwyer *et al.* (2015)	*Competitiveness Review*	Insights into the creation of a successful MNE innovation cluster	Discusses typologies of innovation clusters.
13	Pancholi *et al.* (2015b)	*International Journal of Knowledge-Based Urban Development*	Place-making facilitators of knowledge and innovation spaces: insights from European best practices	Provides insights on typologies and characteristics of the best practice cases from the European context.
14	Diez-Vial & Olmos (2015)	*Journal of Technology Transfer*	Knowledge spill overs in science and technology parks: how can firms benefit most?	Discusses various definitions of innovation districts.
15	Khomsi (2016)	*Technology Innovation Management Review*	The smart city ecosystem as an innovation model: lessons from Montreal	Describes Montreal's innovation and cultural districts.
16	Van Winden & Carvalho (2016)	*Journal of Urban Technology*	Urbanise or perish? assessing the urbanisation of knowledge locations in Europe	Provides insights into typology of knowledge locations from the three case studies: Digital Hub, Kista Science City, and Biocant.

(Continued)

TABLE 9.3 *(Continued)* Classification of Innovation Districts by Functions

No.	Literature	Journal	Title	Relevance
17	Hobbs *et al.* (2017)	*Journal of Technology Transfer*	Science and technology parks: an annotated and analytical literature review	Defines and discusses science and technology parks and their development.
18	Jones (2017)	*Planning Practice and Research*	Regenerating urban water fronts-creating better futures-from commercial and leisure market places to cultural quarters and innovation districts	Provides insights into evolution of a water front development into a cultural and innovation district.
19	Esmaeilpoorarabi *et al.* (2020a)	*Cities*	How can an enhanced community engagement with innovation district be established? Evidence from Sydney, Melbourne and Brisbane	Generates insights into planning, development, features, and typology of innovation districts.

profiles; Quantity and quality of labour pool; Capability of knowledge generation and knowledge spill-over, and; Proficiency of professional networks' (Esmaeilpoorarabi *et al.*, 2018c, p. 735).

As we mentioned earlier, functions of existing innovation districts focus on three sectoral activities: (a) high-technology-intensive activities, (b) creativity-intensive activities, and (c) knowledge-intensive service activities. These are further discussed below.

High-technology-intensive activity: Millar *et al.* (2005) suggested that there are three types of high-technology-intensive innovation districts: (a) market-driven technology districts (e.g., Silicon Valley, USA, and Cambridge Science Park, UK); (b) state-driven technology districts, which target foreign multinational companies (e.g., Sophia Antipolis, France, and One-North, Singapore); and (c) state-driven technology districts, which foster local companies (e.g., Hsinchu Science Park, Taiwan). These districts predominantly encompass ICT or biotechnology industries (Forsyth, 2014). Except for a few privately funded ones (e.g., Silicon Valley, and Cambridge Science Park), most of the existing innovation districts are mandated either by state and/or local governments (Leon, 2008; O'Dwyer *et al.*, 2015).

Creativity-intensive activity: Creativity intensive innovation districts are largely concentrated on cultural knowledge generation such as music, film, and gaming industries (Zheng, 2011). According to Esmaeilpoorarabi *et al.* (2020a), innovation districts' functions have expanded beyond their conventional functions (i.e., a hub for commerce, research, and education) to include the creative activities as well. Location-wise, creative industries are traditionally clustered and practically always located in the core of major urban areas (Baum *et al.*, 2009; Durmaz *et al.*, 2010). Ontario attests to this statement as there are 'more than three quarters of all firms in the film, television, and interactive digital media located in the Toronto metropolitan region' (Davis *et al.*, 2009, p. 203).

Knowledge-intensive service activity (KISA): KISAs cluster around corporate head offices and related activities of transnational corporations, where many of them are located in innovation districts (Pancholi *et al.*, 2018b). KISA companies focus on marketing, auditing, and insurance businesses (Yang & Wang, 2008). Yigitcanlar *et al.* (2008) underline the crucial role of KISAs in innovation districts for knowledge generation and innovation, particularly for R&D businesses as support activities. Many other scholars, such as Holroyd & Coates (2007), also perceive KISAs as a fundamental support element of innovation districts.

The literature highlights the significant differences between high-technology-intensive and creativity-intensive innovation districts (e.g., Canadian technology intensive innovation clusters, and Ontario's screen-based media clusters) in terms of the number of member companies, geographical distribution of customers, and the strength of identity and district/place branding. As identified by Davis *et al.* (2009), whilst creativity-intensive innovation districts primarily orient towards local and regional customers, high-technology-intensive innovation districts are highly extraverted, and work at the global scale. Besides, creativity-intensive innovation districts have a greater number of member companies with stronger identity as compared to high-technology-intensive innovation districts contributing to the district/place branding.

9.4.3 Classification by Feature

Over one-third (36.2%, n = 21) of the reviewed articles concentrated on the features of innovation districts (see Table 9.4). According to these articles, the most common practice in classifying innovation districts is done by focusing on their features (e.g., Feldman, 2014; Kaigorodtsev & Bordiyanu, 2014). These features are grouped under four categories and discussed below: (a) economic, (b) physical, (c) operational, and (d) social.

Economic features: Cities use innovation districts as a vehicle to achieve revitalisation of urban areas and boost their knowledge and economic activities (Benneworth & Ratinho, 2014; Esmaeilpoorarabi *et al.*, 2016). Indicators such as innovation and R&D capital investment, skilled workforce size, productivity, and patents are used to analyse the local innovative production networks and formation of innovation districts (Hu *et al.*, 2005). These indicators are regarded as economic features of innovation districts (Boja, 2011; Jolly & Zhu, 2012). Just to give an example on one of these indicators, Macquarie Park Innovation District (Australia) employs about 45,000 knowledge workers, where this figure is over 250,000 for Suzhou New Innovation District (China), (Jolly & Zhu, 2012).

Physical features: Castells & Halls (1994) and Forsyth & Crewe (2010) classified physical features of innovation districts by their size, location, scale, territorial jurisdiction, and level of physical planning. Forsyth (2014) used these features to identify six typologies of high-technology-intensive innovation districts: (a) corridor, (b) clumps, (c) cores, (d) comprehensive campus, (e) technology sub-divisions, and (f) scattered technology sites. 'The size of these districts ranges in the scale of a few

TABLE 9.4 Classification of Innovation Districts by Features

No	Literature	Journal	Title	Relevance
1	Altenburg & Stamer (1999)	*World Development*	How to promote clusters: policy experiences from Latin America	Focuses on general cluster characteristics in Latin America, and defines clusters and their key features.
2	Hu *et al.* (2005)	*Technovation*	Technology-based regional development strategies and the emergence of technological communities: a case study of the HSIP, Taiwan	Generates insights into the key characteristics of Hsinchu innovation district.
3	Simmie (2005)	*Regional Studies*	Innovation and space: a critical review of the literature	Defines and determines the characteristics of innovative milieus.
4	Battaglia & Tremblay (2011)	*Urban Studies*	22@ and the innovation district at Barcelona and Montreal: a process of clustering development urban regeneration and economic competitiveness	Elaborates the governance strategy used at the 22@Barcelona.
5	Boja (2011)	*International Journal of Economic Practices and Theories*	Clusters models, factors, and characteristics	Provides information on cluster models characteristics and life cycle stages of innovation clusters.
6	Engel & Palacio (2011)	*California Management Review*	Global clusters of innovation: the case of Israel and Silicone Valley	Defines clusters of innovation and reveals the key attributes of Israel and Silicon Valley cases.
7	Foley *et al.* (2011)	*Research-Technology Management*	The greater Philadelphia innovation cluster for energy-efficient buildings: a new model for public-private partnership	Reveals the governance model of the Greater Philadelphia innovation cluster.

(Continued)

TABLE 9.4 *(Continued)* Classification of Innovation Districts by Features

No	Literature	Journal	Title	Relevance
8	Jolly & Zhu (2012)	*Journal of Business Strategy*	Chinese S&T parks: the emergence of a new model	Identifies the different characteristics of Chinese innovation districts compared to North American and other models.
9	Lee & Choi (2013)	*Academy of Entrepreneurship Journal*	Industry-Academia corporation in creative innovation clusters: a comparison of two clusters in Korea	Provides a definition and key characteristics of innovation clusters in Korea.
10	Varga *et al.* (2013)	*International Journal of Public Sector Management*	State aid for innovation clusters in the Republic of Serbia	Focuses on state funding of innovation districts in Serbia and defines and discloses the characteristics of clusters.
11	Benneworth & Ratino (2014)	*Environment and Planning C: Government and Policy*	Reframing the role of knowledge parks and science cities in knowledge-based urban development	Discloses the characteristics of the case study innovation district.
12	Feldman (2014)	*Small Business Economics*	The character of innovation places: entrepreneurial strategy, economic development, and prosperity	Generates insights into the fundamental characteristics of innovation districts.
13	Kaigorodtsev & Bordiyanu (2014)	*Aktual'ni Problemy Ekonomiky*	SWOT-analysis of an innovation cluster creation in East Kazakhstan	Provides insight into characteristics of innovation clusters from East Kazakhstan.
14	Kovacs & Petruska (2014)	*Periodica Polytechnica Social and Management Sciences*	Operational characteristics of Hungarian innovation clusters as reflected by a qualitative research study	Describes the two case study innovation clusters as examples of open and close innovation models.
15	Gabaldon-Estevan & Ybarra (2017)	*European Planning Studies*	Innovative culture in district innovation systems of European ceramics SMEs	Provides information on definition, characteristics and typology of innovation districts.

16	Bank & Sibanda (2018)	*Development Southern Africa*	Universities as city builders: the city-campus development in East London, Buffalo City South Africa	Identifies a governance model between public and higher education institutions to generate a platform for urban renewal through innovation districts.
17	Esmaeilpoorarabi *et al.* (2018a)	*Land Use Policy*	Does place quality matter for innovation districts? Determining the essential characteristics from Brisbane's knowledge precincts.	Generates insights on the essential place characteristics that gives quality to innovation districts.
18	Esmaeilpoorarabi *et al.* (2018b)	*Land Use Policy*	Evaluating place quality in innovation districts: a Delphic hierarchy process approach	Focuses on the place quality and provides insights on characteristics of innovation districts from the selected case innovation districts.
19	Yun *et al.* (2018)	*Sustainability*	Architectural design and open innovation symbiosis: insights from research campuses, manufacturing systems, and innovation districts	Evaluates the role of architectural design in creating an open innovation ambiance, in innovation districts.
20	Cheng *et al.* (2019)	*Information Development*	Triple helix on globalisation: a case study of the China international nanotech innovation cluster	Provides lessons on the practice of an innovation district governance model.
21	Morrison & Bevilacqua (2019)	*Urban Research & Practice*	Balancing gentrification in the knowledge economy: the case of Chattanooga's innovation district	Generates insights into Chattanooga innovation district's characteristics and development story.

hectares to 10,000 s of hectares. They are mostly located in urban or suburban areas, but there exist some in exurban areas. The physical planning level varies among them' (Forsyth, 2014, p. 813). Furthermore, relating to location, 'proximity and connectivity' of actors are considered important locational features. Proximity is also regarded by other scholars (Asheim *et al.*, 2007; Benneworth *et al.*, 2014) as a critically important feature for place specific competitiveness (Coe *et al.*, 2004; Amin & Roberts, 2008).

Operational features: The key operational features include district property and activities management model, operational funding, and business support services. There are different types of management models used in the operation and management of innovation districts. The property and activities management model of a district is generally determined at the planning stage. The management takes place as either a district-level body corporation (e.g., One-North, Singapore), a distributed building-base body corporation (Diamantina Knowledge Precinct, Australia), or without any body corporation involved (Brisbane Technology Park, Australia). In terms of operational funding, innovation districts and their management are usually financially supported by the state and/or local governments either directly providing funds/incentives to the district and companies or indirectly through research and infrastructure grants provided to anchor universities and R&D centres (Battaglia & Tremblay, 2011; Yigitcanlar *et al.*, 2019b). Many innovation districts facilitate the tenant entrepreneurs' access to funding assistance for the traditional venture capital, initial public offering (IPO), and corporate venture capital investment (Feldman, 2014). Business support services are also provided to the companies (Millar & Choi, 2010; Benneworth & Ratinho, 2014).

Social features: Social features of innovation districts comprise the demographic composition and social capital of the labour force. According to Benneworth & Ratinho, (2014), the demographic structure of innovation districts consists of actors and breakup of innovation companies, which can be identified by the following globally connected group of actors: (a) university, (b) local companies, (c) multinational technology companies, (d) technology investors, (e) property investment companies, and (f) service companies. Social capital or the existing knowledge network across innovation districts (Gabaldon-Estevan & Ybarra, 2017) is recognised as the heart of the innovation process (Montresor & Marzetti, 2008). Social capital is a key asset of innovation districts that accumulates benefits for being connected to a network or community (Feldman & Zoller, 2012), and eventually, mutual trust among the actors is established due to their long-term networking—thus, producing intangible knowledge exchange (Lee & Choi, 2013).

9.4.4 Classification by Space-Use

Less than one-third (31%, n = 18) of the reviewed articles concentrate on issues related to space-use in innovation districts (see Table 9.5). Space-use (land and building) mainly focuses on: (a) spatial design and configuration of innovation districts concerning open or close innovation systems, scale, and land use; (b) natural environment and unique surroundings of innovation districts, including both the natural and built environments in and around; and (c) governance model of innovation districts that is determined during the conception of the district.

Spatial design and configuration: Open and green spaces, and architectural and urban design are important in establishing the identity of innovation districts (Loures *et al.*, 2007). Unlike the traditional innovation districts (e.g., science and technology parks), the contemporary examples adopt an open district innovation system model (Chesbrough, 2006) that allows possible involvement of other sectoral activities and territories, especially through established networks. The notion of open innovation is demonstrated in Yun et al.'s (2018) research concerning open architectural design of innovation districts that encourage collaborative and cooperative innovation. Pancholi *et al.* (2018a) underlined that the rise of open innovation, as the increasing returns of knowledge generation—is a sharable commodity. Hence, many innovation districts aspire to become highly networked spaces.

In terms of the scale of development and land-use plans of innovation districts, we observe that new-generation innovation districts are mostly planned to accommodate mixed uses and that the scale of development varies considerably across the global cases. For instance, innovation districts in Australia are typically medium to large in size and mostly planned developments as opposed to spontaneous agglomerations. They are, in general, mixed-use centres of an innovation economy located in the inner city and close-in suburbia (Esmaeilpoorarabi *et al.*, 2018c). However, there are a few single-purpose use innovation districts—e.g., Brisbane Technology Park (Esmaeilpoorarabi *et al.*, 2018c). In the United States and Europe, the general scale of development ranges from medium to large to regional scales (Forsyth, 2014), and are increasingly planned developments

TABLE 9.5 Classification of Innovation Districts by Space-Use

No	Literature	Journal	Title	Relevance
1	Bathelt *et al.* (2004)	*Progress in Human Geography*	Clusters and knowledge: local buzz, global pipelines and the process of knowledge creation	Defines knowledge-based theory of spatial clustering and focuses on the space-use.
2	Van Winden *et al.* (2007)	*Urban Studies*	European cities in the knowledge economy: towards a typology	Focuses on the knowledge city concept and discusses the development concept and definitions of knowledge-based urban development.
3	Smoliński & Pichlak (2009)	*Technology in Society*	Innovation in Polish industry: The cluster applied to clean coal technologies in Silesia	Provides insights into the planning and development process of creating a clean coal technology cluster.
4	Pareja-Eastaway & Piqué (2011)	*Journal of Urban Regeneration and Renewal*	Urban regeneration and the creative knowledge economy: the case of 22@ Barcelona	Generates lessons from the 22@Barcelona's development process, governance, characteristics and typology.
5	Kireeva (2014)	*Economic and Social Changes: Facts, Trends and Forecast*	Theoretical and methodological framework establishing the conditions for the development of innovation clusters	Provides insights into the methodology and prerequisite conditions for formation and development of an ideal innovation district.
6	Pancholi *et al.* (2014)	*Asia Pacific Journal of Innovation and Entrepreneurship*	Urban knowledge and innovation spaces: concepts, conditions and context	Introduces the key concepts, characteristics and conditions that contribute to shaping urban knowledge and innovation spaces.
7	Capdevila (2015)	*International Journal of Innovation Management*	Co-working spaces and the localised dynamics of innovation in Barcelona	Provides definitions of co-working spaces, where they can be regarded as micro-clusters of innovation.

8	Edvardsson *et al.* (2016)	*Knowledge Management Research & Practice*	Knowledge city research and practice under the microscope: a review of empirical findings	Defines the terms such as knowledge-based urban development and knowledge cities and their relevance to innovation districts.
9	Rabelo & Bernus (2015)	*IFAC Papers Online*	A holistic model of building innovation ecosystem	Focuses on the planning and development of innovation ecosystem, which can be applied to an innovation district.
10	Yigitcanlar *et al.* (2016)	*Journal of Urban Technology*	Place-making for knowledge generation and innovation: planning and branding Brisbane's knowledge community precinct	Generates insights from the case studies and covers the definition, features, branding and design of innovation districts.
11	Hawken & Han (2017)	*Journal of Urban Design*	Innovation districts and urban heterogeneity: 3D mapping of industry mix in downtown Sydney	Provides insights into urban design concept of mixed-use development for innovation districts.
12	Esmaeilpoorarabi *et al.* (2018c)	*Cities*	Place quality in innovation clusters: an empirical analysis of global best practices from Singapore, Helsinki, New York and Sydney	Provides information on spatial planning, typology and characteristics of the case study innovation districts.
13	Pancholi *et al.* (2018a)	*Journal of Urban Design*	Attributes of successful place-making in knowledge and innovation spaces: evidence from Brisbane's Diamantina knowledge precinct	Presents the approaches and processes used for place-making of the selected case study innovation district.
14	Pancholi *et al.* (2018b)	*City, Culture, and Society*	Societal integration that matters: place making experience of Macquarie Park Innovation District, Sydney	Provides information on spatial planning, typology and characteristics of the selected case study innovation districts.
15	Yigitcanlar *et al.* (2018)	*Energies*	Towards smart Florianopolis: what does it take to transform a tourist island into an innovation capital?	Reviews the existing strengths and weaknesses of the case study island to identify opportunities and challenges of transforming the island from tourism to innovation district.

(*Continued*)

TABLE 9.5 *(Continued)* Classification of Innovation Districts by Space-Use

No	Literature	Journal	Title	Relevance
16	Pancholi *et al.* (2019)	*Technological Forecasting and Social Change*	Place making for innovation and knowledge-intensive activities: the Australian experience	Discusses the five dimensions of place-making in innovation districts.
17	Valler *et al.* (2019)	*Urban Science*	Science spaces as 'Ethnospaces': identity, perception, and the production of locality	Provides discussion on the spatial planning of the case studies.
18	Esmaeilpoorarabi *et al.* (2020b)	*Sustainable Cities and Society*	How does the public engage with innovation districts? Societal impact assessment of Australian innovation districts	Elaborates the planning and development of the innovation districts from Sydney, Melbourne, and Brisbane.

(e.g., Sophia-Antipolis, France, and Chattanooga Innovation District, USA). Conversely, there exist some districts, which initially emerged from spontaneous agglomeration and later were included in planned developments (e.g., Monash National Employment and Innovation Cluster, Australia). In contrast to the Western countries, the size of innovation districts in Asian countries, especially in China, are typically very large (i.e., regional clusters). On average, Chinese innovation districts house a population of 300,000 employees (Jolly & Zhu, 2012).

Natural environment and unique surroundings: Esmaeilpoorarabi *et al.* (2018c) considered the natural and built environments as key elements of place quality. The natural environment in and around innovation districts enhances the liveability by providing unique natural or human-made surroundings such as waterfront locations, national parks, historic sites, and university campuses. Good practices of urban design connect open spaces (e.g., waterbodies, landscape, parks, streetscapes) with the built structures in a way to create harmony between R&D, office, and education spaces and nature. Kelvin Grove Urban Village (Australia) and Orestad (Denmark) are exemplar cases of using architectural and urban design to increase the appeal of an innovation district (Pancholi *et al.*, 2015a).

Governance model: The most popular governance model used in establishing innovation districts is the triple helix model—an academic-industry-government partnership (Etzkowitz, 2002). Macquarie Park Innovation District (Australia) is an example of this model. The double helix model, also known as a public-private partnership, is used in the Research Triangle Park (USA) (Feldman, 2014). There is also a consolidated triple helix model called the 'novel triple helix' (Cheng *et al.*, 2019). According to this model, independent and interactional relationships among universities, government, and industries are critical and influence innovative output generation positively. An example is China International Nanotech District (Carrillo *et al.*, 2014). Finally, there is a quadruple helix model that adds a fourth strand of society to the standard triple helix model (Marcovich & Shinn, 2011).

9.5 DISCUSSION AND CONCLUSION

The results of our systematic review of the literature revealed that knowledge and innovation spaces are constantly evolving (Huggins, 2008), and accordingly they are being rebranded. As the popular knowledge and innovation spaces of the late 20th century, suburban and exurban science and technology parks (Jolly & Zhu, 2012) have been replaced with urban knowledge (community) precincts (Yigitcanlar *et al.*, 2008) in the early 21st century, and now are rebranded as innovation districts (Esmaeilpoorarabi, 2018a). Innovation districts have become a new type of land use in many cities that have embraced the revitalisation of urban areas by boosting innovation economy activities (Benneworth & Ratinho, 2014). Successful examples include 22@ Barcelona (Spain), Boston (USA), Macquarie Park (Australia), One-North (Singapore), and Toronto (Canada) innovation districts (Katz & Wagner, 2014). Nonetheless, there is limited knowledge on how to create successful urban landscapes to house innovation activities—e.g., innovation districts. In other words, innovation districts are a new land-use type that have not been thoroughly investigated from the planning perspective—that greatly benefit from identifying common typologies (Collier *et al.*, 2012).

Despite being established in various forms and models, innovation districts have some common characteristics specifically in their economic, physical, and networking assets. In contrast, innovation districts are highly distinctive from their predecessors—e.g., science and technology parks—even though their primary goal is to generate new knowledge and innovation. Firstly, their main operational functions show variations. As discussed earlier, innovation districts are mainly host high-technology-intensive activities or creativity-intensive activities—besides, some hybrid models also exist. In both types of districts, knowledge-intensive service activities function as support services (Martinez-Fernandez, 2010). Secondly, innovation districts' economic, physical, and social features have significant differences (Forsyth, 2014). Thirdly, their land and building space-use show significant variations (Loures *et al.*, 2007).

The literature discloses that there is no comprehensive or widely accepted typology framework for innovation districts. Previous research only sheds limited light on possible classifications of innovation districts. The existing innovation districts typology frameworks are not holistic, and hence have limitations in adoption. Markusen (1996), Roelandt (1996), Clark (2010), and Forsyth's (2014) innovation district classifications fell short due to focusing on limited features of innovation districts—e.g., focusing on economic features and neglecting environmental and social ones. While, Katz & Wagner's (2014) general innovation district typology seems to be more holistic, they also failed to capture most indicators under the feature category (Table 9.6).

TABLE 9.6 Indicators of Innovation District Typology Development

Factors	Indicators	Description	Studies				
			Markusen (1996)	Roelandt (1996)	Clark *et al.* (2010)	Forsyth (2014)	Katz & Wagner (2014)
FUNCTION							
High-technology-intensive activities	Market-driven science and technology districts	Private investor- or venture capitalist-led and funded development concentrated on science and technology-related product generation and marketing	+	+	+	–	–
	State-driven science and technology districts aiming to attract foreign companies and direct investment	State-led and funded/incentivised development concentrated on science and technology-related product generation and marketing	+	–	+	–	+
	State-driven technology districts aiming to foster and nurture local companies	State-led and funded/incentivised development concentrated on science and technology-related product generation and marketing	+	–	+	–	+

(*Continued*)

TABLE 9.6 *(Continued)* Indicators of Innovation District Typology Development

Factors	**Indicators**	**Description**	**Studies**				
			Markusen (1996)	**Roelandt (1996)**	**Clark *et al.* (2010)**	**Forsyth (2014)**	**Katz & Wagner (2014)**
Creativity-intensive activities	Cultural and creative industry and business concentrated districts	Mainly state-led and funded/incentivised development concentrated on cultural and creativity-related product generation and marketing	–	–	–	–	+
Knowledge-intensive service activities	Support activities clustered in and around innovation districts, corporate headquarters, and transnational corporations	Mainly private investor-led and funded development concentrated on business support services such as consultancy, marketing, insurance, IT services	+	+	–	–	+
FEATURE							
Economic	Knowledge-based, creative, and innovation companies	Quantity and quality of companies	–	–	–	–	–
	R&D capital investment	Level of state or private sector R&D investment	–	–	–	–	–

	Creative class of knowledge workers	Knowledge worker and community population and density	–	–	–	–	–
	Productivity	Economic outcome and profitability					
	Patents	Patent registry numbers and frequency	–	–	+	–	–
Physical	Size of innovation districts	Small, medium, large	–	–	–	+	–
	Location	Urban, suburban, exurban	–	–	–	+	–
	Territorial jurisdiction	Local, metropolitan, regional and national	–	–	–	–	+
	Level of physical planning	Low, medium, high	–	+	–	+	+
Operational	District property and activities management model	District-level body corporation, distributed building-base body corporation or no management	+	–	+	–	–
	Operational funding	Financed by state or local government or private sector project funds	–	–	–	–	–
	Business support services	High-technology service provision to innovative businesses with the district	+	–	+	–	+

(*Continued*)

TABLE 9.6 *(Continued)* Indicators of Innovation District Typology Development

Factors	Indicators	Description	Studies				
			Markusen (1996)	**Roelandt (1996)**	**Clark *et al.* (2010)**	**Forsyth (2014)**	**Katz & Wagner (2014)**
Social	Composition of actors and stakeholders	University, R&D institution, local companies, multinational companies, investors, property investment, and service companies	–	–	+	–	–
	Type of businesses	IT, biotechnology, business technology, convergence technology, media, film	+	+	–	–	–
	Social capital	Knowledge networks and social activities in and around innovation districts	+	+	+	–	–
SPACE-USE							
Spatial design and configuration	Open design innovation system	Open design innovation system to encourage collaboration with other sectors, activities and territories via established networks	–	–	–	–	–

	Scale of development	Small, medium, large	–	–	+	–	+
	Land use	Mixed-use being the most preferred	–	–	–	–	+
Environment	Natural environment	Green space, climate, air quality, water quality, and noise	–	–	–	–	–
	Unique surroundings	Natural or human-made surroundings such as water front, national parks, historic sites, and anchor universities	–	–	–	–	+
	Urban design	Design of landscapes, parks, streetscapes, open spaces	–	–	–	+	+
	Architectural design	Design of buildings and other physical structures	–	–	–	+	+
Governance	District governance model	Partnership models between sectors to govern the district such as triple or quadruple helix models	+	–	–	–	–

Given the limitations of the existing innovation district typology frameworks, there is room for the development of a holistic framework. This study, hence, proposed an approach to form a typology framework for innovation districts by investigating the key factors of such a framework and their potential indicators. Typology of innovation districts is the result of classification of key characteristics of innovation districts, and can be set up based on a combination of or intersection at the different characteristics of innovation districts. Thus, building on the characteristics of innovation districts identified from the reviewed literature, as listed in Table 9.6, is a logical step to follow to establish a typology framework. Table 9.6 underpins three key factors that can be combined to classify innovation districts—i.e., function, feature, and space-use—that contain 28 indicators.

Nevertheless, the first step of developing a holistic typology framework is to create a conceptual framework with guiding principles, domains, or elements. In light of the review findings, this study also develops a conceptual innovation district typology framework. As illustrated in Figure 9.2, this framework consists of four domains—i.e., context, function, feature, and space-use—that form the cornerstones of innovation district typologies.

The first domain, 'context', includes the regional scale issues that impact, shape up, or restrain the innovation district development—i.e., economic (macroeconomic progress of the city), political (political progress of the city), societal (societal progress of the city), and spatial (city-wide spatio-environmental qualities) systems of the city/region the innovation district is located in.

The second domain, 'function', consists of the key elements that characterise how the innovation district functions—i.e., industry type (dominant business activity operating within the district), investment type (principal support and funding body for the development of the district), and management model (management model of the district's properties and activities).

The third domain, 'feature', contains the essential elements to generate a sense of place for the companies, knowledge workers, and users of the district—i.e., economic scale (skilled employment outcomes of the district's activities), locality setting (location of the district within the metropolitan area), and social activities (public places and socio-cultural activities within the district).

The last domain, 'space-use', consists of built and natural elements that give the innovation district its unique identity—i.e., mixed-use

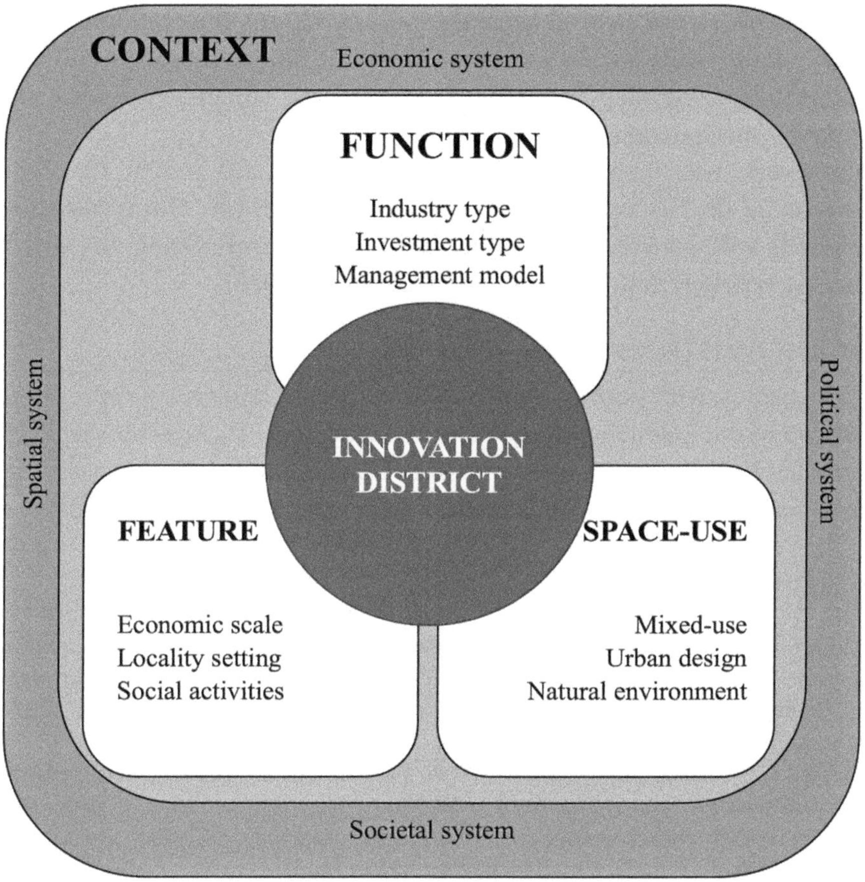

FIGURE 9.2 Conceptual innovation district typology framework.

(the main land-use types of the district), urban design (urban and architectural design encouraging open innovation system within the district), and natural environment (aesthetic qualities of urban green and blue spaces within the district).

The findings of the review helped in partially addressing the research question of how contemporary innovation districts can be classified. The conceptual typology framework brought together the main domains and elements of such classifications. The provided insights into such classifications are invaluable inputs for developing a holistic typology framework for contemporary innovation districts. Subsequently, that framework will be instrumental for innovation district policymakers, planners, designers, developers, and managers in several ways. Firstly, the framework will help in accurately classifying the innovation districts that come in many shapes and sizes. Secondly, this classification will help authorities to develop customised policies and suitable incentive and support mechanisms for each innovation district type. Lastly, the framework will help authorities to determine and match the right innovation district type for the right location to house. Our prospective research will concentrate on the development and testing of such a holistic typology framework (Yigitcanlar *et al.*, 2020).

ACKNOWLEDGEMENTS

This chapter, with permission from the copyright holder, is a reproduced version of the following journal article: Yigitcanlar, T., Adu-McVie, R., & Erol, I., (2020). How can contemporary innovation districts be classified? A systematic review of the literature. Land Use Policy, 95(1), 104595.

REFERENCES

Aldieri, L., & Vinci, C. P. (2017). Innovation, productivity and environmental performance of technology spill overs effects: evidence from European patents within the triad. *Journal of Sustainable Development*, 10(5), 123–129. DOI:10.5539/jsd.v10n5px123

Aldieri, L., Carlucci, F., Vinci, C. P. & Yigitcanlar, T. (2019). Environmental innovation, knowledge spill overs and policy implications: a systematic review of the economic effects literature. *Journal of Cleaner Production*, 239, 118051. DOI:10.1016/j.jclepro.2019.118051

Altenburg, T., & Meyer-Stamer, J. (1999). How to promote clusters: policy experiences from Latin America. *World Development*, 27(9), 1693–1713. DOI:10.1016/S0305-750X(99)00081-9

Amin, A. & Roberts, J. (2008). Knowing in action: beyond communities of practice. *Research Policy*, 37(2), 353–369. DOI:10.1016/j.respol.2007.11.003

Asheim, B., Coenen, L., & Vang, J. (2007). Face-to-face, buzz, and knowledge bases: sociospatial implications for learning, innovation, and innovation policy. *Environment and Planning. C, Government and Policy*, 25(5), 655–670. DOI:10.1068/c0648

Bank, L., & Sibanda, F. (2018). Universities as city-builders: the city-campus development opportunity in East London-Buffalo City, South Africa. *Development Southern Africa (Sandton, South Africa)* 35(5), 701–715. DOI:10.1080/0376835X.2018.1502076

Bask, A., & Rajahonka, M. (2017). The role of environmental sustainability in the freight transport mode choice: a systematic literature review with focus on the EU. *International Journal of Physical Distribution & Logistics Management*, 47(7), 560–602. DOI:10.1108/IJPDLM-03-2017-0127

Bathelt, H., Malmberg, A., & Maskell, P. (2004). Clusters and knowledge: local buzz, global pipelines and the process of knowledge creation. *Progress in Human Geography*, 28(1), 31–56. DOI:10.1191/0309132504ph469oa

Battaglia, A., & Tremblay, D. (2011). 22@ and the Innovation District in Barcelona and Montreal: a process of clustering development between urban regeneration and economic competitiveness. *Urban Studies Research*, 2011, 1–17. DOI:10.1155/2011/568159

Baum, S., O'Connor, K., & Yigitcanlar, T. (2009). The implications of creative industries for regional outcomes. *International Journal of Foresight and Innovation Policy*, 5(1-3), 44–64. DOI:10.1504/IJFIP.2009.022098

Benneworth, P., & Ratinho, T. (2014). Reframing the role of knowledge parks and science cities in knowledge-based urban development. *Environment and Planning. C, Government and Policy*, 32(5), 784–808. DOI:10.1068/c1266r

Benneworth, P., Irawati, D., Rutten, R., & Boekema, F. (2014). The social dynamics of innovation networks: From learning region to learning in socio-spatial context. In *Social dynamics of innovation networks*. Taylor & Francis.

Boix, R., & Galleto, V. (2009). Innovation and industrial districts: a first approach to the measurement and determinants of the I-district effect. *Regional Studies*, 43(9), 1117–1133. DOI:10.1080/00343400801932342

Boja, C. (2011). Clusters models, factors and characteristics. *International Journal of Economic Practices and Theories*, 1(1), 1–10. https://ssrn.com

Capdevila, I. (2015). Co-working spaces and the localized dynamics of innovation in Barcelona. *International Journal of Innovation Management*, 19(03), 1540004. DOI:10.1142/S1363919615400046

Carrillo, F. J., Yigitcanlar, T., García, B., & Lönnqvist, A. (2014). Knowledge and the city: concepts, applications and trends of knowledge-based urban development. *Routledge.*

Castells, M., & Hall, P. (1994). *Technopoles of the world: the making of twenty-first-century industrial complexes.* Routledge.

Clark, J., Huang, H., & Walsh, J. (2010). A typology of innovation districts: what it means for regional resilience. *Cambridge Journal of Regions, Economy and Society*, 3(1), 121–137. DOI:10.1093/cjres/rsp034

Chapple, K., Jackson, S., & Martin, A. (2010). Concentrating creativity: the planning of formal and informal arts districts. *City, Culture and Society*, 1(4), 225–234. DOI:10.1016/j.ccs.2011.01.007

Cheng, Y., Liu, Y., Fan, W., Yan, Z., & Ye, X. (2019). Triple helix on globalization: a case study of the China International Nanotech Innovation Cluster. *Information Development*, 35(2), 272–289. DOI:10.1177/0266666917743050

Chesbrough, H. W. (2006). *Open innovation: the new imperative for creating and profiting from technology*. Harvard Business Press.

Coe, N. M., Hess, M., Yeung, H. W., Dicken, P., & Henderson, J. (2004). Globalising regional a global production networks perspective. *Transactions of the Institute of British Geographers (1965), New Series*, 29(4), 468–484. DOI:10.1111/j.0020-2754.2004.00142.x

Collier, D., LaPorte, J., & Seawright, J. (2012). Putting typologies to work: concept formation, measurement, and analytic rigor. *Political Research Quarterly*, 65(1), 217–232. DOI:10.1177/1065912912437162

Crevoisier, O. (2014). Beyond territorial innovation models: the pertinence of the territorial approach. *Regional Studies*, 48(3), 551–561. DOI:10.1080/00343404.2011.602629

Davis, C., Creutzberg, T., & Arthurs, D. (2009). Applying an innovation cluster framework to a creative industry: the case of screen-based media in Ontario. *Innovation (North Sydney): Management Policy and Practice*, 11(2), 201–214. DOI:10.5172/impp.11.2.201

Díez-Vial, I., & Fernández-Olmos, M. (2015). Knowledge spill overs in science and technology parks: how can firms benefit most? *Journal of Technology Transfer*, 40, 70–84. DOI:10.1007/s10961-013-9329-4

Durmaz, B., Platt, S., & Yigitcanlar, T. (2010). Creativity, culture tourism and place-making: Istanbul and London film industries. *International Journal of Culture, Tourism and Hospitality Research*, 4(3), 198–213. DOI:10.1108/17506181011067592

Engel, J., & Del-Palacio, I. (2011). Global clusters of innovation: the case of Israel and Silicon Valley. *California Management Review*, 53(2), 27–49. DOI:10.1525/cmr.2011.53.2.27

Edvardsson, I., Yigitcanlar, T., & Pancholi, S. (2016). Knowledge city research and practice under the microscope: a review of empirical findings. *Knowledge Management Research & Practice*, 14(4), 537–564. DOI:10.1057/s41275-016-0003-0

Esmaeilpoorarabi, N., Yigitcanlar, T., & Guaralda, M., (2016). Place quality and urban competitiveness symbiosis? *International Journal of Knowledge Development*, 7(1), 4–21. DOI:10.1504/IJKBD.2016.075444

Esmaeilpoorarabi, N., Yigitcanlar, T., Guaralda, M., & Kamruzzaman, M. (2018a). Does place quality matter for innovation districts? Determining the essential place characteristics from Brisbane's knowledge precincts. *Land Use Policy*, 79, 734–747. DOI:10.1016/j.landusepol.2018.09.016

Esmaeilpoorarabi, N., Yigitcanlar, T., Guaralda, M., & Kamruzzaman, M. (2018b). Evaluating place quality in innovation districts: A Delphic hierarchy process approach. *Land Use Policy*, 76, 471–486. DOI:10.1016/j.landusepol.2018.02.027

Esmaeilpoorarabi, N., Yigitcanlar, T., & Guaralda, M. (2018c). Place quality in innovation clusters: An empirical analysis of global best practices from Singapore, Helsinki, New York, and Sydney. *Cities*, 74, 156–168. DOI:10.1016/j.cities.2017.11.017

Esmaeilpoorarabi, N., Yigitcanlar, T., Kamruzzaman, M., & Guaralda, M. (2020a). How can an enhanced community engagement with innovation districts be established? Evidence from Sydney, Melbourne and Brisbane. *Cities*, 96, 102430. DOI:10.1016/j.cities.2019.102430

Esmaeilpoorarabi, N., Yigitcanlar, T., Kamruzzaman, M., & Guaralda, M., (2020b). How does the public engage with innovation districts? Societal impact assessment of Australian innovation districts. *Sustainable Cities and Society*, 52, 101813. DOI:10.1016/j.scs.2019.101813

Esmaeilpoorarabi, N., Yigitcanlar, T., Kamruzzaman, M., & Guaralda, M., (2020c). Conceptual frameworks of innovation district place quality: an opinion paper. *Land Use Policy*, 90, 104166. DOI:10.1016/j.landusepol.2019.104166

Etzkowitz, H. (2002). The triple helix of university-industry-government: implications for policy and evaluation. *Swedish Institute for Studies in Education and Research.*

Evans, G., & Hutton, T. (2009). Creative cities, creative spaces and urban policy. *Urban Studies*, 46(5/6), 1003–1040. DOI:10.1177/0042098009103853

Fan, J. S. (2015). Coming of age: innovation districts and the role of law schools. *Clinical Law Review*, 22(1), 91.

Feldman, M., & Zoller, T. (2012). Dealmakers in place: social capital connections in regional entrepreneurial economies. *Regional Studies*, 46(1), 23–37. DOI:10.1080/00343404.2011.607808

Feldman, M. (2014). The character of innovative places: entrepreneurial strategy, economic development, and prosperity. *Small Business Economics*, 43(1), 9–20. DOI:10.1007/s11187-014-9574-4

Foley, H., Freihaut, J., Hallacher, P., & Knapp, C. (2011). The Greater Philadelphia Innovation Cluster for energy-efficient buildings: a new model for public-private partnerships. *Research-Technology Management*, 54(6), 42–48. DOI:10.5437/08956308X5406014

Forsyth, A. & Crewe, K. (2010). Suburban technopoles as places: the international campus-garden suburb style. *Urban Design International* 15(3), 165–182. DOI:10.1057/udi.2010.15

Forsyth, A. (2014). Alternative forms of the high-technology district: corridors, clumps, cores, campuses, subdivisions, and sites. *Environment and Planning. C, Government and Policy*, 32, 809–823. DOI:10.1068/c1267r

Gabaldon-Estevan, D., & Ybarra, J. A. (2017). Innovative culture in district innovation systems of European ceramics SMEs. *European Planning Studies*, 25(11), 2021–2036. DOI:10.1080/09654313.2017.1353591

Hawken, S., & Hoon Han, J. (2017). Innovation districts and urban heterogeneity: 3D mapping of industry mix in downtown Sydney. *Journal of Urban Design*, 22(5), 568–590. DOI:10.1080/13574809.2017.1301203

Hobbs, K., Link, A., & Scott, J. (2017). Science and technology parks: an annotated and analytical literature review. *Journal of Technology Transfer*, 42(4), 957–976. DOI:10.1007/s10961-016-9522-3

Holroyd, C., & Coates, K. (2007). Japan Tech: the foundations of the innovation revolution. *Journal of Internet Banking & Commerce*, 12(1), 1–6.

Hsieh, H., Hu, T., Chia, P., & Liu, C. (2014). Knowledge patterns and spatial dynamics of industrial districts in knowledge cities: Hsinchu, Taiwan. *Expert Systems with Applications*, 41(12), 5587–5596. DOI:10.1016/j.eswa.2014.02.009

Huggins, R. (2008). The evolution of knowledge clusters: progress and policy. *Economic Development Quarterly*, 22(4), 277–289. DOI:10.1177/08912424 08323196

Hu, T., Lin, C., & Chang, S. (2005). Technology-based regional development strategies and the emergence of technological communities: a case study of HSIP, Taiwan. *Technovation*, 25(4), 367–380. DOI:10.1016/j.technovation.2003.09.002

Jolly, D., & Zhu, F. (2012). Chinese S&T parks: the emergence of a new model. *Journal of Business Strategy*, 33(5), 4–13. DOI:10.1108/02756661211282740

Jones, A. L. (2017). Regenerating urban waterfronts-creating better futures-from commercial and leisure market places to cultural quarters and innovation districts. *Planning Practice & Research*, 32, 333–334. DOI:10.1080/02697459.2016.1222146

Kaigorodtsev, A., & Bordiyanu, I. (2014). SWOT-analysis of an innovation cluster creation in East Kazakhstan. *Aktual'ni Problemy Ekonomiky*, 156, 284–290.

Katz, B., & Wagner, J. (2014). *The rise of innovation districts: a new geography of innovation in America*. Brookings Institution.

Khomsi, M. (2016). The smart city ecosystem as an innovation model: lessons from Montreal. *Technology Innovation Management Review*, 6(11), 26–31. DOI:10.22215/timreview1032

Kireeva, A. (2014). Theoretical and methodological framework establishing the conditions for the development of innovation clusters. *Economic and Social Changes: Facts, Trends, Forecast*, 31(1 (31)), 148–153. DOI:10.15838/esc/2014.1.31.17

Kovács, I., & Petruska, I. (2014). Operational characteristics of Hungarian innovation clusters as reflected by a qualitative research study. *Periodica Polytechnica Social and Management Sciences*, 2(2), 129–139. DOI:10.3311/PPso.7138

Lee, K. (2001). From fragmentation to integration: development process of innovation clusters in Korea. *Science Technology & Society*, 6(2), 305–327. DOI:10.1177/097172180100600202

Lee, W., & Choi, J. (2013). Industry-Academia cooperation in creative innovation clusters: a comparison of two clusters in Korea. *Academy of Entrepreneurship Journal*, 19(3), 79–96.

Leon, N. (2008). Attract and connect: The 22@ Barcelona innovation district and the internationalization of Barcelona business. *Innovation (North Sydney)*, 10, 235–246. DOI:10.5172/imp.453.10.2-3.235

Loures, L., Santos, R., & Panagopoulos, T. (2007). Urban parks and sustainable city planning: the case of Portimão, Portugal. *Population*, 3(10), 171–180.

Marcovich, A., & Shinn, T. (2011). From the triple helix to a quadruple helix? The case of Dip-Pen Nanolithography. *Minerva*, 49(2), 175–190. DOI:10.1007/s11024-011-9169-z

Markusen, A. (1996). Sticky places in slippery space: a typology of industrial districts. *Economic Geography*, 72, 293–313. DOI:10.2307/144402

Martinez-Fernandez, C. (2010). Knowledge-intensive service activities in the success of the Australian mining industry. *The Service Industries Journal*, 30(1), 55–70. DOI:10.1080/02642060802317820

Metaxiotis, K., Carrillo, F. J., & Yigitcanlar, T. (2010). *Knowledge-based development for cities and societies: integrated multi-level approaches*. Hershey, PA: Information Science Reference.

Millar, C., Choi, C., & Chu, R. (2005). The state in science, technology and innovation districts: conceptual models for China. *Technology Analysis & Strategic Management*, 17(3), 367–373. DOI:10.1080/09537320500211722

Millar, C., & Choi, C. (2010). Development and knowledge resources: a conceptual analysis. *Journal of Knowledge Management*, 14(5), 759–776. DOI:10.1108/13673271011074881

Morrison, A., & Bevilacqua, C. (2019). Balancing gentrification in the knowledge economy: the case of Chattanooga's innovation district. *Urban Research and Practice*, 12, 472–492. DOI:10.1080/17535069.2018.1472799

Montresor, S., & Marzetti, G. (2008). Innovation clusters in technological systems: a network analysis of 15 OECD countries for the mid-1990s. *Industry and Innovation*, 15(3), 321–346. DOI:10.1080/13662710802041679

O'Dwyer, M., O'Malley, L., Murphy, S., & Mcnally, R. (2015). Insights into the creation of a successful MNE innovation cluster. *Competitiveness Review*, 25(3), 288–309. DOI:10.1108/CR-08-2014-0026

Pancholi, S., Yigitcanlar, T., & Guaralda, M. (2014). Urban knowledge and innovation spaces: concepts, conditions and context. *Asia Pacific Journal of Innovation and Entrepreneurship*, 8(1), 15–38.

Pancholi, S., Yigitcanlar, T., & Guaralda, M. (2015a). Public space design of knowledge and innovation spaces: learnings from Kelvin Grove Urban Village, Brisbane. *Journal of Open Innovation: Technology, Market, and Complexity*, 1, 13. DOI:10.1186/s40852-015-0015-7

Pancholi, S., Yigitcanlar, T., & Guaralda, M. (2015b). Place making facilitators of knowledge and innovation spaces: insights from European best practices. *International Journal of Knowledge-Based Development*, 6(3), 215–240. DOI:10.1504/IJKBD.2015.072823

Pancholi, S., Yigitcanlar, T., & Guaralda, M. (2018a). Attributes of successful place-making in knowledge and innovation spaces: evidence from Brisbane's Diamantina knowledge precinct. *Journal of Urban Design*, 23, 693–711. DOI:10.1080/13574809.2018.1454259

Pancholi, S., Yigitcanlar, T., & Guaralda, M. (2018b). Societal integration that matters: place making experience of Macquarie Park Innovation District, Sydney. *City, Culture and Society*, 13, 13–21. DOI:10.1016/j.ccs.2017.09.004

Pancholi, S., Yigitcanlar, T., & Guaralda, M. (2019). Place making for innovation and knowledge-intensive activities: the Australian experience. *Technological Forecasting & Social Change*, 146, 616–625. DOI:10.1016/j.techfore.2017.09.014

Pareja-Eastaway, M., & Piqué, J. (2011). Urban regeneration and the creative knowledge economy: the case of 22@ in Barcelona. *Journal of Urban Regeneration and Renewal*, 4(4), 319–327.

Park, S. (2003). Economic spaces in the Pacific Rim: a paradigm shift and new dynamics. *Papers in Regional Science*, 82(2), 223–247. DOI:10.1007/s10110 0200144

Rabelo, R., & Bernus, P. (2015). A holistic model of building innovation ecosystems. *IFAC Papers Online*, 48(3), 2250–2257. DOI:10.1016/j.ifacol.2015.06.423

Roelandt, T. J., & Hertog, P. D. (1996). Assessing the knowledge distribution power of national innovation systems. In OECD conference on new S&T indicators for the knowledge-based economy, OECD.

Ruhlandt, R. W. (2018). The governance of smart cities: a systematic literature review. *Cities*, 81, 1–23. DOI:10.1016/j.cities.2018.02.014

Simmie, J. (2005). Critical surveys edited by Stephen Roper innovation and space: a critical review of the literature. *Regional Studies*, 39(6), 789–804. DOI:10.1080/00343400500213671

Smoliński, A., & Pichlak, M. (2009). Innovation in Polish industry: the cluster concept applied to clean coal technologies in Silesia. *Technology in Society*, 31(4), 356–364. DOI:10.1016/j.techsoc.2009.10.008

Valler, D., Phelps, N., Miao, J. T., Benneworth, P., & Eckardt, F. (2019). Science spaces as ethnoscapes: identity, perception and the production of locality. *Urban Science*, 3(1), 17. DOI:10.3390/urbansci3010017

Van Winden, W., van Den Berg, L., & Pol, P. (2007). European cities in the knowledge economy: towards a typology. *Urban Studies*, 44(3), 525–549. DOI:10.1080/00420980601131886

Van Winden, W., & Carvalho, L. (2016). Urbanize or perish? Assessing the urbanization of knowledge locations in Europe. *Journal of Urban Technology*, 23, 53–70. DOI:10.1080/10630732.2015.1090194

Varga, S., Vujisic, D., & Zdravkovic, M. (2013). State aid for innovation clusters in the Republic of Serbia. *International Journal of Public Sector Management*, 26(2), 102–110. DOI:10.1108/09513551311317960

Yang, T., & Wang, N. (2008). The cultivation of cluster's sustainable competence based on knowledge management. *International Journal of Business Management*, 3, 83–88. DOI:10.5539/ijbm.v3n7p83

Yigitcanlar, T., Velibeyoglu, K., & Martinez-Fernandez, C. (2008). Rising knowledge cities: the role of urban knowledge precincts. *Journal of Knowledge Management*, 12(5), 8–20. DOI:10.1108/13673270810902902

Yigitcanlar, T. (2014). Position paper: benchmarking the performance of global and emerging knowledge cities. *Expert Systems with Applications*, 4(12), 5549–5559. DOI:10.1016/j.eswa.2014.03.032

Yigitcanlar, T., Inkinen, T., & Makkonen, T. (2015). Does size matter? Knowledge-based development of second-order city-regions in Finland. *DISP-The Planning Review*, 51(3), 62–77. DOI:10.1080/02513625.2015.1093352

Yigitcanlar, T., Guaralda, M., Taboada, M., & Pancholi, S. (2016). Place making for knowledge generation and innovation: planning and branding Brisbane's knowledge community precincts. *Journal of Urban Technology*, 23(1), 115–146. DOI:10.1080/10630732.2015.1090198

Yigitcanlar, T., Edvardsson, I. R., Johannesson, H., Kamruzzaman, M., Ioppolo, G. & Pancholi, S. (2017). Knowledge-based development dynamics in less favoured regions. *European Planning Studies* 25(12), 2272–2292. DOI:10.1080/09654313.2017.1358699

Yigitcanlar, T., Sabatini-Marques, J., Lorenzi, C., Bernardinetti, N., Schreiner, T., Fachinelli, A., & Wittmann, T. (2018). Towards smart Florianópolis: what does it take to transform a tourist island into an innovation capital? *Energies*, 11(12), 3265. DOI:10.3390/en11123265

Yigitcanlar, T., & Inkinen, T. (2019). Geographies of disruption: place making for innovation in the age of knowledge economy. *Springer International Publishing*. DOI:10.1007/978-3-030-03207-4

Yigitcanlar, T., Kamruzzaman, M., Foth, M., Sabatini-Marques, J., Da Costa, E., & Ioppolo, G. (2019a). Can cities become smart without being sustainable? A systematic review of the literature. *Sustainable Cities and Society*, 45, 348–365. DOI:10.1016/j.scs.2018.11.033

Yigitcanlar, T., Sabatini-Marques, J., Costa, E., Kamruzzaman, M., & Ioppolo, G. (2019b). Stimulating technological innovation through incentives: perceptions of Australian and Brazilian firms. *Technological Forecasting and Social Change*, 146, 403–412. DOI:10.1016/j.techfore.2017.05.039

Yigitcanlar, T., Adu-McVie, R., & Erol, I., (2020). How can contemporary innovation districts be classified? A systematic review of the literature. *Land Use Policy*, 95(1), 104595. DOI:10.1016/j.landusepol.2020.104595

Yin, R. K. (1994). Discovering the future of the case study: method in evaluation research. *Evaluation Practice*, 15(3), 283–290. DOI:10.1177/109821409401500309

Yin, R. K. (2015). *Qualitative research from start to finish*, Second edition. Guilford Publications.

Yun, J. J., Zhao, X., Yigitcanlar, T., Lee, D., & Ahn, H. (2018). Architectural Design and Open Innovation Symbiosis: Insights from Research Campuses, Manufacturing Systems, and Innovation Districts. *Sustainability*, 10(12), 4495. DOI:10.3390/su10124495

Zheng, J. (2011). Creative industry clusters and the entrepreneurial city of Shanghai. *Urban Studies*, 48(16), 3561–3582. DOI:10.1177/0042098011399593

CHAPTER 10

Innovation District Classification

10.1 INTRODUCTION

Across the globe, many cities have been developing policies to prioritise and incentivise the clustering of knowledge and innovation activities (Yigitcanlar & Inkinen, 2019). These policy efforts have become the springboard for the formation of urban knowledge and innovation spaces. As a result of these policy efforts, a new land-use type, so-called 'innovation districts' (Esmaeilpoorarabi *et al.*, 2018a), has emerged (Metaxiotis *et al.*, 2010; Morisson & Bevilacqua, 2019). Innovation districts are defined as 'geographic zones that cluster and connect leading-edge anchor organisations (universities, R&D centres) and innovative firms with supporting and spin-off entities, business incubators, mixed-use housing, office and retail space, high-tech amenities, and high-quality public transportation, among other perks' (Katz & Wagner, 2014, p. 1).

Owing to the local contextual factors, innovation districts differ in terms of their features, functions, and spatial characteristics. This is to say, there exists a rich variety of innovation districts throughout the globe. The existence of such variety makes it harder for urban administrations to decide on the kind of innovation district to invest in (Pancholi *et al.*, 2020). This calls for a holistic classification that details the key attributes or characteristics of innovation districts.

So far, several scholars have attempted to classify them (see Table 10.1 for the list of these attempts). Nonetheless, these classifications were based on only limited features, functions, or spatial characteristics. Among these classifications, the most popular one was developed by Markusen (1996). She classified innovation districts as follows: '(a) Marshallian district; (b) Hub-and-spoke district; (c) Satellite platform

DOI: 10.1201/9781032657431-14

TABLE 10.1 Innovation District Types and Classification Categories, Derived from Yigitcanlar *et al.* (2020)

Study	Type	Classification Category	Factor
Roelandt *et al.* (1996)	a. Industrial clusters based on their specialisation patterns	Function	Hard
	b. Industrial clusters based on their innovation characteristics	Characteristics of knowledge activities	Hard
		Formation process	Hard
		Behaviour (i.e., competition and collaboration)	Soft
Markusen (1996)	a. Marshallian district	Firm configuration	Hard
	b. Hub-and-spoke district	Internal versus external orientation	Hard
	c. Satellite platform	Governance structure	Hard
	d. State-anchored districts		
Clark *et al.* (2010)	a. Type 1: Marshallian innovation districts	Patent data	Soft
	b. Type 2a: Hub-and-spoke innovation district	Regional resilience	Hard
	c. Type 2b: Satellite platform innovation district		
Forsyth (2014)	a. Corridors	Location	Hard
	b. Clumps	Physical scale of development	Hard
	c. Cores	Level of physical planning, and urban design	Hard
	d. Comprehensive campus		
	e. Technology sub-divisions		
	f. Scattered technology sites		
Katz & Wagner (2014)	a. Anchor plus	General observations	Hard
	b. Re-imagined urban areas		
	c. Urbanised science park models		
NSW-IPC (2018)	a. Health and education innovation district	Sectors	Hard
	b. Innovation precincts around universities	Locality setting	Hard
	c. Innovation precincts around a major asset		
	d. Inner-city innovation locations		

(*Continued*)

TABLE 10.1 *(Continued)* Innovation District Types and Classification Categories, Derived from Yigitcanlar *et al.* (2020)

Study	Type	Classification Category	Factor
SGS (2020)	a. Services innovation district	Sectors	Hard
	b. Multi-sector design-driven innovation district	Business activity type	Hard
	c. Science innovation district		
	d. Manufacturing innovation district		
	e. Regional resource innovation district		

district, and; (d) State-anchored district' (Yigitcanlar *et al.*, 2020, p. 2). This classification was based on 'firm configuration', 'internal or external orientation', and 'governance structure', which only partially covered feature and function attributes, and totally excluded spatial attributes.

A thorough review of the literature, by Yigitcanlar *et al.* (2020), confirmed that there is no innovation district classification framework that holistically covers features, functions, and spatial characteristics. This limitation prompts the question of: How can innovation districts be holistically classified by considering their multidimensional characteristics? This study, hence, aims to contribute to the efforts in classifying innovation districts holistically through a multidimensional framework—by developing and validating a holistic classification framework based on the key features, functions, and spatial characteristics of innovation districts.

The study first identified the potential attributes—i.e., dimensions, indicators, and measures—of a classification framework through a comprehensive review of the literature. Then, the Delphi study method was employed to determine the adequacy and accuracy of the proposed attributes of the framework. The significant output of the Delphi study, in which a total of 32 international multidisciplinary experts participated, is the multidimensional classification framework of innovation districts. As a classification tool, the developed framework will contribute to our understanding on how innovation districts can holistically be classified.

Following this introduction, the rest of the chapter is organised as follows. Section 10.2 provides a review of the literature concerning innovation districts and their classification attempts. Section 10.3 presents the methodological approach of the study including the conceptual framework developed by Yigitcanlar *et al.* (2020), its expansion into a fully fledged operational framework, and validation of the framework by an international panel of 32 Delphi experts. Section 10.4 reveals the results of the Delphi study, and shows the revised and finalised framework. Section 10.5 closes the chapter with a discussion and final remarks.

10.2 LITERATURE BACKGROUND

An innovation district is an emerging land-use type, also referred to as an urban model of innovation (Millar & Choi, 2010; Wagner *et al.*, 2019) that has become a global phenomenon for many cities in recent years, primarily due to the agglomeration benefits attached with it. The term 'innovation district' is used interchangeably with 'high-technology district' (Forsyth, 2014), 'science and technology park' (Díez-Vial & Fernández-Olmos, 2015), 'knowledge community precinct' (Esmaeilpoorarabi *et al.*, 2020b), 'innovation and cultural districts' (Jones, 2017), 'innovation precincts' (Esmaeilpoorarabi *et al.*, 2018b), 'knowledge and innovation spaces' (Pancholi *et al.*, 2019), and the likes—that are mostly inner-city and suburban mixed-function land uses (Yigitcanlar *et al.*, 2020). In a nutshell, an innovation district is the nexus of knowledge-based urban development (Yigitcanlar & Dur, 2013; Yigitcanlar & Inkinen, 2019) that promotes sustainable innovation.

Classic examples of innovation districts include the Silicon Valley in the United States and Sophia-Antipolis in France (Pancholi *et al.*, 2015; Esmaeilpoorarabi *et al.*, 2020a). The modern examples are Singapore's One-North, and Spain's 22@ Barcelona Innovation District. Whilst the former innovation districts were developed for single-purpose use within enclosed district walls based on closed innovation systems design (Yigitcanlar *et al.*, 2016), the more contemporary ones are designed and developed as boundaryless environments and mixed land uses encouraging open innovation systems with strong social networks (Van Winden & Carvalho, 2016; Jones, 2017; Wagner *et al.*, 2019; Yigitcanlar *et al.*, 2020).

The new-generation innovation districts prosper as the growth nodes for their host cities to achieve the promised agglomeration benefits that come in the form of economic, technological, sociocultural, and environmental outcomes (Yigitcanlar *et al.*, 2017; Pancholi *et al.*, 2018). They also provide a mixed-use cyber environment for knowledge workers and other users within the district (Yigitcanlar *et al.*, 2015; Pancholi *et al.*, 2019), which encourages networking and collaboration amongst the users, and hence contributes to the success of innovation activities (Kovács & Petruska, 2014; Wagner *et al.*, 2019).

There is evidence in the literature on the contributions that both sustainable innovation and knowledge-based development bring to smart places (i.e., cities, districts, neighbourhood, ecosystem). The contributions include, but are not limited to, environmental innovation (i.e., innovations focused on environmental goals and motivations such as facilitating sustainable development); each firm's productivity is positively affected by environmental knowledge (Aldieri *et al.*, 2020). Similarly, 'digitalisation of systems of innovation makes an open system of innovation result in the creation of cyber-physical systems that collaboration networks, platforms, data and analytics sustain innovation processes, capabilities and performance' (Panori *et al.*, 2021, p. 2).

Cities develop innovation districts primarily for the agglomeration benefits that come in the forms of economic, technological, sociocultural, and environmental outcomes (Yigitcanlar *et al.*, 2017; Pancholi, *et al.*, 2018). Despite their popularity, not all innovation districts are successful in delivering the expected agglomeration benefits (Yigitcanlar & Inkinen, 2019). This may be due to the lack of the state government's early interest and participation (O'Mara, 2005), low level of private-sector research and development, and lack of collaboration amongst firms (Dodgson *et al.*, 2011; Yigitcanlar & Bulu, 2016; Yigitcanlar & Inkinen, 2019). These reasons, coupled with excessive changes in the form of emergence of new key players (knowledge and creative workers), population movements, firms clustering patterns, and job creation taking place in cities over time continually challenge policymakers to provide solutions (Carrillo *et al.*, 2014). In this context, a potential solution is to identify the main characteristics of existing innovation districts and holistically classify them. Such classification will inform related authorities to decide on which type of innovation district to develop in which location (Yigitcanlar *et al.*, 2020).

Despite being the nexus of knowledge and innovation economy, the key functional and spatial characteristics of innovation districts vastly vary due to their differring local contextual factors. Consequently, we observe a rich variety of innovation districts (Forsyth, 2014; Hsieh *et al.*, 2014; Hawken & Hoon Han, 2017). This makes it difficult to holistically classify them with existing approaches. Most innovation districts have some common characteristics—in terms of general economic, spatial, social networking assets, governance, and funding support (Katz & Wagner, 2014; Wagner *et al.*, 2019); they are distinctive in possessing specific functions, features, and spatial qualities. Table 10.1 lists studies on the common types of innovation districts and their classification categories. All classification categories listed in Table 10.1 concern either hard (tangible) or soft (intangible) factors. While hard factors are related to 'place focus', soft factors cover 'people focus' (Esmaeilpoorarabi *et al.*, 2018a). Although both hard and soft factors play a fundamental role in the classification of innovation districts, previous research has dominantly focused on hard factors.

In one of the earlier studies, Roelandt *et al.* (1996) used function, characteristics of knowledge activities, formation process, behaviour in competition, and collaboration as the main classification categories and identified two innovation clusters based on either their specialisation patterns or innovation characteristics. In the same year, Markusen (1996) used the classification categories of firm configurations, internal versus external orientation, and governance to identify four types of innovation districts: (a) Marshallian, (b) hub-and-spoke, (c) satellite platform, and (d) state-anchored. After a while, patent data and regional resilience were added to Markusen's categories by Clark *et al.* (2010), which resulted in the rebranding of the innovation districts. Four years later, Forsyth (2014) employed location, level of physical planning, and urban design to classify innovation districts. The study identified six types of innovation districts, as presented in Table 10.1. Since 2014, researchers have been using sectors, business activity types, and locality as the predominant classification categories (e.g., NSW-IPC, 2018; SGS, 2020). Evidently, the classification categories have evolved over time, but, none of these studies have attempted to holistically classify these districts.

Recently, Yigitcanlar *et al.* (2020) offered a conceptual framework for classification of innovation districts, as illustrated in Figure 10.1. This framework was developed based on a comprehensive literature review on how to develop a classification framework guideline (Collier *et al.*, 2012) and the main characteristics of innovation districts identified in the literature—i.e., function, feature, space-use. and context. These four dimensions, together with their indicators, form the cornerstone of developing a classification framework for innovation districts. The specifics and expansion of this conceptual framework are presented in the next section.

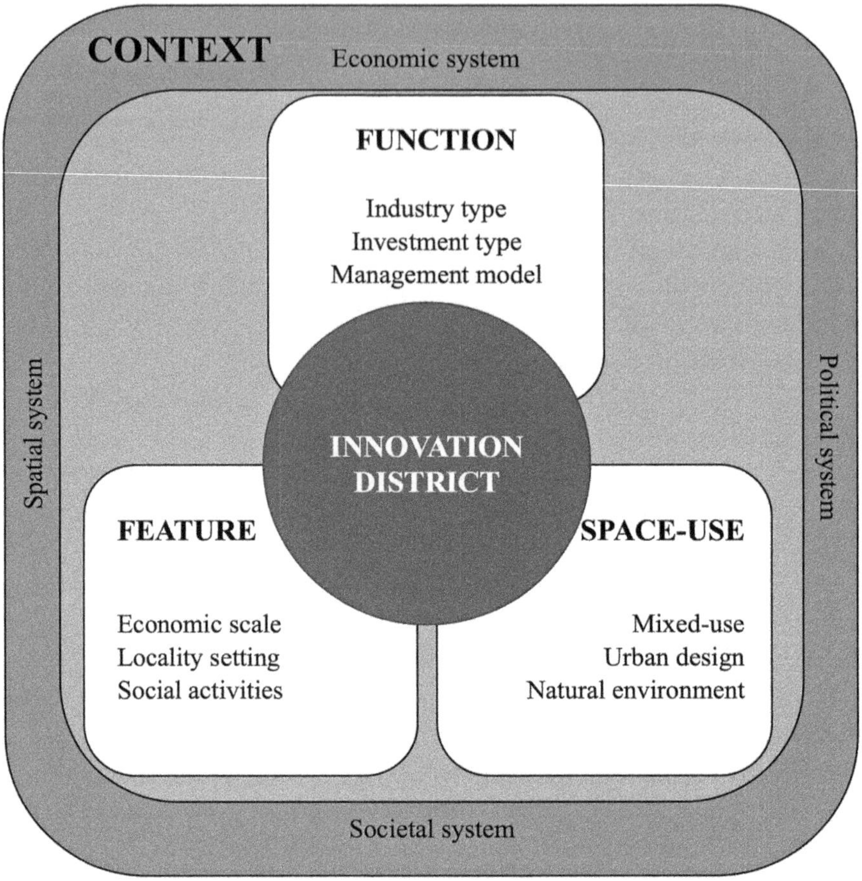

FIGURE 10.1 Conceptual framework of innovation district classification (initial) (Yigitcanlar *et al.*, 2020).

10.3 EMPIRICAL INVESTIGATION

10.3.1 Methodology

This study adopts the conceptual framework (Figure 10.1) developed by Yigitcanlar *et al.* (2020), which is based on the review of 58 scholarly articles to identify the most cited indicators relating to the classification of innovation districts. The study then expands the conceptual framework into the proposed classification attributes (presented in Table 10.2) after additional peer-reviewed articles and other relevant sources are consulted for appropriate measures and parameters to use. A variety of approaches can be used to identify initial attributes (i.e., dimensions, indicators, measures) for developing a classification framework. This

TABLE 10.2 Multidimensional Innovation District Classification Framework (Initial)

Dimension	Indicator	Description	Measure
Context	Economic system	Macroeconomic progress of the city (e.g., monetary and fiscal performance to maintain stability of economic growth)	• High-performance economic system • Mid-performance economic system • Low-performance economic system
	Political system	Political progress of the city (e.g., political institution effectiveness, accountability, transparency, participation)	• High-level governance effectiveness • Mid-level governance effectiveness • Low-level governance effectiveness
	Societal system	Societal progress of the city (e.g., equality, age structure, participation in cultural/community activities, tolerance, diversity)	• High-level societal equality • Mid-level societal equality • Low-level societal equality
	Spatial system	City-wide spatial-environmental qualities (e.g., physical environment, spatial conditions, physical urban development)	• High-quality spatial environment • Mid-quality spatial environment • Low-quality spatial environment
Function	Industry type	Dominant business activity operating within the innovation district	• High-tech business intensive • Creative business intensive • Business support service intensive
	Investment type	Principal support and funding body for the development of the district	• Public-private partnership driven • Private sector driven • Public sector driven
	Management model	Management model of the innovation district's properties and activities	• District-wide body corporate • Building-base body corporate • No body corporate

(*Continued*)

TABLE 10.2 *(Continued)* Multidimensional Innovation District Classification Framework (Initial)

Dimension	Indicator	Description	Measure
Feature	Economic scale	Skilled employment outcome of the district activities	• High-level skilled employment • Mid-level skilled employment • Low-level skilled employment
	Locality setting	Location of the district within the metropolitan area	• Urban • Suburban • Ex-urban
	Sociocultural places/activities	Public places and socio-cultural activities within the innovation district	• High-quality public/ sociocultural places • Mid-quality public/ sociocultural places • Low-quality public/ socio cultural places
Space Use	Land use	Main land use types of the innovation district	• For work-learn-play-live uses • For work-learn-play uses • For work use only
	Built environment	Urban and architectural design encouraging open innovation system within the innovation district	• High-level design qualities (e.g., open design) • Mid-level design qualities (e.g., semi open design • Low-level design qualities (e.g., close design)
	Natural environment	Aesthetic qualities of urban green and blue spaces within the district; significant natural features (e.g., >50% water coverage, >50% tree cover, good view/vista points)	• High presence of green and blue spaces • Moderate presence of green and blue space • Low presence of green and blue space

study employed the most popular approach—the literature review method, and then made recommendation to the experts for the evaluation of each attribute's suitability and adequacy (Ameyaw & Chan, 2015; Kiba-Janiak, 2016; Esmaeilpoorarabi *et al.*, 2018a).

Table 10.2 provides a detailed descriptions of the potential attributes to develop a holistic classification framework for innovation districts, where 'indicators' are the key measurable elements selected for each dimension (e.g., investment type, management model, locality setting), and the 'measures' describe each of the indicator's performance to classify innovation districts. Particularly, the top indicator for 'function' dimension is 'industry type', which identifies the dominant business activity within an innovation district. The measures defined for industry type are high-tech business intensive, creative business intensive, and business support service intensive. At this stage, the recommended attributes are identified as potentials; therefore, the experts were encouraged to suggest any additional or replacement attributes that they deemed important and need to be included or otherwise excluded in the classification framework.

Following other similar studies (Von der Gracht, 2012; Mafi *et al.*, 2015; Esmaeilpoorarabi *et al.*, 2018a; Perveen *et al.*, 2018), a Delphi study method was employed to validate the proposed classification attributes in Table 10.2. Experts with multidisciplinary backgrounds from both Australia and overseas were consulted to validate the proposed dimensions, indicators, and measures. Subsequently, the study developed a multidimensional operational classification framework, which can holistically classify the variety of innovation districts. The rationale for employing Delphi study as the validation method for the proposed classification attributes is as follows. First, the previous research confirms that there is limited empirical research on investigating and developing a holistic classification framework. Second, the Delphi study is suitable for circumstances where there is limited resources and documents (Ruppert & Duncan, 2017; Esmaeilpoorarabi *et al.*, 2018a).

10.3.2 Delphi Method

The Delphi method is widely used and accepted by researchers for obtaining experts' opinions on a topic within their domain of expertise. The method was introduced by the Rand Corporation in 1950 (Dalkey & Helmer, 1963; He *et al.*, 2016). 'The technique is designed as a group communication process, which aims to achieve a convergence of opinion on a specific real-world issue' (Hsu & Sandford, 2007, p. 1). Scholars, including Ruppert & Duncan (2017) and Rust (2017), refer to this technique as 'a reiterative systematic policymaking process, which utilises a series of anonymous questionnaires to collect expert opinions' (Esmaeilpoorarabi *et al.*, 2018a, p. 473). The Delphi method has four distinct features: (a) anonymity—a group of experts (panellists) are selected to participate on an online questionnaire about a specific research topic. The process is anonymous to avoid social pressure and potential bias in responses; (b) iteration—the process comprised of several rounds of enquiry which subsequent rounds are designed as feedback process; (c) controlled feedback—a group summary of responses for each round is presented to the experts in the next round to allow and encourage revisions of their initial judgements until consensus is achieved; and (d) statistical group response—the Delphi method produces two outcomes, namely the analytical statistics and the consensus levels (von der Gracht, 2012; Barnes & Mattsson, 2016).

Our study executed the Delphi method in the following manner.

Selection of the experts: Three main principles are followed in selecting experts for our Delphi study. First, the experts are selected from both the academic and professional sectors to ensure both theory- and practice-oriented views are gathered. Second, the experts are selected from different geographical locations, including Europe, North America, Latin America, Asiatic region, Pacific region, and the Middle East, to ensure wider coverage/validity of opinions (Ruppert & Duncan, 2017). Third, the experts are selected from a diverse, but related disciplinary areas 'to provide a heterogeneous landscape to the research' (Esmaeilpoorarabi *et al.*, 2018a, p. 475). The disciplinary areas include architecture and urban design; economics and business; communication and information technology; sustainability; geography, planning, and development (specifically innovation districts); creative industries and cultural policies; property and real estate; and public policy and administration. Furthermore, two key eligibility criteria are employed in the selection of the Delphi survey experts: (a) academics must be employed in an academic institution, and have publications on innovation district or related topic in international peer-reviewed journals in the past five years (Meijering *et al.*, 2015); and (b) professionals must be employed in either a public or private organisation, and have been actively involved in the planning, design, development, or management of an innovation district during the past five years. These eligibility criteria ensured the quality of the sample pool and reliability of expert inputs (Esmaeilpoorarabi *et al.*, 2018a).

Expert profiles: After checking the experts' profiles regarding their disciplinary areas and geographical locations, it is observed that the two most prevalent groups of experts are specialised in urban planning and real estate (41%), and architecture and urban design (22%) disciplines. The prevalent groups are actively participating in the design, planning, development, and management of innovation districts. On the lower end, social sciences, business, and communication studies equally share the remaining 38%. The lower representation is because they have a focus on limited aspects of the innovation districts (Esmaeilpoorarabi *et al.*, 2018a). In terms of geographical distribution, 31% of the experts were from Europe; 22% from the Middle East; those from Latin America and Pacific region have equal shares of 19%; 6% from the Asiatic region; and the remaining 3% from North America (3%). Hence, a heterogeneous sample of experts is assembled that represents a rich variety of views. The experts' invaluable inputs provided critical insights in the selection of the dimensions, indicators, measures, and parameters to finalise the classification framework. Furthermore, having an adequate number of experts in the study is equally important. For homogenous samples, 10–15 experts are said to be reasonably adequate; nonetheless, for heterogeneous samples 30–50 experts are required (Ameyaw & Chan, 2015; Mafi *et al.*, 2015; Nourouzian-Maleki *et al.*, 2015; Alawadi & Dooling, 2016). As the present study required a heterogeneous sample, we targeted a minimum of 30 experts to participate, and invited a total of 113 experts to ensure the minimum target of 30 is achieved.

Number of rounds: To date, there has been no consensus among scholars regarding the number of rounds required to reach a consensus in the survey. Instead, the number of rounds depends on when consensus is reached by the participants (Esmaeilpoorarabi *et al.*, 2018a). Whilst some studies recommend two rounds (Gigovic *et al.*, 2016; Soria-Lara & Banister, 2017), or three rounds (Jordan & Javernick-Will, 2013; Singhal *et al.*, 2013), others recommend more than three rounds until a consensus is reached (Ruppert & Duncan, 2017). Our study conducted the Delphi survey in two rounds as that is when the consensus was achieved.

Response rate: In the first round, 32 international experts of multi-disciplinary areas validated the proposed dimensions, indicators, measures, and their parameters. At the end of Round 1, completed questionnaires are returned to the researchers to collate and edit and then results are summarised and incorporated into the questionnaire for the next round of the survey. By doing so, each participant was informed of the general viewpoints and underlying reasons. Thus, the feedback process allowed and encouraged experts to revise their initial judgements (Esmaeilpoorarabi *et al.*, 2018a). In the second round, only 17 of the 32 experts from Round 1 participated, in which a similar process was followed in collating and editing of the completed questionnaires.

Both qualitative and quantitative analyses were carried out using the data extracted from the experts' responses. Whilst the qualitative analysis employed the experts' suggestions and comments for possible revision of the proposed frameworks, the quantitative analysis employed the use of statistical analysis to determine the central tendency and dispersion measures, to evaluate reliability of the questionnaire, internal homogeneity, and consistency of opinion among experts. In addition, the consensus level among experts was determined using the report generated from the Key Survey Tool, which is an enterprise survey platform.

10.3.3 Delphi Survey

The proposed classification attributes, presented in Table 10.2, formed the Delphi survey questionnaire. As the survey's aim was to collect both quantitative and qualitative data, Likert-scale and open-ended questions were used. 'Likert-scale questions were used to measure the suitability/adequacy of the recommended dimensions, indicators, and measures; to assess the level of consensus among experts; and generate the mean weightings of expert's scores, whilst the open-ended questions allowed the experts to provide rationales for their scores' and make suggestions for any renamed or new dimensions, indicators, measures, and parameters (Esmaeilpoorarabi *et al.*, 2018a, p. 475). Following the other relevant studies (Kiba-Janiak, 2016; Soria-Lara & Banister, 2017), our study applied a 11-point scale (from 0–no, 1–strongly disagree, to 10–strongly agree) on the following bases. Firstly, it provides respondents more selection options for rating then a limited lower scale (e.g., a 5-point scale). Secondly, it provides multi-categories for calculating consensus levels—i.e., 0–2 (strongly disagree), 3–4 (disagree), 5 (neutral), 6–7 (agree), 8–10 (strongly agree)—and makes reporting easier.

The Delphi survey process commenced Round 1 with an invitation email sent initially to 78 potential experts and then biweekly reminder emails were sent out. After two weeks, only 29% (n = 23) experts completed the survey, which was below this study's minimum sample size target of 30. A third reminder was sent to experts yet to complete, and new invitation emails were sent to an additional 35 experts with the intention to achieve the sample target of 30. This increased our potential participants from 78 to 113 experts. The survey expiry date was then extended for another week to achieve more responses; accordingly, the survey was completed with a total of 32 experts. As the number of responses (n = 32) was within the range of an acceptable sample size of

TABLE 10.3 Sample Size Used in Delphi Studies

References	Field of Study	Round 1 Sample Size	Round 2 Sample Size	Response Rate (%) between Rounds
Hayati *et al.* (2013)	Land use and transportation	9	9	100
Spickermann *et al.* (2014)	Urban planning	57	39	68
Musa *et al.* (2015)	Urban sustainability	34	31	91
Kaufmann (2016)	Land use	18	10	56
Howell *et al.* (2017)	Healthcare	30	26	87
Perveen *et al.* (2018)	Urban sustainability	29	29	100
Esmaeilpoorarabi *et al.* (2018a)	Innovation district	43	34	79

30–50 participants (Ameyaw & Chan, 2015; Mafi *et al.*, 2015; Nourouzian-Maleki *et al.*, 2015; Alawadi & Dooling, 2016), Round 1 was then closed. In Round 2, email invitations were sent to the 32 experts who completed Round 1 survey. Out of the 32 experts invited to participate in Round 2, only 17 (or 53%) completed the survey. According to the authors, this is an acceptable response rate on the premise that the response is over half (>50%) of the total invited experts and consistent with the response rates of similar studies listed in Table 10.3.

10.3.3.1 Selection of Indicators in Round 1

The questionnaire was distributed through the online Key Survey tool, which comprised of an introduction of the project, research aims and objectives, the conceptual framework (Figure 10.1), and the proposed classification framework (Table 10.2) and questions. Descriptions of the recommended indicators and their measures were also provided to avoid misunderstandings among experts that may affect their scoring in the survey. The experts were required to score the importance of 4 dimensions, 13 indicators, and 39 measures in the classification of innovation districts. For example, the experts were asked to score between the four context indicators and which indicators they thought were more important in classifying innovation districts. In total, 13 open-ended questions were included to obtain experts' opinions on the adequacy and accuracy of the recommended dimensions, indicators, and measures. After Round 1, both the conceptual and the proposed indicator frameworks were modified based on the results of the qualitative and quantitative analysis. The duration of the survey was approximately three months (September–November 2020) for both rounds.

10.3.3.2 Selection of Indicators in Round 2

In Round 2, a similar process employed in the previous round was followed to distribute the questionnaire. However, the content of the questionnaire was reformatted to include instructions for the Round 2 requirement, the revised conceptual framework (see Figure 10.2), and the classification attributes (see Table 10.4) based on Round 1 feedback. As both categories of 'dimension' and 'indicators' achieved consensus in Round 1, the researchers agreed that it was not necessary for the experts

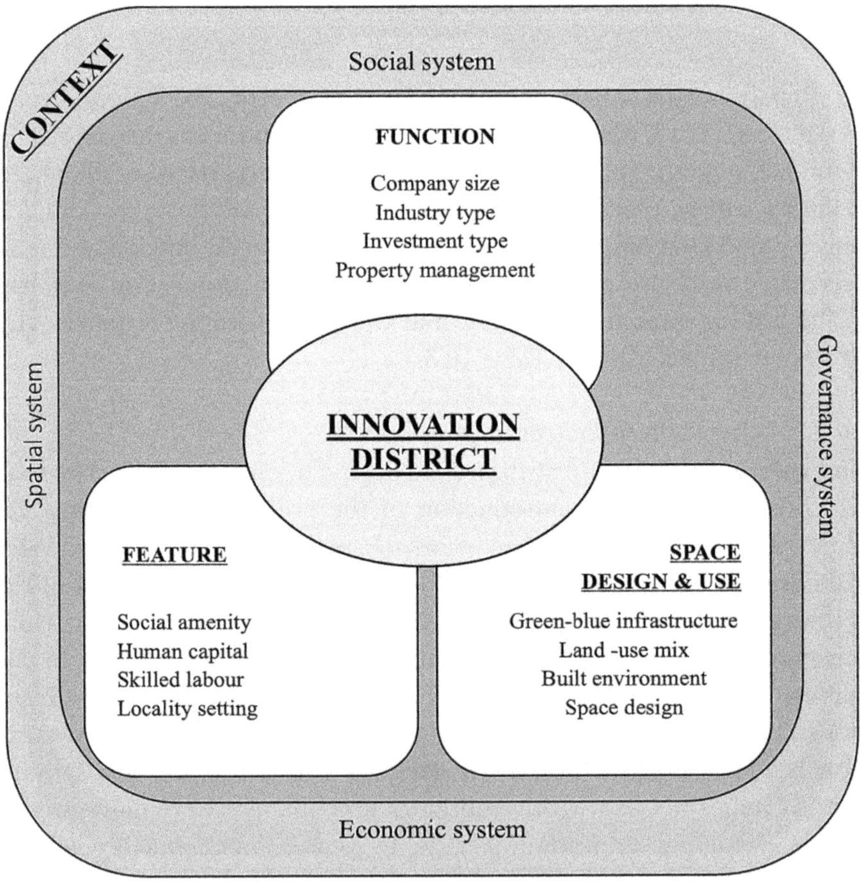

FIGURE 10.2 Conceptual framework of innovation district classification (interim).

TABLE 10.4 Multidimensional Innovation District Classification Framework (Delphi Round 1)

Dimension	Indicator	Description	Measure
Context	Economic system	Macroeconomic progress of the city (e.g., monetary, and fiscal performance to maintain stability of economic growth)	• Leading economic performance • Moderate economic performance • Low economic performance
	Political system	Political progress of the city (e.g., political institution effectiveness, accountability, transparency, participation)	• Leading governance effectiveness • Moderate governance effectiveness • Low governance effectiveness
	Societal system	Societal progress of the city (e.g., diversity, tolerance, equality, age structure, participation in cultural/community activities)	• Leading social assets • Moderate social assets • Low social assets
	Spatial system	City-wide spatial layout and architecture qualities (e.g., physical environment, spatial conditions, physical urban development)	• High-quality spatial design • Moderate-quality spatial design • Low-quality spatial design
Function	Industry type	Dominant business activity operating within the innovation district	• High-technology intensive businesses • Creativity intensive businesses • Business support services
	Investment type	Principal support and funding body for the development of the innovation districts	• Public-private partnership driven • Private sector driven • Public sector driven • Public-private-community partnership driven
	Management model	Management model of the innovation district's properties and activities	• District-wide body corporate • Building-base body corporate • No management

(*Continued*)

TABLE 10.4 *(Continued)* Multidimensional Innovation District Classification Framework (Delphi Round 1)

Dimension	Indicator	Description	Measure
	Firm size classification	Relative size of the firms within the innovation district (i.e., SME dominated, MNE anchored)	• Multinational enterprise (MNE) anchored • Small and medium enterprise (SME) dominated
Feature	Economic scale	Skilled employment outcome of the innovation district activities	• High-level skilled employment • Moderate-level skilled employment • Low-level skilled employment
	Human capital	Inventory of skilled people (i.e., information about the education and skill levels of the population and the potential stock of qualified people)	• High-level human capital • Moderate-level human capital • Low-level human capital
	Locality setting	Location of the district within the metropolitan area	• Urban setting • Suburban setting • Ex-urban setting
	Sociocultural setting	Presence or availability of social amenities for public use within the innovation district	• High presence of social amenities • Moderate presence of social amenities • Low presence of social amenities
Space Design and Use	Space design	Spatial layouts design encouraging open innovation system within the innovation district	• Open layout design • Part open layout design • Close layout design
	Land use	Main land use types within the innovation district	• Work only • Work-learn-play • Work-learn-live • Work-learn-play-live
	Built environment	Architectural design of built forms and functions encouraging open innovation systems, connectivity,	• High-level design qualities (i.e., built form, function, and connectivity) • Mid-level design qualities (i.e., built

(*Continued*)

TABLE 10.4 *(Continued)* Multidimensional Innovation District Classification Framework (Delphi Round 1)

Dimension	Indicator	Description	Measure
		and mobility within the innovation districts	form, function, and connectivity) • Low-level design qualities (i.e., built form, function, and connectivity)
	Green-blue infrastructure	Aesthetic qualities of urban green and blue infrastructure within the innovation (i.e., all natural and semi natural landscape elements that form a green-blue network)	• High-level presence of green or blue infrastructure • Mid-level presence of green or blue infrastructure • Low-level presence of green or blue infrastructure

to reassess their initial scores, except for the renamed or new dimension, indicators, and measures. The attribute that did not achieve consensus in Round 1 was 'measures'; thus, all the measures needed reassessment by the experts in Round 2. Furthermore, the experts strongly recommended replacing the subjective measures with objective measures and including parameters for measures. We incorporated these suggestions in the formulation of Round 2 questions.

A total of seven questions were formulated, three relating to rating the importance of the attributes of concern (i.e., those which fail to achieve consensus, renamed or new addition) in the classification of innovation districts. For example, the experts were asked to score on a 11-point Likert-scale (0–10) how much they agree on the name change of 'space-use' to 'form' under the dimension category. Similar question types were used for the 'indicator' category. However, the question for 'measures' category was slightly different as the experts were required to review all their initial ratings in Round 1 (including any name change or new additions) by indicating the importance of these measures on a 11-point Likert scale (0–10). Also, a summary table of consensus achieved/not achieved in Round 1 was provided for the experts' information. The other three questions are relative to the former questions, which required the experts to provide an overall rationale for their scores and the final question wass for the experts to make any general comments.

After a revision, 'firm size classification' was added as the fourth indicator to the function dimension with appropriate measures. Likewise, 'human capital' was added to the feature dimension and 'space design' and 'urban green-blue infrastructure' (renamed for natural environment) were added to space design and use dimension. The revised conceptual framework displayed in Figure 10.2 maintained the initial four dimensions, but the number of indicators increased from 13 in Round 1 to 16 in Round 2, and similarly the number of measures increased from 39 in Round 1 to 48 in Round 2. Detailed descriptions of these attributes are illustrated in Table 10.4.

10.3.4 Analysis

The Delphi method produces two outcomes; namely the analytical statistics and the consensus levels. The most common analytical statistics are the 'central tendency measure' (i.e., mean, median, mode) and the 'dispersion measures' (i.e., standard deviation, interquartile range), whilst the most common definition for consensus level is 'percentage of agreement' (Diamond *et al.*, 2014; Esmaeilpoorarabi *et al.*, 2018a). This study used mean values to calculate the level of importance of each of the 'dimensions', 'indicators', and 'measures' within their categories, which is appropriate for receiving feedback and calculating weights (Holey *et al.*, 2007; Jordan & Javernick-Will, 2013; Singhal *et al.*, 2013), and standard deviation to evaluate dispersion measures. A lower level of standard deviation (SD) and a higher mean value indicate that there is a stronger agreement among experts.

Some studies suggested that the SD of experts mean scores below 1 in a 5-point Likert-scale questionnaire is accepted as a strong agreement amongst experts (Julsrud & Priya-Uteng, 2015; Perveen *et al.*, 2018). Others suggested an SD below 2 is reasonable for a 4–5 Likert-scale questionnaire (West & Cannon, 1988; Rogers & Lopez, 2002). In addition, relating to a 10- to 11-point Likert-scale, Schmiedel *et al.* (2013) and Esmaeilpoorarabi *et al.* (2018a) suggested that the SD of experts' mean scores below 2 is reasonable. As the present study employed an interval 11-point Likert-scale (grouped into five agreement levels), it is expected that such a multi-level of agreement/disagreement will cause a high dispersion level amongst the experts' mean scores. Thus, it is reasonable for the SD to be over 2 points. We applied the rule of thumb suggested in the mathematics and statistics literature to determine the appropriate SD point threshold.

The rule of thumb for SD is 'the maximum standard deviation to minimum standard deviation should be about 2:1 ratio. If the item (in this case attributes) does not fulfil the rule, it needs to be standardised to align with the scale' (Othman *et al.*, 2011, p. 12). Our quantitative results in Rounds 1 and 2 (see Appendices 10A and 10B) revealed the maximum SD of experts' mean scores is 3.46 and minimum SD is 1.21, which does not fulfil the rule of thumb because the maximum SD of 3.46 is almost three times the minimum SD. To determine an ideal maximum SD of experts' mean scores, a simple calculation is done by multiplying the minimum SD by two (i.e., 1.21 × 2), which resulted in 2.42 points. Thus, a maximum SD of 2.42 and minimum 1.21 would meet the acceptable ratio of 2:1. Hence, we suggest the SD point threshold for experts' mean scores for this Delphi study is 2.42. Hence, SDs below 2.42 points are considered an indication of stronger agreement amongst experts.

Another important analysis that must be done prior to determining the level of agreements is the measure of Cronbach's alpha (α), which examines 'the reliability of the questionnaire, the internal homogeneity and consistency of opinion among experts' (Esmaeilpoorarabi *et al.*, 2018a, p. 476). Typically, Cronbach's α value is between 0 and 1. According to the previous research, α values above 0.7 indicate that the ratings are strongly associated and a value lower than 0.7 shows that they are unrelated (Hassanzadeh *et al.*, 2014; Mafi *et al.*, 2015). The measure of the consensus level (amongst experts) evaluates the levels of agreement based on the 11-point Likert scale. More specifically, two levels of agreement are calculated in each round.

First, the overall agreement which is the sum of the percentage of scores for 'agree' and 'strongly agree' (Ruppert & Duncan, 2017; Sutterluty *et al.*, 2017). Second, the specific agreement which is the percentage of scores for 'agree' and 'strongly agree' calculated separately. If the overall agreement for dimensions, indicators, and measures achieves more than the majority scores (>60%), then these should be retained in the classification framework. On the other hand, the specific agreement level indicates priorities in each category. If the strongly agree scores for a dimension, indicator, or measure achieve more than majority votes (>60%), then these are ranked as high importance in the related category (Kaufmann, 2016; Perveen *et al.*, 2018).

Lastly, stability tests on experts' responses between the two rounds were carried out using both Kendall's coefficient of concordance (W) and the coefficient of variation (CV) (He *et al.*, 2016; Kiba-Janiak, 2016; Esmaeilpoorarabi *et al.*, 2018a; Perveen *et al.*, 2018). Kendall's W calculates only the 'mean values and percentage of overall agreement among continuous rounds' excluding the 'level of agreements between participants'; hence, Kendall's W > 0.5 indicates 'there is stability of mean scores and consistency of expert's judgements between the survey rounds' (Esmaeilpoorarabi *et al.*, 2018a, p. 476). Any further rounds will not make a difference to the stability results already reached. Likewise, changes in the CV value between two survey rounds can be used to measure stability (Dajani et al., 1979). If the percentage change in CV value between two rounds is less than 15%, the stability of consensus is achieved and the Delphi survey is completed (Scheibe *et al.*, 2002).

10.3.4.1 Qualitative Analysis

Round 1: In the qualitative analysis, each of the 32 experts rated 4 dimensions, 13 indicators, and 39 measures, resulting in 56 responses overall. There were more than 300 comments made by the experts through open-ended questions, which mainly focused on (a) expert's rationale for the scores and recommendations for additional, new, or replacement of the dimensions, indicators, and measures; and (b) comments on which of the dimensions they think is important in classifying innovation districts. Using the 'eyeball' technique to summarise the experts' comments, the majority (74%) of the experts think the dimensions 'function, 'context', and 'feature' are equally important to 'space-design' (12%), while only 14% say all four dimensions are equally important in the classification of innovation districts. As for indicators, one expert suggested to rename 'industry type' as 'creative industry type'; another suggested 'social/societal asset' to replace 'societal equality'. However, the most noteworthy comment was on 'measures', where the majority of the experts preferred using objective measures than the subjective ones recommended. In general, the experts preferred additional information or elaboration on the definition of the measures for further clarification. After a thorough consideration of the experts' inputs, both the conceptual and classification frameworks for the innovation districts were revised. Figure 10.2 and Table 10.4 present the qualitative analysis of Round 1.

Round 2: A similar process applied in Round 1 for qualitative analysis was also followed in Round 2. From the 32 experts invited, only 17 experts responded and re-assessed 1 dimension, 5 indicators, and 43 measures, resulting in a total of 49 responses. More than 90 comments were made by the experts through open-ended questions, which focused again on their rationale for scores given and any further recommendations and comments. Some of the experts reiterated their initial suggestions to use objective instead of subjective measures. There were also additional suggestions for some renamed and new indicators, measures, and parameters that the researchers agreed to accept only the most relevant ones instead of conducting a Round 3 survey. Significant changes were made to all the 'high-mid-low measures' (subjective measures) to 'strong-moderate-weak' (objective measures). Indeed, the decision to not conduct a Round 3 survey was also supported and confirmed by Round 2 quantitative analysis results (as will be discussed below). Consequently, the classification attributes, particularly the 'measures' category, was further revised as illustrated in Table 10.5.

10.3.4.2 Quantitative Analysis

Round 1: The quantitative analysis is based on the expert's rating on an 11-point Likert scale (i.e., 0–10) in Round 1 of the importance of the recommended attributes to classify innovation districts. First, analysis was done to test the reliability of the quantitative data derived from the experts' responses. The calculated Cronbach's alpha (α) was 0.962, which is above the minimum value of 0.7 and indicates there is internal consistency and reliability of data collected. Second, analysis of mean values, SDs, and levels of agreements within each category were calculated, as discussed below. An aggregate summary of mean value, SD, and overall and specific agreements calculations are provided in Appendix 10.A. The SDs for experts' mean scores on all the 'dimensions' and 'indicators' calculated are below the threshold of 2.42 points, which suggest that there is convergence and reliability in responses for the proposed dimensions and indicators. Nevertheless, almost 54% (30/56) of the recommended measures (in italics in Appendix 10.A) had SDs above 2.42 points, indicating a weaker agreement amongst the experts.

In terms of overall agreement, all the proposed dimensions and indicators reached a consensus with an overall agreement of more than 60%, which indicated that dimensions and indicators are crucial for forming the innovation district classification framework and these were maintained to be used in the Round 2 survey. Although the literature suggested that 50% is the minimum acceptable consensus level (Zeeman *et al.*, 2016; Esmaeilpoorarabi *et al.*, 2018a), this study adopted a higher consensus level of 60%. The highest overall agreement (in bold; see Appendix 10.A) was in the 80–94% range and the lowest agreement ranged from 60% to 78%.

TABLE 10.5 Multidimensional Innovation District Classification Framework (Delphi Round 2)

Dimension	Indicators	Description	Measure
Context	Economic system	Macroeconomic progress of the city (e.g., monetary, and fiscal performance to maintain stability of economic growth)	• Strong economic performance • Moderate economic performance • Weak economic performance
	Governance system	Political progress of the city (e.g., political institution effectiveness, accountability, transparency, participation)	• Strong governance effectiveness • Moderate governance effectiveness • Weak governance effectiveness
	Societal system	Societal progress of the city (e.g., diversity, tolerance, equality, age structure, participation in cultural/community activities)	• Strong social assets • Moderate social assets • Weak social assets
	Spatial system	City-wide spatial layout and architecture qualities (e.g., physical environment, spatial conditions, physical urban development)	• Strong spatial design • Moderate spatial design • Weak spatial design
Function	Industry type	Dominant business activity operating within the innovation district	• Technology intensive businesses • Creativity intensive businesses • Business support services
	Investment type	Principal support and funding body for the development of the innovation districts	• Multi-sectors • Two sectors • Single sectors
	Property management	Management model of the innovation district's properties and activities	• District-wide body corporate • Building-base body corporate • None
	Company size	Relative size of the firms within the innovation district (i.e., SME	• Multinational enterprise (MNE) anchored

(*Continued*)

TABLE 10.5 *(Continued)* Multidimensional Innovation District Classification Framework (Delphi Round 2)

Dimension	Indicators	Description	Measure
		dominated, MNE anchored)	• Large national enterprise (LNE) dominated • Small and medium enterprise (SME) dominated
Feature	Skilled labour	Skilled employment outcome of the innovation district activities	• Strong skilled employment • Moderate skilled employment • Weak skilled employment
	Human capital	Inventory of skilled people (i.e., information about the education and skill levels of the population and the potential stock of qualified people)	• Strong human capital • Moderate human capital • Weak human capital
	Locality setting	Location of the district within the metropolitan area	• Inner city setting • Suburban setting • Regional setting
	Social amenity	Presence or availability of social amenities for public use within the innovation district	• Strong presence of social amenities • Moderate presence of social amenities • Weak presence of social amenities
Space Design and Use	Space design	Spatial layouts design encouraging open innovation system within the innovation district	• Open layout design • Semi-open layout design • Close layout design
	Land -use mix	Main land use types within the innovation district	• Complex mix • Mixed use • Single use
	Built environment	Architectural design of built forms and functions encouraging open innovation systems, connectivity, and mobility within the innovation districts	• Strong design qualities (i.e., built form, function, and connectivity) • Moderate design qualities (i.e., built

(Continued)

TABLE 10.5 *(Continued)* Multidimensional Innovation District Classification Framework (Delphi Round 2)

Dimension	Indicators	Description	Measure
			form, function, and connectivity) • Weak design qualities (i.e., built form, function, and connectivity)
	Green-blue infrastructure	Aesthetic qualities of urban green and blue infrastructure within the innovation (i.e., all natural and semi natural landscape elements that form a green-blue network)	• Strong presence of green or blue infrastructure • Moderate presence of green or blue infrastructure • Weak presence of green or blue infrastructure

In terms of specific agreement (highlighted in grey in Appendix 10.A), 43% (24/56) of the overall dimensions, indicators, and measures achieved consensus. The weak consensus among the experts could possibly be due to the missing additional information on parameters and measures as pointed out by some of the experts in the comments section. The second-round survey, therefore, aimed to improve consensus for both the overall agreement and the specific agreement for all the dimensions, indicators, and measures that are scored below the consensus level of 60% in Round 1. Generally, improvements in both agreement levels are expected to positively improve the relative SDs of experts' mean scores.

Round 2: Discussion on Round 2 quantitative analysis hereafter is focused on the changes between the two survey rounds in terms of: (a) the overall agreement and (b) the SD of experts' mean scores. In terms of percentage change in the number of 'agreements' for all attributes, the specific agreement increased by nine percentage points, from 42.86% in Round 1 to 52.17% in Round 2, while the overall agreement increased by almost 14 percentage points, from 73.21% in Round 1 to 86.96% in Round 2. Overall, the percentage for 'specific agreement' highlighted in light blue was slightly more than half (52%) of the total dimensions, indicators, and measures that is acceptable. However, despite 87% of all the categories achieved an overall agreement, it appeared that there is still inconsistency in experts' opinions, as some categories, such as 'Form', achieved an overall agreement, yet had a SD over the threshold of 2.42 points. Further analysis confirmed that such a case (as in 'form') will not negatively affect finalising of the classification framework as the calculated value of Cronbach's α for Round 2 was 0.956, indicating that there was a good overall consistency in expert opinions.

Concerning the SDs in experts' mean scores, the results revealed that the number of attributes with SDs higher than the threshold of 2.42 points reduced significantly by 27 percentage points, from 54% in Round 1 to 27% in Round 2, confirming that a considerably lower SD increased the overall level of consensus in the Round 2 survey (see Appendix 10.B for the calculations). Lastly, the tests for stability between the two survey rounds revealed Kendall's W was 0.785 for the mean and 0.919 for overall agreement. Both are above the 0.5 cutoff mark. Further, CV calculation changes between the two rounds was lower than 15%. These results confirmed stability and consistency of experts' judgement between the two survey rounds. Consequently, as these results met the above-mentioned stop criteria for Delphi studies, the researchers decided to conclude the survey at the end of Round 2.

10.4 RESULTS

This study was designed to validate identified attributes associated with comprehensive classification of innovation districts and develop a multidimensional classification framework. The Delphi study method (discussed in Section 10.3) was employed as the validation method. The respondent feedbacks were analysed using both the survey platform and the statistical package, SPSS 26.0. The final multidimensional classification framework for innovation districts is presented in Table 10.6, where the calculated mean scores reflect the experts' opinion about the level of importance of each dimension, indicators, and measures contribute toward developing a robust classification framework. The validated conceptual framework is shown in Figure 10.3.

The validated framework comprises 4 dimensions, 16 indicators, and 48 measures. Although all these attributes were considered important for the classification of innovation districts, the 'Feature' dimension has the highest importance, followed by 'Context', 'Function', and 'Form', respectively. Within the feature dimension, 'social amenity' is found to be the most important indicator, followed by both 'human capital and 'skilled labour' as the second-most important, and 'locality setting' as the

TABLE 10.6 Multidimensional Innovation District Classification Framework (Final)

Dimension	Mean Score	Indicator	Description	Mean Score	Measure	Mean Score
Feature	8.38	Social amenity	Presence or availability of social amenities for public use within the innovation district	8.81	Strong presence of social amenities	7.31
					Moderate presence of social amenities	6.56
					Weak presence of social amenities	6.00
		Human capital	Inventory of skilled people (i.e., information about the education and skill levels of the population and the potential stock of qualified people)	8.19	Strong human capital	8.06
					Moderate human capital	7.31
					Weak human capital	6.38
		Skilled labour	Skilled employment outcome of the innovation district activities	8.19	Strong skilled employment	8.00
					Moderate skilled employment	7.25
					Weak skilled employment	6.25
		Locality setting	Location of the district within the metropolitan area	8.13	Inner city setting	7.75
					Suburban setting	7.13
					Regional setting	6.13
Context	8.00	Spatial system	City-wide spatial layout and architecture qualities (e.g., physical environment, spatial conditions, physical urban development)	8.38	Strong spatial design	8.06
					Moderate spatial design	7.13
					Weak spatial design	5.63
		Societal system	Societal progress of the city (e.g., diversity, tolerance, equality, age structure, participation in cultural/community activities)	8.19	Strong social assets	7.56
					Moderate social assets	6.56
					Weak social assets	4.94

		Governance system	Political progress of the city (e.g., political institution effectiveness, accountability, transparency, participation)	8.06	Strong governance effectiveness	8.44
					Moderate governance effectiveness	7.50
					Weak governance effectiveness	5.75
		Economic system	Macroeconomic progress of the city (e.g., monetary, and fiscal performance to maintain stability of economic growth)	7.50	Strong economic performance	8.06
					Moderate economic performance	7.13
					Weak economic performance	6.06
Function	7.81	Company size	Relative size of the firms within the innovation district (i.e., SME dominated, LNE dominated or MNE anchored)	8.06	Small and medium enterprise (SME) dominated	8.19
					Large national enterprise (LNE) dominated	8.13
					Multinational enterprise (MNE) anchored	8.06
		Industry type	Dominant business activity operating within the innovation district	7.63	Creativity intensive businesses	8.69
					Technology intensive business	8.56
					Business support services	8.44
		Investment type	Principal support and funding body for the development of the innovation districts	7.31	Public-private-community partnership-driven	8.69
					Public-private partnership-driven	8.25
					Public or private sector driven	7.10
		Property management	Management model of the innovation district's properties and activities	7.13	Building-based body corporate	7.50
					District-wide body corporate	7.13
					None	7.06
Form	6.38	Green-blue infrastructure	Aesthetic qualities of urban green and blue infrastructure within the innovation district (i.e., all	8.06	Strong presence of ecosystem services	7.63
					Moderate presence of ecosystem services	6.75

(*Continued*)

TABLE 10.6 *(Continued)* Multidimensional Innovation District Classification Framework (Final)

Dimension	Mean Score	Indicator	Description	Mean Score	Measure	Mean Score
			natural and seminatural landscape elements that form a green-blue network)		Weak presence of ecosystem services	5.69
		Land-use mix	Main land use types within the innovation district	7.94	Complex mix	8.44
					Mixed use	7.88
					Single use	7.13
		Built environment	Architectural design of built forms and functions encouraging open innovation systems, connectivity, and mobility within the innovation districts	7.94	Strong internal connectivity	7.50
					Moderate internal connectivity	6.50
					Weak internal connectivity	5.75
		Space design	Spatial layouts design encouraging open innovation system within the innovation district	7.69	Open layout plan	7.56
					Semi-open layout plan	7.13
					Close layout plan	6.31

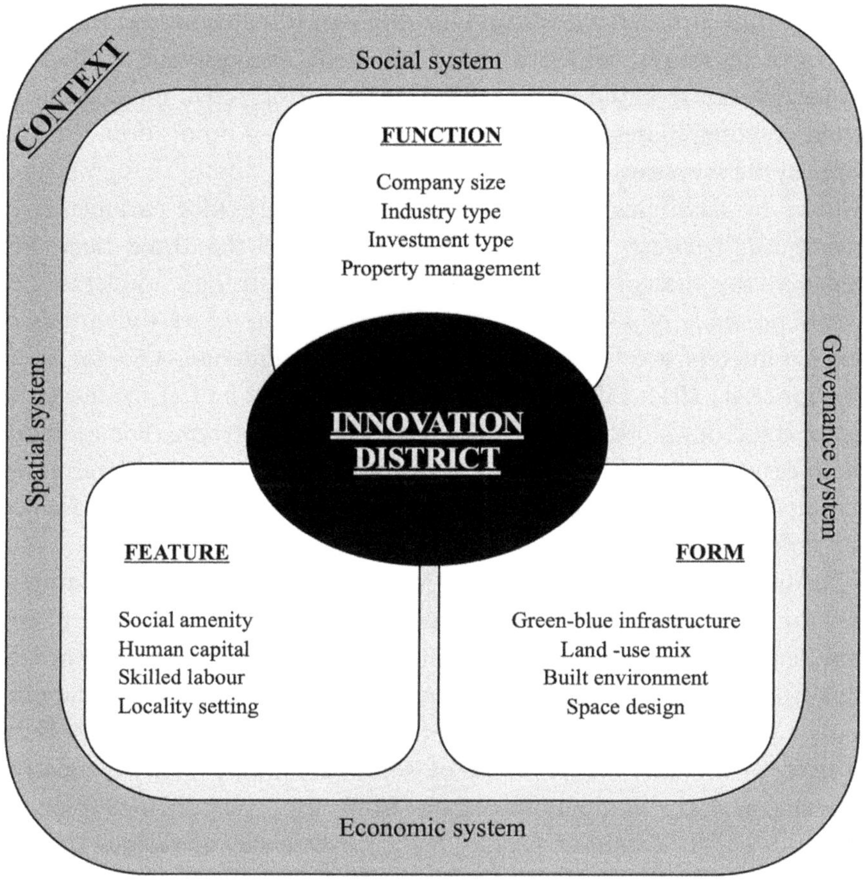

FIGURE 10.3 Conceptual framework of innovation district classification (final).

least important. In addition, the feature dimension has a balanced mixture of both hard and soft factors, such as 'locality setting' (hard factor) and 'human capital' (soft factor), while the other dimensions consist of only hard factors, as discussed below. Within the context dimension, 'spatial system' was rated of higher importance than 'societal system', 'governance system', and 'economic system'; within the function dimension, 'company size' leads the importance list, followed by 'industry type', 'investment type', and 'property management'; and within the form dimension, 'green-blue infrastructure' leads the importance list, followed by 'land-use mix', 'built environment', and 'space design'. It is noteworthy that in the final framework, the name of the fourth dimension was changed from 'space design and use' to 'form' following the experts' recommendations.

Each indicator defined within four different dimensions has three-tier objective measures, which are derived from the relevant multidisciplinary literature-based parameters. The use of objective measures is to avoid potential biases in the classification process of innovation districts. Most of the measures describe each indicator's conditions or significance relative to classifying innovation districts and provides parameters to distinguish between the thresholds for each of the three tiers. For instance, the measures for 'social amenity' are 'strong', 'moderate', or 'weak' presence of social amenities. The composite score weightings of the parameters are: >60 for strong, >50 for moderate, <50 for weak (Taylor et al., 2011; Edwards et al., 2013). The other half of the measures use specific descriptions depending on the indicator type. For example, the measures for locality setting comprise 'inner city setting', 'suburban setting', and 'regional setting' (Van Winden & Carvalho, 2016; Moonen & Clark, 2017; NSW-IPC, 2018).

Additionally, the majority of the indicators employ different parameters for their measures, except for 'social amenity'; 'societal system' and 'built environment' employ composite score weightings. For instance, to measure the 'social amenities' of innovation districts, relevant mapping tools such as Google Earth, Google My Maps, and Nearmap will be utilised to identify the presence of social amenities. And to measure 'human capital', demographic data from the Australian Bureau of Statistics (ABS), company profiles from the websites of various innovation districts, and business directories such as Dunn & Bradstreet will be accessed to identify the number of knowledge workers with a bachelor's degree or higher and the total number of workers employed within the innovation districts, respectively. The percentage of knowledge workers is calculated as the total number of knowledge workers divided by the total employment population of the innovation district.

In sum, the validated multidimensional classification framework is dominated by hard factors, including locality setting, firm size classification, industry type, green-blue infrastructure, and built environment indicator, so as to provide a more quantitative, and hence objective, classification and evaluation of innovative districts.

10.5 DISCUSSION AND CONCLUSION

Innovation districts are a new land-use type that started to appear in cities as their development has become a highly popular urban policy. There is, however, limited information available to assist urban administrations to determine what type of innovation district is the right one for them. Particularly, there is a lack of holistic frameworks that can be used for classifying innovation districts, where such classification provides opportunity for identifying the most suitable type. This study focused on developing and validating such a framework as it is invaluable for urban administrators, policymakers, and planners in understanding what works in certain locations and what does not, and informs their decisions in investing the type of innovation districts suitable for their local circumstances.

The multidimensional innovation district classification framework, the study developed, comprises four dimensions, 'Context', 'Feature', 'Function', and 'Space use and design', 16 indicators (four indicators for each dimension) and 48 measures (three measures for each indicator). The Delphi study findings confirmed that the framework is robust. 'Feature', 'Function', and 'Space use and design' are identified as primary classification dimensions, where 'Context' is seen as a secondary classification dimension as it generates city- or regional-level supporting information to be considered in decisions (Esmaeilpoorarabi *et al.*, 2018a). The context indicators are kept in the framework as contextual or background information for the policymakers to consider.

Out of 16 indicators, 2 represent soft factors—i.e., 'human capital' and 'skilled labour' indicators of the 'Feature' dimension—and the rest represent hard factors. Despite only a fraction of indicators covering the soft factors, they are listed among the top-priority indicators. This finding not only suggests that the soft factors are equally important for the classification of innovation districts, but also shows that the inclusion of soft factors is an important obligation to be successful in the knowledge and innovation economy (Florida, 2005; Yigitcanlar *et al.*, 2007; Alfken *et al.*, 2015).

The high-priority hard factors are identified as 'social amenity' (Feature), 'spatial system' (Context), 'company size' (Function), and 'green-blue infrastructure'(Form) indicators. This indicates that the hard factors continue to play a leading role in the classification of innovation districts as they traditionally have been. For example, Forsyth's (2014) classification framework has focused on hard factors of 'location', 'physical scale of development', 'level of physical planning', and 'urban design'. Likewise, hard factors are critical for the knowledge and innovation economy. For instance, the 'social amenity' indicator focuses on classifying innovation districts by determining the presence and availability of the social amenities for public use within the innovation districts. This indicator aligns with the knowledge and innovation economy's socio-cultural development perspective (Yigitcanlar & Lönnqvist, 2013; Katz & Wagner, 2014). In other words, both the soft and hard indicators have critical roles in the classification of innovation districts and all the indicators comply with the requirements of the knowledge and innovation economy.

This study assembled a framework and thus provided invaluable insights for urban administrators and planners for the planning and development of innovation districts in their cities. Particularly, we envisaged innovation districts to be classified into typologies based on the indicator's level of condition, significance, and specific descriptions. However, developing typologies is beyond the scope of the study at hand. Nonetheless, our prospective studies will focus on developing generic typologies based on the presented framework through empirical studies of innovation districts in Australia and overseas.

Nevertheless, for the sake of giving an example, for instance, Typology A may comprise innovation districts that have all their indicators at the first tier of measures with the following characteristics: strong social amenities, human capital, and green-blue infrastructure; strong skilled labour and built environments and located in the inner cities; dominated by small and medium size enterprises in the line of creativity intensive businesses; and funded by multiple sectors and managed by a building-based body corporate. This type of innovation districts is designed for complex mixed-use developments and encourages open innovation system through their open layout plans. The other typologies may compose of innovation districts that have all their indicators at the second-tier of measures or a mixture of the first, second, and third tier of measures. Again, developing these innovation district typologies will form the core of our prospective research.

Lastly, it should be noted that this study did not follow the 'rule of thumb' for Delphi studies, where at the end of the final Delphi round, those attributes still below the consensus level are to be excluded from the final framework. Instead, the most affected ones, for example, 'weak presence of social amenities' and 'weak skilled employment' with the rest of the lower- or third-tier measures retained and included in the final framework. This is because the authors envisage that not all existing innovation district indicators will be rated on the first- and second-tier measures. There may be some whose indicators will fall in the lower-tier measure. Thus, it is necessary to include lower-tier measures in the classification framework to cater for such innovation districts. On this basis, this study developed a classification framework that has a three-tier measures—e.g., strong (first tier), moderate (second tier), and weak (third tier) (Adu-McVie *et al.*, 2021).

ACKNOWLEDGEMENTS

This chapter, with permission from the copyright holder, is a reproduced version of the following journal article: Adu-McVie, R., Yigitcanlar, T., Erol, I., & Xia, B., (2021). Classifying innovation districts: Delphi validation of a multidimensional framework. Land Use Policy, 111(1), 105779.

REFERENCES

Adu-McVie, R., Yigitcanlar, T., Erol, I., & Xia, B., (2021). Classifying innovation districts: Delphi validation of a multidimensional framework. *Land Use Policy*, 111(1), 105779. DOI: 10.1016/j.landusepol.2021.105779

Alawadi, K., & Dooling, S. (2016). Challenges and opportunities for participatory planning approaches within Dubai's urban context. *Journal of Urbanism*, 9(3), 276–301. DOI:10.1080/17549175.2015.1045924

Aldieri, L., Makkonen, T., & Paolo Vinci, C. (2020). Environmental knowledge spill overs and productivity: A patent analysis for large international firms in the energy, water and land resources fields. *Resources Policy*, 69, 101877. DOI:10.1016/j.resourpol.2020.101877

Alfken, C., Broekel, T., & Sternberg, R. (2015). Factors explaining the spatial agglomeration of the creative class: empirical evidence for German artists. *European Planning Studies*, 23(12), 2438–2463. DOI:10.1080/09654313.2014.979767

Ameyaw, E.E., Chan, A.P., & Anna Walker, Dr. (2015). Evaluating key risks factors for PPP water projects in Ghana: a Delphi study. *Journal of Facilities Management*, 13(2), 133–155. DOI:10.1108/JFM-10-2013-0051

Barnes, S.J., & Mattsson, J. (2016). Understanding current and future issues in collaboration consumption: a four-stage Delphi study. *Technological Forecasting & Social Change*, 104, 200–211. DOI:10.1016/j.techfore.2016.01.006

Carrillo, F.J., Yigitcanlar, T., García, B., & Lönnqvist, A. (2014). *Knowledge and the city: concepts, applications and trends of knowledge-based urban development*. Routledge.

Clark, J., Huang, H., & Walsh, J. (2010). A typology of innovation districts: what it means for regional resilience. *Cambridge Journal of Regions, Economy and Society*, 3(1), 121–137. DOI:10.1093/cjres/rsp034

Collier, D., LaPorte, J., & Seawright, J. (2012). Putting typologies to work concept formation, measurement, and analytic rigor. *Political Research Quarterly*, 65(1), 217–232. DOI:10.1177/1065912912437162

Dajani, J. S., Sincoff, M. Z., & Talley, W. K. (1979). Stability and agreement criteria for the termination of Delphi studies. *Technological Forecasting and Social Change*, 13, 83–90. DOI: 10.1016/0040-1625(79)90007-6

Dalkey, M. & Helmer, O. (1963). An experimental application of the DELPHI method to the use of experts. *Management Science*, 9(3), 458–467. DOI:10.1287/mnsc.9.3.458

Diamond, I.R., Grant, R.C., Feldman, B.M., Pencharz, P.B., Ling, S.C., Moore, A.M., & Wales, P.W. (2014). Defining consensus: a systematic review recommends methodologic criteria for reporting of Delphi studies. *Journal of Clinical Epidemiology*, 67(4), 401, 409. DOI:10.1016/j.jclinepi.2013.12.002

Díez-Vial, I., & Fernández-Olmos, M. (2015). Knowledge spill overs in science and technology parks: how can firms benefit most? *Journal of Technology Transfer*, 40(1), 70–84. DOI:10.1007/s10961-013-9329-4

Dodgson, M., Hughes, A., Foster, J., & Metcalfe, S. (2011). Systems thinking, market failure, and the development of innovation policy: The case of Australia. *Research Policy*, 40(9), 1145–1156. DOI:10.1016/j.respol.2011.05.015

Edwards, N., Hooper, P., Trapp, G. S.A., Bull, F., Boruff, B., & Giles-Corti, B. (2013). Development of a Public Open Space Desktop Auditing Tool (POSDAT): A remote sensing approach. *Applied Geography*, 38, 22–30. DOI: 10.1016/j.apgeog.2012.11.010

Esmaeilpoorarabi, N., Yigitcanlar, T., Guaralda, M., & Kamruzzaman, M., (2018a). Evaluating place quality in innovation districts: a Delphic hierarchy process approach. *Land Use Policy*, 76, 471–486. DOI:10.1016/j.landusepol.2018.02.027

Esmaeilpoorarabi, N., Yigitcanlar, T., Guaralda, M., & Kamruzzaman, M. (2018b). Does place quality matter for innovation districts? Determining the essential place characteristics from Brisbane's knowledge precincts. *Land Use Policy*, 79, 734–747. DOI:10.1016/j.landusepol.2018.09.016

Esmaeilpoorarabi, N., Yigitcanlar, T., Kamruzzaman, M., & Guaralda, M. (2020a). How can an enhanced community engagement with innovation districts be established? Evidence from Sydney, Melbourne, and Brisbane. *Cities*, 96, 102430. DOI:10.1016/j.cities.2019.102430

Esmaeilpoorarabi, N., Yigitcanlar, T., Kamruzzaman, M., & Guaralda, M. (2020b). Conceptual frameworks of innovation district place quality: an opinion paper. *Land Use Policy*, 90, 104166. DOI:10.1016/j.landusepol.2019.104166

Florida, R. (2005). *The flight of the creative class: the new global competition for talent*. Harper Collins, London.

Forsyth, A. (2014). Alternative forms of the high-technology district: corridors, clumps, cores, campuses, subdivisions, and sites. *Environment and Planning. C, Government and Policy*, 32(5), 809–823. DOI:10.1068/c1267r

Gigovic, L., Pamucar, D., Lukic, D., & Markovic, S. (2016). GIS-fuzzy DEMATEL MCDA model for the evaluation of the sites for ecotourism development. *Land Use Policy*, 58, 348–365. DOI:10.1016/j.landusepol.2016.07.030

Hassanzadeh, A., Namdarian, L., Elahi, S.B., & Majidpour, M. (2014). Impact of technology foresight on the policy-making process in Iran. *Science,*

Technology & Society (New-Delhi, India), 19(3), 275–304. DOI:10.1177/0971721814548112

Hayati, E., Majnounian, B., Abdi, E., Sessions, J., & Makhdoum, M. (2013). An expert-based approach to forest road network planning by combining Delphi and spatial multi-criteria evaluation. *Environmental Monitoring and Assessment*, 185(2), 1767–1776. DOI:10.1007/s10661-012-2666-1

He, S.Y., Kuwo, Y.H., & Wu, D. (2016). Incorporating institutional and spatial factors in the selection of the optimal locations of public electric vehicle charging facilities: a case study of Beijing, China. *Transportation Research Part C, Emerging Technologies*, 67, 131–148. DOI:10.1016/j.trc.2016.02.003

Hawken, S., & Hoon Han, J. (2017). Innovation districts and urban heterogeneity: 3D mapping of industry mix in downtown Sydney. *Journal of Urban Design*, 22(5), 568–590. DOI:10.1080/13574809.2017.1301203

Holey, E.A., Feeley, J.L., Dixon, J., & Whittaker, V.J. (2007). An exploration of the use of simple statistics to measure consensus and stability in Delphi studies. *BMC Medical Research Methodology*, 7(1), 52- 52. DOI:10.1186/1471-2288-7-52

Howell, A.-M., Burns, E. M., Hull, L., Mayer, E., Sevdalis, N., & Darzi, A. (2017). International recommendations for national patient safety incident reporting systems: an expert Delphi consensus-building process. *BMJ Quality & Safety*, 26(2), 150–163. DOI:10.1136/bmjqs-2015-004456

Hsieh, H., Hu, T., Chia, P., & Liu, C. (2014). Knowledge patterns and spatial dynamics of industrial districts in knowledge cities: Hsinchu, Taiwan. *Expert Systems with Applications*, 41(12), 5587–5596. DOI:10.1016/j.eswa.2014.02.009

Hsu, C.C., & Sandford, B.A. (2007). The Delhi technique: making sense of consensus. *Practical Assessment Research & Evaluation*, 12(1), 1–8. DOI:10.7275/pdz9-th90

Jones, A.L. (2017). Regenerating urban waterfronts-creating better futures-from commercial and leisure marketplace to cultural quarters and innovation districts. *Planning Practice & Research*, 32(3), 333–334. DOI:10.1080/02697459.2016.1222146

Jordan, E., & Javernick-Will, A. (2013). Indicators of community recovery: content analysis and Delphi approach. *Natural Hazards Review*, 14(1), 21–28. DOI:10.1061/(ASCE)NH.1527-6996.0000087

Julsrud, T.E., & Priya-Uteng, T. (2015). Technopolis, shared resources or controlled mobility? A net-based Delphi study to explore visions of future urban daily mobility in Norway. *European Journal of Future Research*, 3(1), 1–13. DOI:10.1007/s40309-015-0069-6

Katz, B., & Wagner, J. (2014). *The rise of innovation districts: a new geography of innovation in America*. Brookings Institute.

Kaufmann, R.P. (2016). Integrating factor analysis and the Delphi method in scenario development: a case study of Dalmatia, Croatia. *Applied Geography*, 71, 56–58. DOI:10.1016/j.apgeog.2016.04.007

Kiba-Janiak, M. (2016). Key success factors for city logistics from the perspective of various group of stakeholders. *Transportation Research Procedia*, 12, 557–569. DOI:10.1016/j.trpro.2016.02.011

Kovács, I., & Petruska, I. (2014). Operational characteristics of Hungarian innovation clusters as reflected by a qualitative research study. *Periodica Polytechnica Social and Management Sciences*, 22(2), 129–139. DOI:10.3311/PPso.7138

Mafi, D., Feghhi, J., Danehkar, A., Yarali, N. (2015). Classification and prioritization of negative factors affecting on mangrove forests using Delphi method. *Advances in Bioresearch*, 6(3), 78–92. DOI: 10.15515/abr.0976-4585.6.3.7892

Markusen, A. (1996). Sticky places in slippery space: a typology of industrial districts. *Economic Geography*, 72(3), 293–313. DOI: 10.2307/144402

Meijering, J.V., Tobi, H., Van Dan Brink, A., Morris, F., & Bruns, D. (2015). Exploring research priorities in landscape architecture. *Landscape and Urban Planning*, 137, 85–94. DOI:10.1016/j.landurbplan.2015.01.002

Metaxiotis, K., Carrillo, F. J., & Yigitcanlar, T. (2010). *Knowledge-based development for cities and societies: integrated multi-level approaches*. Hershey, PA: Information Science Reference.

Millar, C., & Choi, C. (2010). Development and knowledge resources: a conceptual analysis. *Journal of Knowledge Management*, 14(5), 759–776. DOI:10.1108/13673271011074881

Moonen, T. & Clark, G. (2017). The logic of innovation locations: understanding the drivers that enables cities to host innovation economies. A Report by The Business of Cities and Future Cities Catapult.

Morisson, A., & Bevilacqua, C. (2019). Balancing gentrification in the knowledge economy: the case of Chattanooga's innovation district. *Urban Research & Practice*, 12(4), 472–492. DOI:10.1080/17535069.2018.1472799

Musa, H. D., Yacob, M. R., Abdullah, A. M., & Ishak, M. Y. (2015). Delphi Method of Developing Environmental Well-being Indicators for the Evaluation of Urban Sustainability in Malaysia. *Procedia Environmental Sciences*, 30, 244–249. DOI:10.1016/j.proenv.2015.10.044

NSW-IPC (2018). *NSW innovation precincts: lessons from international experience*. NSW Government, Sydney.

Nourouzian-Maleki, S., Bell, S., Hosseini, S.B., & Faizi, M. (2015). Developing and testing the framework for the assessment of neighbourhood liveability in two contrasting countries: Iran and Estonia. *Ecological Indicators*, 48, 263–271. DOI:10.1016/j.ecolind.2014.07.033

O'Mara, M. (2005). *Cities of knowledge: Cold war science and the search for the next Silicon Valley*. Princeton, NJ: Princeton University Press.

Othman, A., Yin, T., Sulaiman, S., Mohamed, M., & Razha-Rashid, M. (2011). Application of mean and standard deviation in questionnaire surveys. *Menemui Matematik (Discovering Mathematics)*, 33(1),11–22. https://myjms.mohe.gov.my/index.php/dismath/article/view/13707

Pancholi, S., Yigitcanlar, T., & Guaralda, M. (2015). Public space design of knowledge and innovation spaces: learnings from Kelvin Grove Urban Village, Brisbane. *Journal of Open Innovation*, 1, 1–17. DOI:10.1186/s40852-015-0015-7

Pancholi, S., Yigitcanlar, T., & Guaralda, M. (2018). Attributes of successful place-making in knowledge and innovation spaces: evidence from Brisbane's Diamantina knowledge precinct. *Journal of Urban Design*, 23(5), 693–711. DOI:10.1080/13574809.2018.1454259

Pancholi, S., Yigitcanlar, T., & Guaralda, M. (2019). Place making for innovation and knowledge-intensive activities: the Australian experience. *Technological Forecasting & Social Change*, 146, 616–625. DOI:10.1016/j.techfore.2017.09.014

Pancholi, S., Yigitcanlar, T., Guaralda, M., Mayere, S., Caldwell, G.A., & Medland, R. (2020). University and innovation district symbiosis in the context of placemaking: Insights from Australian cities. *Land Use Policy*, 99, 105109. DOI:10.1016/j.landusepol.2020.105109

Panori, A., Kakderi, C., Komninos, N., Fellnhofer, K., Reid, A., & Mora, L. (2021). Smart systems of innovation for smart places: Challenges in deploying digital platforms for co-creation and data-intelligence. *Land Use Policy*, 104631. DOI:10.1016/j.landusepol.2020.104631

Perveen, S., Kamruzzaman, M., Yigitcanlar, T., (2018). Developing policy scenarios for sustainable urban growth management: a Delphi approach. *Sustainability (Basel, Switzerland)*, 9(10),1787. DOI:10.3390/su9101787

Roelandt, T.J., & Hertog, P.D. (1996). Assessing the knowledge distribution power of national innovation systems. In OECD conference on new S&T indicators for the knowledge-based economy, OECD.

Rogers, M.R., & Lopez, E.C. (2002). Identifying critical cross-cultural school psychology competencies. *Journal of School Psychology*, 40(2), 115–141. DOI:10.1016/S0022-4405(02)00093-6

Ruppert, J., & Duncan, R.G. (2017). Defining and characterising ecosystem services for education: a Delphi study. *Journal of Research in Science Teaching*, 54(6), 737–763. DOI:10.1002/tea.21384

Rust, N.A. (2017). Can stakeholders agree on how to reduce human-carnivore conflict on Namibian livestock farms? *Oryx*, 51(2), 339–346. DOI:10.1017/S0030605315001179

Scheibe, M., Skutsch, M., & Schofer, J. (2002). IV.C. Experiments in Delphi methodology. In *The Delphi method: techniques and applications*, 257–281.

Schmiedel, T., Vom Brocke, J., & Recker, J. (2013). Which cultural values matter to business process management? Results from a global Delphi study. *Business Process Management Journal*, 19 (2), 292–317. DOI:10.1108/14637151311308321

SGS (2020). *From alleys to valleys: creating innovation precincts through inclusive policy*. SGS Economics & Planning, Sydney.

Singhal, S., Mcgreal, S., & Berry, J. (2013). An evaluative model for city competitiveness: application to UK cities. *Land Use Policy*, 30(1), 214–222. DOI:10.1016/j.landusepol.2012.03.018

Soria-Lara, J.A., Julio, A., & Banister, D. (2017). Participatory visioning in transport back casting studies: Methodological lessons from Andalusia (Spain). *Journal of Transport Geography*, 58, 113–126. DOI:10.1016/j.jtrangeo.2016.11.012

Spickermann, A., Grienitz, V., & Von Der Gracht, H. A. (2014). Heading towards a multimodal city of the future: Multi-stakeholder scenarios for urban mobility. *Technological Forecasting & Social Change*, 89, 201–221. DOI:10.1016/j.techfore.2013.08.036

Sutterluty, A., Hesser, F., Schwarzbauer, P., Schuster, K.C., Windsperger, A., & Stern, T. (2017). A Delphi approach to understanding varying expert viewpoints in sustainability communication: the case of water footprints of bio-based fibre resources. *Journal of Industrial Ecology*, 21, 412–422. DOI: 10.1111/jiec.12427

Taylor, B. T., Fernando, P., Bauman, A. E., Williamson, A., Craig, J. C., & Redman, S. (2011). Measuring the Quality of Public Open Space Using Google Earth. *American Journal of Preventive Medicine*, 40, 105–112. DOI: 10.1016/j.amepre.2010.10.024

Van Winden, W., & Carvalho, L. (2016). Urbanize or perish? Assessing the urbanization of knowledge locations in Europe. *Journal of Urban Technology*, 23(1), 53–70. DOI:10.1080/10630732.2015.1090194

Von der Gracht, H.A. (2012). Consensus measurement in Delphi studies. *Technological Forecasting and Social Change*, 79(8), 1525–1536. DOI:10.1016/j.techfore.2012.04.013

Wagner, J., Katz, B., Osha, T. (2019). *The evolution of innovation districts: the new geography of global innovation*. The Global Institute of Innovation Districts. New York.

West, J.F., & Cannon, G.S. (1988). Essential collaborative consultation competencies for regular and special educators. *Journal of Learning Disabilities*, 21(1), 56–63. DOI:10.1177/002221948802100111

Yigitcanlar, T., Baum, S., & Horton, S. (2007). Attracting and retaining knowledge workers in knowledge cities. *Journal of Knowledge Management*, 11(5), 6–17. DOI:10.1108/13673270710819762

Yigitcanlar, T., & Dur, F. (2013). Making space and place for knowledge communities: lessons for Australian practice. *Australasian Journal of Regional Studies*, 19(1), 36–63.

Yigitcanlar, T & Lönnqvist, A. (2013). Benchmarking knowledge-based urban development performance: results from the international comparison of Helsinki. *Cities*, 31, 357–369. DOI:10.1016/j.cities.2012.11.005

Yigitcanlar, T., Inkinen, T., & Makkonen, T. (2015). Does size matter? Knowledge-based development of second-order city-regions in Finland. *DISP-The Planning Review*, 51(3), 62–77. DOI:10.1080/02513625.2015.1093352

Yigitcanlar, T., & Bulu, M. (2016). Urban knowledge and innovation spaces. *Journal of Urban Technology*, 23(1), 1–9. DOI:10.1080/10630732.2016.1164443

Yigitcanlar, T., Guaralda, M., Taboada, M., & Pancholi, S. (2016). Place making for knowledge generation and innovation: planning and branding

Brisbane's knowledge community precincts. *Journal of Urban Technology*, 23(1), 115–146. DOI:10.1080/10630732.2015.1090198

Yigitcanlar, T., Edvardsson, I. R., Johannesson, H., Kamruzzaman, M., Ioppolo, G. & Pancholi, S. (2017). Knowledge-based development dynamics in less favoured regions. *European Planning Studies*. 25(12), 2272–2292. DOI:10.1080/09654313.2017.1358699

Yigitcanlar, T., & Inkinen, T. (2019). *Geographies of disruption: place making for innovation in the age of knowledge economy*. Springer International Publishing. DOI:10.1007/978-3-030-03207-4

Yigitcanlar, T., Adu-McVie, R., & Erol, I. (2020). How can contemporary innovation districts be classified? A systematic review of the literature. *Journal of Land Use Policy*, 95, 104595. DOI:10.1016/j.landusepol.2020.104595

Zeeman, H., Wright, C.J., & Hellyer, T. (2016). Developing design guidelines for inclusive housing: a multi- stakeholder approach using a Delphi method. *Journal of Housing and Built Environment*, 31(4), 761–772. DOI:10.1007/s10901-016-9499-0

APPENDIX 10.A ROUND 1 DELPHI SURVEY RESULTS

Dimension	M	SD	A	SA	OA	Indicator	M	SD	A	SA	OA	Measure	M	SD	A	SA	OA
Context	8.28	*1.89*	12.51	78.13	**90.64**	Economic system	8.17	*2.00*	9.38	78.14	**87.52**	High-performance economic system	6.66	*3.12*	9.38	56.26	**65.64**
												Mid-performance economic system	5.90	*2.77*	9.68	38.71	**48.39**
												Low-performance economic system	5.17	*3.28*	9.68	32.26	**41.94**
						Political system	8.28	*1.91*	9.38	78.13	**87.51**	High-level governance effectiveness	7.00	*3.26*	12.51	62.51	**75.02**
												Mid-level governance effectiveness	6.38	*2.98*	9.68	51.61	**61.29**
												Low-level governance effectiveness	5.55	*3.46*	6.45	41.94	**48.39**
						Social system	8.10	*1.90*	12.50	75.01	**87.51**	High-level societal equality	6.69	*3.32*	6.26	62.50	**68.76**
												Mid-level societal equality	6.00	*3.13*	13.33	43.33	**56.66**
												Low-level societal equality	5.17	*3.32*	16.13	32.25	**48.38**

						Spatial system	8.00	*1.60*	18.76	71.88	**90.64**	High-quality spatial environment	7.00	*3.04*	15.63	59.38	**75.01**
												Mid-quality spatial environment	6.52	*2.87*	19.35	48.38	**67.73**
												Low-quality spatial environment	5.72	*3.23*	19.36	35.48	**54.84**
Function	8.17	*1.34*	21.88	71.88	**93.76**	Industry type	7.90	*1.54*	25.01	65.63	**90.64**	High-tech business intensive	8.28	*1.79*	15.63	71.89	**87.52**
												Creative business intensive	8.45	*1.84*	9.38	78.13	**87.51**
												Business support service intensive	8.00	*1.81*	15.63	68.76	**84.39**
						Investment type	7.62	*1.57*	37.50	53.13	**90.63**	Public-private partnership -driven	8.38	*1.88*	9.38	75.01	**84.39**
												Private-sector -driven	7.69	*1.83*	9.38	68.76	**78.14**
												Public sector-driven	7.41	*1.96*	18.76	59.38	**78.14**
						Management type	7.55	*1.55*	37.50	53.13	**90.63**	District-wide body corporate	7.55	*2.57*	6.26	71.88	**78.14**
												Building -base body corporate	6.90	*2.19*	18.76	50.01	**68.77**
												No body corporate	6.69	*2.58*	15.63	50.00	**65.63**
Feature	8.10	*1.82*	25	65.63	**90.63**	Economic scale	8.31	*1.63*	12.51	78.13	**90.64**	High-level skilled employment	7.76	*2.86*	9.38	68.76	**78.14**
												Mid-level skilled employment	7.03	*2.68*	12.51	56.26	**68.77**

(Continued)

Dimension	M	SD	A	SA	OA	Indicator	M	SD	A	SA	OA	Measure	M	SD	A	SA	OA
												Low-level skilled employment	6.07	*3.21*	9.38	43.76	**53.14**
						Locality setting	8.31	*1.37*	12.50	78.13	**90.63**	Urban (i.e., inner-city)	8.03	*2.38*	12.51	75.00	**87.51**
												Suburban (i.e., suburban area)	7.28	*2.36*	25.00	56.25	**81.25**
												Ex-urban (i.e., outside of suburban area	6.38	*3.03*	12.51	46.88	**59.39**
						Sociocult-ural setting	8.41	*1.66*	9.38	81.26	**90.64**	High- quality public social/ cultural places	7.55	*2.57*	9.38	62.50	**71.88**
												Mid-quality public social/ cultural places	6.59	*2.82*	9.38	50.01	**59.39**
												Low-quality public social/ cultural places	5.59	*3.43*	6.25	40.64	**46.89**
Space- use	7.76	*1.86*	18.76	65.63	84.39	Land use form	7.17	*2.14*	15.63	59.39	**75.02**	For work-learn-play-live uses	7.41	*2.82*	9.38	59.38	**68.76**
												For-work-learn-play uses	7.03	*2.69*	3.13	59.38	**62.51**
												For work use only	5.97	*2.98*	9.38	37.51	**46.89**
						Built environ-ment	7.59	*1.90*	21.88	62.51	**84.39**	High-level design qualities (e.g., open design)	6.90	*3.19*	6.26	56.26	**62.52**
												Mid-level design qualities (e.g., semi-open design)	6.00	*2.73*	15.63	40.63	**56.26**

						Low-level design qualities (e.g., close design)	4.97	*3.01*	9.38	25.01	**34.39**
Natural environment	7.48	*2.01*	31.26	50.01	**81.27**	High-level presence of green/blue spaces	6.69	*3.22*	6.25	56.26	**62.51**
						Mid-level presence of green/blue spaces	6.24	*3.17*	18.76	40.63	**59.39**
						Low-level presence of green/blue spaces	5.10	*3.45*	12.51	31.26	**43.77**

SD = Standard deviation, A = Agree (6&7), SA = Strongly agree (8,9 &10), OA = Overall agreement (sum of A + SA). Bold = Overall agreement >60%, grey shade = Specific agreement >60% (separate calculations of A and SA), Italics = SD >2.42 (high dispersion level-lesser agreement).

Reliability Statistics

C Cronbach's Alpha	Cronbach's Alpha Based on Standardised Items	N of Items
0.962	0.959	56

Case Processing Summary

		N	%
Cases	Valid	29	90.6
	Excluded[a]	3	9.4
	Total	32	100.0

a. Listwise deletion based on all variables in the procedure.

Consensus level cut off mark = 60%

Agreement level summary	N=56	%
Overall agreement	41	**73.21**
Specific agreement	24	46.86
Lesser agreement	30	*53.57*

APPENDIX 10.B ROUND 2 DELPHI SURVEY RESULTS

Dimension	M	SD	A	SA	OA	Indicator	M	SD	A	SA	OA	Measure	M	SD	A	SA	OA
Context	8.0	*2.03*	12.51	78.13	**90.64**	Economic system	7.50	*2.34*	9.38	78.14	**87.52**	Leading economic performance	8.06	*2.14*	23.54	64.70	**88.24**
												Moderate economic performance	7.13	*2.09*	47.06	35.30	**82.36**
												Low economic performance	6.06	*2.91*	35.30	25.53	**58.83**
						Political system	8.06	*2.24*	9.38	78.13	**87.51**	Leading governance effectiveness	8.44	*1.67*	29.41	64.71	**94.12**
												Moderate governance effectiveness	7.50	*1.59*	47.06	41.19	**88.25**
												Low governance effectiveness	5.75	2.82	29.42	23.53	**52.95**
						Social system	8.19	*2.10*	12.50	75.01	**87.51**	Leading social assets	7.56	3.05	17.65	58.83	**76.48**
												Moderate social assets	6.56	2.73	35.30	35.31	**70.61**
												Low social assets	4.94	3.23	17.65	23.53	**41.18**
						Spatial system	8.38	*1.59*	18.76	71.88	**90.64**	High quality spatial design	8.06	2.26	17.65	64.71	**82.36**
												Moderate quality spatial design	7.13	2.22	35.30	41.18	**76.48**
												Low quality spatial design	5.63	3.36	11.76	35.30	**47.06**

Function	7.81	*1.42*	21.88	71.88	**93.76**	Industry type	7.63	*1.71*	25.01	65.63	**90.64**	High-technology intensive business	8.56	*1.90*	11.77	76.48	**88.25**
												Creative intensive business	8.69	*1.58*	11.77	82.35	**94.12**
												Business support services	8.44	*1.21*	29.41	70.59	**100**
						Investment type	7.31	*1.82*	37.50	53.13	**90.63**	Public-private partnership -driven	8.25	*2.14*	9.38	75.01	**84.39**
												Private sector -driven	7.31	*2.06*	9.38	68.76	**78.14**
												Public sector-driven	6.88	*2.28*	18.76	59.38	**78.14**
												Public-private-community partnership-driven	8.69	*1.25*	23.53	76.47	**100**
						Management type	7.13	*1.67*	37.50	53.13	**90.63**	District-wide body corporate	7.13	*2.47*	6.26	71.88	**78.14**
												Building -base body corporate	7.50	*2.10*	23.53	52.95	**76.48**
												No management	7.06	*2.57*	23.54	52.94	**76.48**
						Firm size classification	8.06	*1.81*	23.53	70.59	**94.12**	Multinational enterprise (MNE) anchored	8.06	*1.45*	23.53	70.59	**94.12**
												Small and medium enterprise (SME) dominated	8.19	*1.56*	23.53	70.59	**94.12**
Feature	8.38	*1.41*	25	65.63	**90.63**	Economic scale	8.19	*1.22*	12.51	78.13	**90.64**	High-level skilled employment	8.00	*2.76*	9.38	68.76	**78.14**

(Continued)

Dimension	M	SD	A	SA	OA	Indicator	M	SD	A	SA	OA	Measure	M	SD	A	SA	OA
												Moderate-level skilled employment	7.25	*1.57*	35.30	41.18	**76.48**
												Low-level skilled employment	6.25	*3.21*	9.38	43.76	**53.14**
						Human capital	8.19	*1.22*	23.53	76.47	**100**	High-level human capital	8.06	*2.44*	23.54	64.71	**88.25**
												Moderate- level human capital	7.31	*2.13*	35.30	52.94	**88.24**
												Low-level human capital	6.38	*2.28*	29.42	35.29	**64.71**
						Locality setting	8.13	*1.67*	12.50	78.13	**90.63**	Urban setting	7.75	*1.88*	29.41	52.94	**82.35**
												Suburban setting	7.13	*1.45*	35.30	41.19	**76.49**
												Ex-urban setting	6.13	*2.45*	29.42	29.41	**58.83**
						Sociocult-ural setting	8.81	*1.28*	9.38	81.26	**90.64**	High presence of social amenities	7.31	*2.24*	23.54	52.94	**76.48**
												Moderate presence of social amenities	6.56	*2.09*	41.18	29.42	**70.60**
												Low presence of social amenities	6.00	*2.63*	29.42	29.41	**58.83**
Form	6.38	*3.22*	25.53	47.07	**70.60**	Space design	7.69	*2.68*	29.41	58.83	**88.24**	Open layout design	7.56	*2.16*	23.53	58.83	**82.36**
												Part open layout design	7.13	*2.22*	11.77	58.82	**70.59**
												Close layout design	6.31	*2.85*	5.88	47.06	**52.94**
						Land use	7.94	*2.08*	23.53	64.70	**88.23**	Work only	7.13	*1.86*	47.06	35.31	**82.37**
												Work-learn-play	7.88	*1.93*	17.65	70.59	**88.24**
												Work-learn-live	7.88	*2.09*	17.65	64.71	**82.36**
												Work-learn-play-live	8.44	*2.19*	11.77	76.47	**88.24**

Built environment	7.94	*1.77*	21.88	62.51	**84.39**	High-level design qualities (i.e., built form, function, and connectivity)	7.50	*2.31*	17.65	64.71	**82.36**
						Mid-level design qualities (i.e., built form, function, and connectivity)	6.50	*2.07*	35.30	35.29	**70.59**
						Low-level design qualities (i.e., built form, function, and connectivity)	5.75	*2.82*	41.17	29.42	**70.59**
Urban green-blue-infrastructure	8.06	*1.95*	23.54	70.59	**94.13**	High-level presence of green or blue spaces	7.63	*2.63*	17.65	64.71	**82.36**
						Mid-level presence of green or blue spaces	6.75	*2.41*	35.30	41.18	**76.48**
						Low-level presence of green or blue spaces	5.69	*3.09*	17.65	35.30	**52.95**

Notes: SD = Standard deviation, A = Agree (6&7), SA = Strongly agree (8,9 &10), OA = Overall agreement (sum of A + SA). Bold = Overall agreement >60% grey shade = Specific agreement >60% (separate calculations of A and SA), Italics =SD>2.42 (high dispersion level-lessor agreement).

Reliability Statistics		
C Cronbach's Alpha Items	Cronbach's Alpha Based on Standardised Items	N of
0.956	0.963	69

Case Processing Summary			
		N	%
Cases	Valid	16	94.1
	Excluded[a]	1	5.9
	Total	17	100.0

a. Listwise deletion based on all variables in the procedure.

Consensus level cut off mark = 60		
Agreement level summary	**N=69%**	**%**
Overall agreement	60	**88.96**
Specific agreement	36	52.17
Lesser agreement	19	*27.54*

Test Statistics - ΣMean (R1-R2)	
N	2
Kendall's W[a]	**0.785**
Chi-Square	106.818
df	68
Asymp. Sig.	0.002

a. Kendall's Coefficient of Concordance

Test Statistics- Σ Overall Agreement (R1-R2)	
N	2
Kendall's W[a]	**0.919**
Chi-Square	125.018
df	68
Asymp. Sig.	0.000

a. Kendall's Coefficient of Concordance

CHAPTER 11

Innovation District Typologies

11.1 INTRODUCTION

During the last several decades, urban locations have started to heavily dominate the hosting of innovation activities (Yigitcanlar & Inkinen, 2019; Yigitcanlar *et al.*, 2021). In line with such domination, the concept of innovation districts has globally been recognized and accepted as the new urban land-use type by many cities for the economic, social, and spatial benefits they offer to the host city (Lawrence *et al.*, 2019; Morawska *et al.*, 2021; Cohendet *et al.*, 2022; Kayanan, 2022). However, developing innovation districts is a risky investment with a high cost; hence, there is a need for the holistic assessment of their performance to inform relevant stakeholders involved in the planning, development, and management of these districts (Morisson & Bevilacqua, 2019; Sisi, 2019; Morisson, 2020; Spirou, 2021). Such an assessment will identify areas with the most-needed interventions as well as policy and investment decisions on developing innovation district types with characteristics that are most suitable for the specific locations (Borsi & Schulte, 2018; Pancholi *et al.*, 2019; Wong, 2022), and, consequently, contributing to the success of innovation districts. This calls for the employment of a holistic assessment framework, which the literature confirms is lacking (Adu-McVie *et al.*, 2021).

Typologies are analytical tools that are employed by diverse disciplines including urban planning and development and related fields for classification analysis, which involves the process of performance assessment and the creation of typologies. So far, there are limited studies on the classification of innovation districts, including Markusen (1996), who identified four typologies and SGS (2020)'s five typologies.

DOI: 10.1201/9781032657431-15

However, these classification studies only focus on partial dimensions—e.g., ‘firm configuration’ and ‘industry type’ (Markusen, 1996; SGS, 2020)—and exclude other important spatial attributes; hence, they do not holistically classify their case study innovation districts. A notable reason is due to the lack of suitable performance frameworks. To date, and to the best of our knowledge, only one multidimensional performance framework has been designed by the authors in their previous study (Adu-McVie *et al.*, 2021) to holistically classify innovation districts.

The study aims to contribute to the body of knowledge by adopting the multidimensional innovation district performance framework and testing its practicality through a three-step process (Pancholi *et al.*, 2019), as illustrated in Figure 11.1, on three selected innovation districts representing Australia’s largest cities, i.e., Sydney, Melbourne, and Brisbane. This study focuses on producing: (a) a ranking of innovation districts based on their performances, (b) a performance matrix that categorizes innovation districts into clusters, (c) a typology matrix that describes the character of innovation districts, and (d) subsequently a better understanding on the holistic assessment of innovation district performances.

This chapter is structured as follows. Section 11.2 presents the literature review concerning the various name tags for innovation districts; typology as an analytical tool and its use not only in social science, but by diverse disciplines including urban planning and development; the evolution of innovation district classification; and the multidimensional innovation

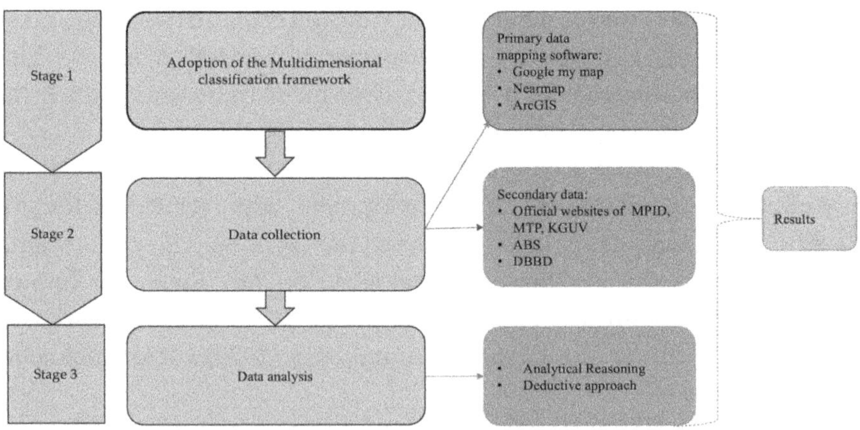

FIGURE 11.1 Stages of the research.

district performance framework. Section 11.3 presents the research design, introduces the case study, and then discusses the methods the study utilized in the data collection and analysis. Section 11.4 presents the study's findings. Section 11.5 discusses the findings. Section 11.6 highlights the implications and concludes the chapter.

11.2 LITERATURE BACKGROUND

The term 'innovation district' is described by Adu-McVie *et al.* (2021) as an expansive term that includes similar developments such as high-technology districts (Forsyth, 2014), science and technology parks (Díez-Vial & Fern'andez-Olmos, 2015), innovation and cultural districts (Jones, 2017), innovation precincts (Esmaeilpoorarabi *et al.*, 2018), knowledge and innovation spaces (Pancholi *et al.*, 2019), and knowledge community precincts (Esmaeilpoorarabi *et al.*, 2020). Some of the world's most renowned innovation districts include the United States' (US) Silicone Valley, France's Sophia Antipolis, Spain's 22@ Barcelona, Singapore's One North, and Australia's Macquarie Park Innovation District. Despite the fact that these districts come in various sizes, forms, and characteristics, they all have a common purpose, i.e., to achieve agglomeration benefits in terms of economic, technological, sociocultural, and environmental outcomes for their host cities (Adu-McVie *et al.*, 2021).

The term 'typology' is defined as an organized system of types and is a well-established analytical tool in the social sciences. Typologies facilitate 'diverse analytical tasks forming and refining concepts, drawing out underlying dimensions, creating categories for classification and measurement, and sorting cases' (Collier, 2012). The use of typologies as analytical tools dates to the early 1970s and includes Dahl's (1971) analysis and identification of 'polyarchies', 'competitive oligarchies', 'inclusive hegemonies', and 'closed hegemonies'. To date, typologies are still employed for classification analysis by diverse disciplines including climatology (Koc *et al.*, 2016), architectural and building sciences (Klein, 2014; Pozas & Gonzalez, 2018), urban studies (Yigitcanlar et al., 2007; Benefoh *et al.*, 2018), and ecosystems studies (De Groot *et al.*, 2002). Typology outcomes are typically illustrated in a matrix form, an approach seen as a method that uses a system of different criteria for classifying elements (Klein, 2014).

Attempts to classify industrial districts and clusters, which are accepted as the predecessors of innovation districts, date back to 1996

(Markusen, 1996; Roelandt, 1996; Boix & Galletto, 2009; Metaxiotis *et al.*, 2010). From the start of the millennium, most of these industrial clusters were transformed into innovation districts (Leon, 2008; Turkina & Van Assche, 2018; Wu *et al.*, 2020), and in recent years, there has been an increasing interest in classifying innovation districts or the new 'innovation spaces' (Lee, 2001; Heaphy & Wiig, 2020; Hansasooksin & Tontisirin, 2021; Komninos *et al.*, 2022). These attempts have resulted in the emergence of some famous innovation district typologies including the anchor plus, re-imagined urban areas, and urbanized science parks (Katz & Wagner, 2014; Yigitcanlar *et al.*, 2019); and corridors, clumps, cores, and comprehensive campuses (Forsyth, 2014). Whilst these studies identified some typologies, none of them holistically classified innovation districts due to the lack of a comprehensive performance framework (Adu-McVie *et al.*, 2021).

Despite the urgent call for a suitable innovation district performance framework, so far, to our best knowledge, there is only one such framework in the current body of knowledge, which was developed recently by Adu-McVie *et al.* (2021): the multidimensional innovation district performance framework (see Table 11.1). The framework was validated by 32 international experts through a two-round Delphi survey in 2020. The multidimensional performance framework comprises four key dimensions of context, feature, function, and form, with 16 indicators including social amenity, industry type, and land-use mix and 48 measures comprising three tiers (e.g., strong, moderate, weak). It uses multidisciplinary objective methods for measuring indicators. The framework's main purpose is to classify the typologies of innovation districts based on their characteristics and performance. Hence, the multidimensional framework serves a dual purpose: first as a performance framework and second as a classification framework.

11.3 METHODOLOGY

This study applied a case study method to test the above-mentioned multidimensional performance framework on three internationally recognized Australian innovation districts and classify these districts based on their performance. The case study method is used by diverse disciplines as a 'robust research method particularly when a holistic, in-depth investigation is required' (Zainal, 2007; Rashid *et al.*, 2019). The case studies are identified as a qualitative form of research design. Their subjects for analysis usually include persons, social communities, and

TABLE 11.1 Multidimensional Innovation District Performance Framework

Dimension	Indicator	Description	Measure
Context	Spatial system	City-wide spatial layout and architecture qualities, including physical environment	Strong Moderate Weak
	Societal system	Societal progress of the city, including diversity and age structure	Strong Moderate Weak
	Governance system	Political progress of the city, including political institution effectiveness, transparency, and accountability	Strong Moderate Weak
	Economic system	Macroeconomic progress of the city, including monetary and fiscal performance	Strong Moderate Weak
Feature	Social amenity	Presence/availability of social amenities for public use within the innovation district	Strong Moderate Weak
	Human capital	Inventory of skilled people	Strong Moderate Weak
	Skilled labor	Skilled employment outcome of the innovation district activities	Strong Weak Moderate
	Locality setting	Location of the district within the metropolitan area	Inner city Suburban Regional
Function	Company size	Relative size of the firms within the innovation district	Multinational enterprises (MNE) anchored Large national enterprises (LNE) dominated Small and medium enterprises (SME) dominated
	Industry type	Dominant business activity operating within the innovation district	Creativity-intensive businesses Technology-intensive businesses Business support services

(Continued)

TABLE 11.1 *(Continued)* Multidimensional Innovation District Performance Framework

Dimension	Indicator	Description	Measure
	Investment type	Principle support and funding body for the development of the innovation district	Public–private–community partnership Public–private partnership Public or private sector driven
	Property management	Management model of the innovation districts' properties and activities	District-wide body corporate Building-based body corporate None
Form	Green–blue infrastructure	Aesthetic qualities of urban green and blue	Strong Moderate Weak
	Land-use mix	Main land-use types within the innovation district	Complex mix Mixed-use Single use
	Built environment	Architectural design of built forms and functions encouraging open innovation systems, connectivity, and mobility within the innovation district	Strong Moderate Weak
	Space design	Spatial layout design encouraging open innovation system within the innovation district	Open layout Semi-open layout Closed layout

organizations (Cacar & Aykol, 2020). The present study's aim falls within the above-stated requirement; hence, a case study approach is suitable. Information on the case study is provided in Section 3.1. To apply the framework, the study followed Pancholi *et al.* (2019)'s three-step process (Figure 11.1), which is briefly discussed in Section 3.2.

11.3.1 Case Studies

The following three leading Australian innovation districts are selected for the study: Macquarie Park Innovation District (MPID) in Sydney, Monash Technology Precinct (MTP) in Melbourne, and Kelvin Grove Urban Village (KGUV) in Brisbane. From these three cases, MPID is recognized as the largest and most mature innovation district in Australia, MTP is seen as the leading non-central business district hub for employment, economic growth, and innovation in Victoria (DIIS, 2019; MTP, 2022), and KGUV is one of the best known 'new urbanism' projects in Australia formed as a mixed-use master planned community (Wardner & Hefferan, 2015). Figures 11.2 through 11.4 show the location of the case study innovation districts—licensed Nearmap software was used to create these figures.

MPID is an 'industry-led initiative' that was re-established in 2015 (as an extension of Macquarie Park that was established in 1999) by key stakeholders to strengthen 'their competitive advantage in life sciences, health and pharmaceuticals, digital and biotechnology' (NSW-IPC, 2018). MTP was established in 1984, based on a public–university–private initiative. It hosts

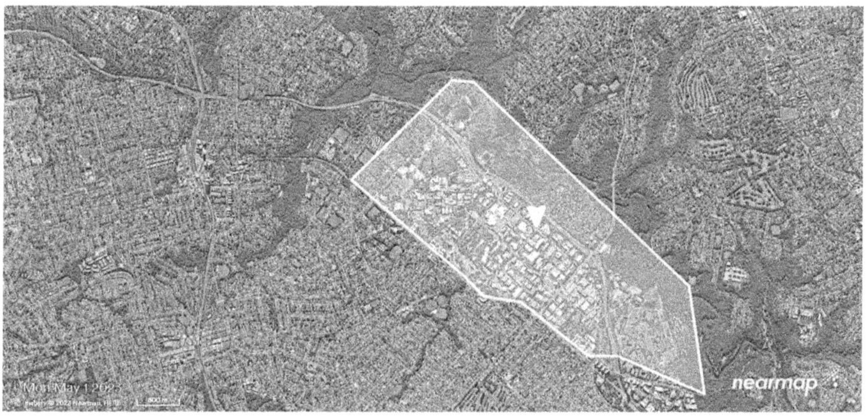

FIGURE 11.2 Location of Macquarie Park Innovation District.

FIGURE 11.3 Location of Monash Technology Precinct.

FIGURE 11.4 Location of Kelvin Grove Urban Village.

leading institutions in education, health, research, and innovation business activities. Compared to the other two case study innovation districts, KGUV is a medium-size district in terms of land area and was established in 2003 under a public–private–university–community investment partnership. Additionally, KGUV is Australia's first mixed-used master planned urban village and university precinct (Pancholi *et al.*, 2015; Esmaeilpoorarabi *et al.*, 2018). It is also one of Brisbane's eight global precincts (BSB, 2017; BCC, 2019; Advance Queensland, 2021), which hosts health and education research, creativity, residential housing, and retail activities.

Each of the selected innovation districts is strategically located in their respective state's economic corridors, which obviously contributed to their rapid and successful growth. Located in Sydney's northern district and along the 'Global Economic Corridor of Sydney', MPID's suburban area is approximately 12 km from the CBD and is adjacent to Lane Cove National Park, which provides green and blue spaces for leisure and physical exercise activities (Pancholi *et al.*, 2019). Likewise, MTP is also a suburban area in Clayton (Melbourne's southeast), which is 23 km from Melbourne's CBD. It sits at the crossroads of Clayton, Wellington, Blackburn, and Dandenong roads, and the Monash Freeway. On the other hand, KGUV is an inner-city suburb located less than 3 km from Brisbane's CBD. It sits amidst the Kelvin Grove and Victoria Park roads and is surrounded by major arterial roads to the north and south and is accessible from the Inner-City Bypass, Airport Link, CLEM7, and Legacy Way tunnels (BCC, 2020). Regarding the size, MTP and MPID have larger land areas of 6.9 km^2 and 6.8 km^2, respectively, whilst KGUV is only 0.37 km^2 (ABS, 2016).

All three case districts contain top universities, such as the Macquarie University, Monash University, and Queensland University of Technology, as well as research and training hospitals, e.g., the Macquarie University Hospital (MPID), Monash Medical Trial Center at Monash Health (MTP), and Kelvin Grove Medical Centre as the main anchors. In terms of the workforce and number of existing businesses, MPID has more than 380 businesses employing over 45,000 people and has some of Australia's top 100 companies, including Cochlear, Sony Australia, Foxtel, Microsoft, and Johnson & Johnson, with their head offices located within the district (MPID, 2022). MTP, on the other hand, is referred to as an ecosystem because it hosts 13,000 businesses employing over 94,000 employees (MTP, 2022). The top organizations and companies located in this district include CSIRO, ANSTO Synchrotron, Agilent, Bosch, and Johnson & Johnson. In contrast, KGUV only has around 50 businesses employing approximately 4,000 people. Besides its anchors and the Queensland Academy of Creative Industries, the other existing businesses are mainly in the health care (e.g., medical centers), retail, and food service (e.g., restaurants) sectors.

In terms of economic growth, MPID and MTP contribute approximately AUD 9 billion to their state's economic output. Whilst there are no publicly available data on KGUV's economic contribution, it is believed that KGUV (as one of Brisbane's eight global precincts) has contributed significantly to the Brisbane region's AUD 181 billion gross regional product for 2020–2021 (BCC, 2022). Demographically, all three districts have a large proportion of their populations with high education degrees—i.e., in MPID, 78% of the district population has a minimum of a bachelor's or higher degree (BA+) qualification, and this ratio is 93% for KGUV and 61% for MTP. Considering the potential labor force, 67% of MTP's, 70% of MPID's, and 68% of KGUV's respective Statistical Area 2 (SA2) population fall into the age category of 15 to 44 years with a graduate certificate and bachelor's or further qualifications (ABS, 2016); thus, indicating the strong presence of a young workforce.

11.3.2 Data Collection

The present study employed Pancholi *et al.* (2019)'s three-step process (Figure 11.1) to test the application of the multidimensional framework on three selected innovation districts. Firstly, it involved the adoption of the framework to guide the remaining two stages of data collection and analysis, respectively. Secondly, the study applied desktop audits using mapping and GIS software, specifically Google My Map, Nearmap, and ArcGIS, to collect the primary data, whereas the secondary data were collected from the official websites of the case innovation districts, the Australian Bureau of Statistics (ABS), and the Dun & Bradstreet Business Directory (DBBD). Thirdly, the study employed an analytical reasoning method and a deductive approach for the data analysis. The validated multidimensional framework was employed in the deductive approach to test each of the indicators for the three innovation districts.

Primary and secondary data were collected and used to assess the performance of the selected case districts. The primary data included spatial data such as land-use mix and space design and descriptive data of social amenities, and the secondary data included demographic statistics on human capital and skilled labor, financial data on tenant companies' annual income, and descriptive information on types of investment. The audits on MPID, MTP, and KGUV were all conducted remotely on a desktop utilizing the mapping software. The legal boundaries of each of the case districts were first identified and marked to ensure the audits were done within the boundary of the case innovation districts. These layout maps also acted as the base maps for further spatial and descriptive analysis.

The three main types of activity that typically exist in any innovation district are 'technology-intensive businesses' that are composed of firms involved in information communications and technology services, biotechnology, or use high technology and knowledge for the production of goods and services, and carry out research to generate knowledge and innovation (Forsyth, 2014; Baum *et al.*, 2007); 'business support services', composed of firms that provide services such as marketing, auditing, and insurance (Yang & Wang, 2008; Cravo & Piza, 2019) to tenants within the innovation districts; and 'creativity-intensive businesses' that are mainly involved in music, film, and gaming industries to generate cultural knowledge (Durmaz *et al.*, 2010; Zheng, 2011; Schmahmann, 2019).

To determine an accurate estimate of the total business population of the case districts, the study adopted Yigitcanlar *et al.* (2020)'s definition of 'technology-intensive businesses', 'creativity-intensive business categories', and 'business support services', wherein the first one is expanded to include other healthcare facilities besides hospitals (e.g., private surgeries and dentistry services), all manufacturing activities, and mechanical and engineering workshops. The second contains businesses involved in music, film, entertainment, cultural, and gaming industries.

The last category is composed of services, except for retail, that do not fall into either of the technology-intensive businesses and creativity-intensive business categories, including real estate, wholesalers, consultancy services in the built environment, engineering and financial services, and community services (i.e., senior and childcare services).

The next section discusses the various audit tools employed and describes the process followed to collect the required data. The discussion is presented under the subsections of 'context', 'form', 'feature', and 'function'. It is noteworthy that previous related studies recommended excluding the 'context' dimension when assessing the performance of innovation districts, due to its broadness and suitability for regional- or city-level assessments (Esmaeilpoorarabi *et al.*, 2018; Adu-McVie *et al.*, 2021). However, the present study includes it for the following two reasons. First, three case innovation districts are selected from three major capital cities (i.e., Sydney, Melbourne, and Brisbane) of Australia, respectively; hence, it is useful to identify the influence of cities on these case study districts. Second, it would be more proper to assess all four dimensions, since this is a study to test the practicality of the newly validated 'multidimensional innovation district performance framework'.

11.3.2.1 Context

The secondary data for the indicators of context dimension, namely, 'economic', 'societal', 'spatial', and 'governance' systems, are sourced from the ABS; official websites of Sydney, Melbourne, and Brisbane City Councils; and other relevant reports and related studies. It should be noted that the investigation or audit referred to in this section focuses on the three hosting cities of the case innovation districts.

The audit of the economic system aims to investigate the macroeconomic progress of the case cities by considering their five-year (2016–2020) trend in terms of attributes: gross net debt (GND), gross state product (GSP) growth, inflation, and unemployment rates. The raw data obtained for each city are already in percentage points and therefore need to be normalized. We opt to compute the median rather than mean scores specifically to avoid skewed results due to the extremely small and large values when calculating the mean (Watson, 2020). The individual median score for each attribute over the five-year period is computed before calculating a composite median. Finally, we employ Excel's quartile function in the spreadsheet to run the three cities' composite median scores to determine the minimum, median, and maximum values. Note that the composite median scores include attributes GND, inflation, and unemployment rate, which have negative impact; hence, the minimum, median, and maximum values are interpreted in reverse, i.e., minimum = strong, medium = moderate, maximum = weak. These values inform the weighting value range, which defines the cities' overall economic performance.

For the societal system, the audit aims to investigate the societal progress of each city in terms of 'diversity', 'tolerance', 'equality', 'age structure', and 'participation in cultural and community activities'. However, due to the availability of data and lengthy processing time, the study only selects 'diversity' and employed Brookings Institute's audit guide to investigate each of the city's baseline measures of diversity, specifically 'workers by ethnicity and gender', 'graduate students by ethnicity and gender', and 'residents by race/gender and foreign-born status'. The raw data obtained are used to compute individual percentage scores for the population by place of birth (POB), and then the composite mean scores are used to determine the ratio of the dominant to minority population by POB. For this attribute, we opt for composite mean because it has a small disparity in percentage scores. Individual mean scores for each POB category per ABS classification are computed. The mean scores are then recategorized into two classes: a dominant class comprising Oceania/Antarctica (Australia included), and a minority class (i.e., all other POBs not included in the former class). The composite mean scores are converted into a minimum, medium, and maximum value, which define the level of diversity of each city.

For the 'spatial system', the audit aims to investigate each city's spatial layout and architecture qualities in terms of the physical environment (i.e., public green space and playgrounds), unique natural conditions (i.e., waterfronts), unique man-made built environment (i.e., landmarks, heritage preservation, and 'image of the city'), and quality of physical patterns (i.e., characters of the public built environment include basic infrastructure, transportation, and mobility). The study employs the direct method of age standardization—rate per 1,000 population (Esmaeilpoorarabi *et al.*, 2016)—to individually measure most of the spatial attributes except for the city image, where average scores are used, and for basic infrastructure and transportation, where composite scores are employed. To standardize the raw scores, weighting values are given to each attribute before computing the composite and mean scores. Then, quartile values of the mean scores are computed to determine the minimum, median, and maximum value, which define each city's overall spatial system performance.

Finally, for the 'governance system', the audit aims to assess the city's political progress in terms of political institution effectiveness through public services, accountability and transparency, and participation. Using an audit checklist, the stated governance indicators are primarily assessed from descriptive reports of each city's official annual reports and plans. After the individual scores are given within the range of 0 = absent, 0.5 = limited, and 1 = unlimited, the composite scores and mean are calculated. Likewise, quartile values of the mean scores are then computed to determine the minimum to maximum value, which defines the overall performance level of each of the three cities' governance system.

11.3.2.2 Form

The three case districts are virtually visited through Google My Map and Nearmap to obtain primary data on the following indicators: 'land-use mix', 'space design', 'built environment', and 'urban green and blue infrastructure'. Audits on land-use mix aim to identify the type of existing land use within the legal boundary of the case districts, such as a complex mixed use (i.e., work–learn–live–play), mixed use (i.e., work–learn–live or play), or single use (work or learn). Descriptive and spatial analysis is done through an aerial view of each of the innovation districts. An innovation district with a clear aerial image of complex mixed use is rated 'strong', mixed use 'moderate', and single use 'weak'. Evidence of land-use types is captured by screenshots and recorded for reference.

The audit on space design is guided by Yun *et al.*'s (2018) three-element cluster model to identify whether the case districts' spatial layout design encourages knowledge generation. The model is used to determine whether the innovation district is an open, semi-open, or closed layout innovation system. It regards three land-use zonings of R&D (university), house (market or consumer), and park/entertainment facilities (museum) based on three theories. First, an open innovation system in a three-factor cluster takes place when those from the 'house zone' must pass through the 'R&D (university) zone' to get to the 'park (museum) zone'. Second, a semi-open innovation system is when those from the 'house zone' can go directly to the 'park zone' without passing through the 'R&D zone'; however, the R&D is nearby to incite their interest to contact the R&D. Finally, a closed innovation system is when those from the house go directly to the park without passing the R&D and have minimal contact with the R&D. In sum, the measure is about people's permeability to the innovation district. Similar analysis performed on land-use mix is repeated here. An innovation district with a clear aerial image showing evidence of an open layout design is rated 'strong', a semi-open design is 'moderate', and a closed layout design is 'weak'. Evidence of space design for innovation districts is captured by screenshots and recorded for reference.

Furthermore, an audit on built environment features aims to investigate whether the case districts' architectural design, built form, and function encourage an open innovation system, connectivity, and mobility within its legal boundary.

The audit sheet employed was derived from the Healthy Build Environment Checklist (NSW Health, 2021) and an urban design for walking checklist (Victoria Walks, 2021), comprising seven themes, including 'street connectivity and smaller block sizes' and 'number of local living destinations (e.g., transit stations, grocery shops, hospitals, education institutes, etc.) within walking or cycling destinations within 1 km radius of the case district'. A similar analysis process employed for the above form indicators is followed for the built environment. The auditor uses aerial images to identify the presence of the attributes within a 1 km radius of the innovation districts. For instance, if there is at least one each of the living destinations within the 1 km radius, a score of 10 (unlimited) is given, and vice versa. In total, seven attributes are audited, and individual scores are given as 0 = No, 5 = Limited, and 10 = Unlimited. The composite scores derived from the audit are computed to mean scores to define the innovation districts' overall performance for the built environment.

The audits on urban green and blue infrastructure aim to investigate any presence of urban green and blue ecosystem services for aesthetic qualities both within (green) and without (blue) the legal boundary of the case districts. The audit sheet employed to obtain data is based on the design principles for green–blue infrastructure (Gehrels *et al.*, 2016), which has two main parts. The first part concerns green ecosystem services at the innovation district or cluster level and has seven themes, including mitigating heat stress, noise reduction, and air quality regulation. The second part focuses on blue ecosystem services at the city level with eight themes, including green–blue corridors, places to meet, swimming, and playing.

Furthermore, the audit sheet has a guideline to assist the auditor(s) in their ratings between various innovation district sizes. For instance, a small innovation district that has fewer than two trees with large crowns for 'mitigating heat stress' is rated 'limited presence' (0.5 score) and those districts with more than two trees with large crowns are considered 'unlimited' (1 score). In terms of the blue ecosystem services, if the district's host city has fewer than 10 places to meet, e.g., restaurants/eateries along seashores and riverbanks, it is rated 'limited' and more than 10 is 'unlimited'. Scores are given as 0 = Absent, 0.5 = Limited, and 1 = Unlimited. Together, 23 attributes are audited for each case district.

Overall, the composite scores of green ecosystems and blue ecosystems are combined and computed to a percentage score to define the innovation district's performance—i.e., strong, moderate, or weak.

11.3.2.3 Feature

The indicators of 'human capital', 'skilled labor', and locality setting require secondary data, whilst 'social amenity' requires primary data. The audit on human capital aims to identify the number of potential workforces with BA+ qualifications in the surrounding suburbs of the case district, which are identified from the ABS Statistical Area 2 (SA2). First, we identify the suburbs in the same SA2 as the one that the case innovation district is in, and then we use the ABS table builder to generate a report on these suburbs' labor force status. Only data for the qualifications of bachelor's degree and above are extracted and a percentage score of the population is computed. The rating value for this attribute is >50% strong, <50% moderate, and <25% weak. On the other hand, an audit on skilled labor aims to identify the number of people in the workforce or knowledge workers with BA+ qualifications within the case district. As for human capital, a similar analysis process is carried out, except the data analyzed concern the number of employees within the innovation district.

Furthermore, the information on the estimated or actual number of employees is generated by running the tenant company names on the DBBD as well as from the tenant company's official website. In addition, these data sources provide other required information, including the actual or estimated annual income of the companies, the company nationality status, and the nature of the business, which facilitates determining the 'company size' and 'industry type', respectively. These characteristics will be discussed further in the subsequent section for function dimension.

As for the locality setting, the audits aim to identify whether the case district is in an 'inner city', 'suburban', or 'regional' area. The localities are identified from the SA2 areas. In terms of the audit on 'social amenity', it aims to investigate any presence and availability of amenities for public use within the legal boundaries of the case district. The audit tool employed is derived from Taylor *et al.'s* (2011) 'public open space tool' (POST), which was used to assess the quality of public parks and open space in their study. The POST focuses on the following themes, namely, 'activities', 'centrality/locality', 'environment quality', 'amenity', and 'safety'. The present study not only adopts the POST, but also expands this measurement tool to include restaurants and cafes, cultural and entertainment facilities, public parks/open space, and playgrounds.

The audit sheet is designed with a key question (for each theme), which guides the auditor to answer either by checking options 'No', 'Yes', or 'N/A', and Likert scale scores ranging from 0 to 5. The auditor virtually performs the desktop assessment. For instance, under the theme 'environment quality', question 4 is 'Is there clear presence of outdoor dining and coffee shop? (Check an answer below)'. The auditor searches the aerial image of the restaurant or coffee shops for evidence of outdoor tables and chairs to confirm, and checks an answer (i.e., Yes = 1, No = 0, N/A= 2) accordingly. On the other hand, question 6(b) for the theme 'environment quality' is 'Is there shade along paths (check one only)'. This question is only relevant for parks, open spaces, and outdoor ground-level social amenities; hence, indoor social amenities above ground level are checked as N/A. The auditor checks the selected Likert scale score ranging from 0 to 5 based on the aerial observation of the tree canopies, guided by the given parameters of 6(b): the scores given for 'Yes', 'No', or 'N/A' range from 0 to 2, where No = 0, Yes = 1, and N/A = 2, whilst the Likert scores range from 0 to 5 where 0 = Very poor (little or no shade), 1 = Poor (canopies of trees do not touch and trees are spread apart), 2 = Medium (canopies of trees do not touch, but trees are close together), 3 = Good (canopies of some trees touch), 4 = Very good (canopies of many trees touch), and 5 = Not applicable (as there are no paths).

The descriptive scores derived above are converted to values and transferred to an Excel spreadsheet to formulate a 'master scoresheet'. The overall composite score is then normalized to obtain a composite mean score, which defines the innovation district's performance—i.e., >50 strong, >40 moderate, <40 weak. Overall, 44 social amenity attributes are audited for each innovation district.

11.3.2.4 Function

The data obtained for the indicators of 'investment type', 'industry type', 'company size', and 'property management' are all from secondary sources such as DBBD and the official websites of tenant companies. The audit on investment type focused on identifying whether the principal support and funding body of the district is a 'multiple-sector' (public–university–private–community) investment, a 'two-sector' (public and private) investment, or a 'single-sector' (public or private) investment. The auditor identifies each case district's investment type from their history or background information sourced from their official websites. Multiple-sector investment is the most preferred type; hence, it is rated as the top-tier measure for this attribute. Additionally, the audit on industry type aims to identify the dominant industry activity within the case district, i.e., 'technology-intensive business', 'creativity-intensive business', or 'business support services', where the industry type that has a score over 50% becomes the dominant one.

The auditor identifies existing tenants (at the time of audit) and the business type, and then categorizes them into three classes, as per the business types mentioned above. Following this, the percentage of each class is computed to determine the most dominant industry. A similar process is followed in auditing the company size, wherein the audit aims to identify the relative size of the tenant companies. For example, a company is regarded as being multinational if it meets the condition-matching criteria (i.e., ≥AUD 1 million, ≥50 employees, Australia-based company with branches overseas or vice versa); a large national enterprise (≥AUD 1 million, ≥50 employees, Australia-based), or small and medium enterprise (<AUD 1 million, <50 employees, Australia-based). After categorizing the companies into the three classes above, their percentage to total tenant companies within the innovation district is computed. A company size with more than 50% is the dominant one, but the most preferred company size is the multinational enterprise, which is rated as the top-tier measure for this attribute.

Finally, the audit on 'property management' assists in identifying the type of operations and asset management practices in the case district. While a 'district-wide management' has an organization body that supervises or manages the overall operation and assets of the district, a 'building level management' does not have an independent district management, having individual buildings managed separately. The third management type is defined as one in which there is no form of management. Each innovation district's property management type is identified from the official website's 'about' or 'who we are' information. The most preferred management type is 'district-wide management', which is rated as the top-tier measure for this attribute.

All raw data obtained are filtered and normalized to avoid any potential bias in the analysis stage (Morais & Camanho, 2011; Audretsch & Belitski, 2022). For instance, those tenant companies within the districts that have missing values for 'number of employees' and 'annual income' are excluded from the data analysis. However, those with at least one value are retained, and average estimates are adopted to replace the missing values. The study opts for an average value instead of a minimum value (Morais & Camanho, 2011), mainly to ensure that the concern variable is not unnecessarily replaced with a lower value, which may contribute negatively toward the performance of the studied case district. Finally, all of the standardized scores for each dimension above are converted to descriptive values, as illustrated in Table 11.2, which defines the innovation districts' overall performance.

11.4 ANALYSIS AND RESULTS

The results for each case innovation district are presented in Table 11.2 in a descriptive form. Both primary and secondary data obtained are analyzed using both qualitative and quantitative analysis methods, which are utilized to examine all secondary data and compute percentage scores by simple math calculation.

For ease of reference, the descriptive data from the preliminary findings in Table 11.2 are converted into a 'case study matrix', which provides a brief description of each case district. As illustrated in Table 11.3, the matrix comprises three vertical levels of A, B, and C representing the three tier measures (e.g., strong, moderate, weak, or inner city, suburban, regional) in the framework and 16 horizontal levels (i.e., o1–o4; e1–e4; u1–u4; c1–c4) representing 16 indicators under the 4 dimensions of feature, function, form, and context.

Table 11.3 is an example of the 'case study matrix' showing the audit results on Monash Technology Precinct (MTP), presented in the shaded area, which can be interpreted as per type (o1A, o2B, o3A, o4A) + (e1A, e2A, e3B, e4B) + (u1A, u2A, u3B, u4A) + (c1A, c2A, c3A, c4A) and described as a complex mixed-use semi-open design innovation district with strong features in the forms of built environment and urban green

TABLE 11.2 Case Study Results

Dimension	Indicator	Description	Parameter	Measure	MPID (Sydney)	MTP (Melbourne)	KGUV (Brisbane)
Feature	Social amenity	Presence or availability of social amenities for public use	Strong presence of social amenities Moderate presence of social amenities Weak presence of social amenities	Mean composite score Strong > 50, Moderate > 40, Weak < 40	Moderate	Moderate	Moderate
	Human capital	Inventory of skilled people (i.e., information about education and skilled level of the population and potential stock of qualified people)	Strong human capital Moderate human capital Weak human capital	Percentage of knowledge workers with BA or higher qualification Strong > 50%, Moderate > 25%, Weak < 25%	Strong	Strong	Strong
	Skilled labor	Skilled employment outcome of the innovation district activities	Strong skilled employment Moderate skilled employment Weak skilled employment	Ratio of knowledge worker jobs to total innovation district jobs Strong > 50%, Moderate > 25%, Weak < 25%	Strong	Strong	Strong
	Locality setting	Location of the district metropolitan area	Inner city Suburban Regional	Location of the innovation district Inner city, Suburban, Regional	Suburban	Suburban	Inner city
Function	Company size	Relative size of the firms within the innovation district	MNE anchored LNE dominated SME dominated	Ratio of number of firm types to total firms within the innovation district MNE anchored if > 50%, LNE dominated > 50%, SME dominated if > 50%	MNE	LNE	SME

(Continued)

TABLE 11.2 *(Continued)* Case Study Results

Dimension	Indicator	Description	Parameter	Measure	MPID (Sydney)	MTP (Melbourne)	KGUV (Brisbane)
	Industry type	Dominant business activity operating within the innovation district	Technology-intensive business Creativity-intensive business Business support services	Dominant business activity of the district Technology-intensive if > 50%, Creativity-intensive if > 50%, Business support services if > 50%	Technology-intensive business	Technology-intensive business	Business support services
	Investment type	Principle support and funding body for the development of the innovation district	Public–private–community partnership Public–private partnership Public or private sector	Multiple sectors Two sectors Single sector	Multiple sector	Multiple sector	Multiple sector
	Property management	Management model of the innovation district's properties and activities	Managed by a district management Managed by a building management No form of management	District wide Building level None	District wide	District wide	District wide
Form	Urban green and blue infrastructure	Aesthetic qualities of urban green and blue infrastructure within/out the innovation district (i.e., all natural and semi-natural landscape elements that form a green–blue network)	Strong presence of ecosystem services Moderate presence of ecosystem services Weak presence of ecosystem services	Based on design principles of green–blue infrastructure innovation district level (green infrastructure), city level (blue infrastructure) Strong > 50%, Moderate > 25%, Weak, 25%	Strong	Strong	Strong

	Land-use mix	Main land-use types within the innovation districts	Complex mixed Mixed use Single use	Work–learn–live–play Work–learn, live or play Work or learn	Complex mixed	Complex mixed	Complex mixed
	Built environment	Architectural designs of built forms encouraging connectivity and mobility within the innovation districts	Strong internal connectivity Moderate internal connectivity Weak internal connectivity	Based on healthy built environment guidelines Composite scores Strong > 60, Moderate > 50, Weak < 50	Strong	Strong	Strong
	Space design	Spatial layout design encouraging open innovation system within the innovation districts	Open layout Semi-open layout Closed layout	Based on three element cluster model to determine whether the spatial design encourages knowledge generation within the innovation district Open layout, semi-open layout, closed layout	Semi-open layout	Semi-open layout	Open layout
Context	Spatial system	City-wide spatial layout and architectural qualities (i.e., physical environment, spatial conditions, and urban development)	Strong spatial design Moderate spatial design Weak spatial design	Composite index of quality of physical environment, unique natural environment, and physical patterns based on mean value Strong >1.8, Moderate <1.7, Weak < 1.7	Strong	Strong	Strong
	Societal system	Societal progress of the city (i.e., diversity, tolerance, equality, age structure, and participation in cultural/community activities)	Strong social assets Moderate social assets Weak social assets	Diversity and inclusiveness measured by composite scores based on Brookings's audit guide Strong > 70, Moderate > 66, Weak < 66	Strong	Strong	Moderate

(*Continued*)

TABLE 11.2 *(Continued)* Case Study Results

Dimension	Indicator	Description	Parameter	Measure	MPID (Sydney)	MTP (Melbourne)	KGUV (Brisbane)
	Governance system	Political progress of the city (i.e., political institution effectiveness, accountability, transparency, and participation)	Strong governance effectiveness Moderate governance effectiveness Weak governance effectiveness	Composite index of quality public services, civil services, quality of policy formulation and implementation, and credibility of the government's commitment to such policies of the city Strong > 0.75, Moderate < 0.75, Weak < 0.7	Strong	Strong	Strong
	Economy system	Macroeconomic progress of the city (i.e., monetary, and fiscal performance to maintain stability of economic growth)	Strong economic performance Moderate economic performance Weak economic performance	Composite index of governance net debt, real GDP growth, inflation rate, and unemployment rate of the city based on composite median scores Strong < 8.46, Moderate > 8.46, Weak > 9.22	Strong	Strong	Moderate

TABLE 11.3 An Exemplar Case Study Matrix of MTP

		Form		
		A	B	C
Land-use mix	o1	Complex mixed	Mixed use	Single use
Space design	o2	Open	Semi-open	Closed
Built environment	o3	Strong	Moderate	Weak
Urban green and blue infrastructure	o4	Strong	Moderate	Weak
		Feature		
		A	B	C
Human capital	e1	Strong	Moderate	Weak
Skilled labor	e2	Strong	Moderate	Weak
Social amenity	e3	Strong	Moderate	Weak
Locality setting	e4	Inner city	Suburban	Regional
		Function		
		A	B	C
Investment type	u1	Multiple sectors	Two sectors	Single sector
Industry type	u2	Technology-intensive	Creativity-intensive	Business support
Company size	u3	Multinational	Large national	Small and medium
Property management	u4	District wide	Building level	None
		Context		
		A	B	C
Economic system	c1	Strong	Moderate	Weak
Societal system	c2	Strong	Moderate	Weak
Spatial system	c3	Strong	Moderate	Weak
Governance system	c4	Strong	Moderate	Weak

and blue infrastructure, located in a suburban area with moderate social amenity, strong human capital, and skilled labor. It is funded by a multiple-sector investment partnership and is dominated by large, national technology-intensive businesses under district-wide management. MTP is surrounded by strong systems of economic, societal, spatial, and governance.

To compute the overall performance scores, the descriptive measures, for example, Strong, Moderate, Weak, or Open, Semi-open, Closed, are assigned following the categorical values of Strong = A, Moderate = B, and Weak = C; Open = A, Semi-open = B, and Closed = C. Hence, converting MTP's descriptive measures in Table 11.3 results in the following categorical values (see Table 11.4): Land-use mix = A, Space design = B, Built environment = A, Urban green and blue infrastructure = A, Human capital = A, Skilled labor = A, Social amenity = B, Locality setting = B, Investment type = A, Industry type = A, Company size = B, Property management = A, Economic system = A, Societal system = A, Spatial system = A, and Governance system = A. A similar conversion of descriptive measures to categorical values is repeated for the other two case districts. The categorical scores are further analyzed to compute the 'net scores', also known as 'net performance scores', using the following simple formula. The 'net scores' indicate each case district's overall performance.

$$\textit{Net score} = \textit{Percentage of A dimensions} - \textit{Percentage of C dimensions} \quad (11.1)$$

For example, MTP has 12 As, 0 Cs, and 4 Bs; therefore, the net score is 75. The maximum net score is achieved if a district receives all As, as the net score would be 100, whereas a district with all Cs will have a net score of −100. In other words, innovation districts with more As have more positive net scores than those with more Cs that will have negative scores. According to the rule of calculating net scores, the B scores are excluded because they are 'passive' scores (Baehre *et al.*, 2022). Furthermore, the percentages of both As and Cs are expressed as a percentile score, whilst the net scores are expressed in metrics. The analyzed performance scores for all case districts are presented in Table 11.4. For ease of reference, each indicator is color coded according to their performance.

The study considered two potential criteria rating systems, firstly, the Australia Green Star rating system—an internationally recognized Australian sustainability rating and certification system by the Green Building Council of Australia (GBCA). The Green Star rating system has four tools, of which the 'Green Star—Communities' and 'Green Star—Performance' are considered the most suitable for this study. 'Green Star—Communities' assesses the development stages of large-scale development projects of a precinct or community scale, and 'Green Star—Performance' assesses the operational performance of existing buildings, which covers broad environmental issues related with the building development process (Zuo *et al.*, 2016; GBCA, 2021). Green Star's ranking criteria range from 1 Star to 6 Star, where 1 Star = 10

TABLE 11.4 Case Study Categorical Performance Scores

Dimension	Category	Indicator	MPID (Sydney)	MTP (Melbourne)	KGUV (Brisbane)
Form	Complexity and Layout	Land-use mix	A	A	A
		Space design	B	B	A
	Connectivity and Space Design	Built environment	A	A	A
		Urban green–blue infrastructure	A	A	A
Feature	Centrality and Amenity	Social amenity	B	B	B
		Locality setting	B	B	A
	Intelligence and Concentration	Human capital	A	A	A
		Skilled labor	A	A	A
Function	Specialization and Diversity	Investment type	A	A	A
		Industry type	A	A	C
	Scale and Support	Company size	A	B	C
		Property management	A	A	A
Context	Social and Economic	Economic system	A	A	B
		Societal system	A	A	B
	Spatial and Governance	Spatial system	A	A	A
		Governance system	A	A	A
	Total number of categorical values		16	16	16
	Percentage of 'A' dimensions	(n = 13, n = 12, n = 11)	81%	75%	69%
	Percentage of 'C' dimensions	(n = 0, n = 0, n = 2)	0	0	13%
	Net score		81	75	56

Note: Dark grey = Desired performance; Pale grey = Acceptable performance; Light grey = Unsavory performance.

points, indicating minimum practice, and 2–4 Star = 20–40 points, representing average to best practice, respectively. A 5-Star rating of 60 points or above represents Australia's best practice, and a 6-Star rating (75-point score or above) represents world leadership.

Secondly, the 'net promoter score' (NPS) 'is a summary statistic commonly used in commercial survey research to estimate the propensity of business customers to exhibit desirable behaviors' (Reichheld, 2003; Rocks, 2016). The NPS typically uses a marketing accountability metric known as a 'likelihood-to-recommend' (LTR) question to obtain responses from customers. A 0–10 Likert scale is presented in a single questionnaire for customer ratings (Baehre *et al.*, 2022). The NPS ranking criteria range from −100 to 0 (Needs improvement), 0–30 (Good), 30–70 (Great), and 70–100 (Excellent) (Retently, 2022). Note that the criterion range can also be in decimal form (i.e., −1.0–0.00, 0–0.3, 0.3–0.7, 0.7–1.0).

The study adopts a combination of Green Star's ranking system and the NPS's operational method of analyzing the audit scores. The reason being that the Green Star rating system is most relevant to the study as it concerns the performance assessment of built environment and land development projects, unlike NPS, which concerns customer recommendations for businesses. However, whilst we chose not to use NPS's ranking system, its operational method of calculating the net scores is statistically sound for the analysis of the categorical variables and hence adopted (see Table 11.4).

The study generally adopts the Green Star rating system with modified ranking criteria, as follows: scores from 0 to 30 are defined as unsavory (light grey color); scores from 30 to 60 are defined as acceptable (pale grey color); and scores up to 60+ are defined as desired (dark grey color).

11.5 FINDINGS AND DISCUSSION

This section uses a comparison approach to present the key findings of applying the multidimensional innovation district performance framework to the MPID (Sydney), MTP (Melbourne), and KGUV (Brisbane). The findings for each of the case districts are presented separately through a 'performance matrix' (Figure 11.5), and a 'typology matrix' (Table 11.5). Our discussion on the findings is presented for the context, feature, function, and form.

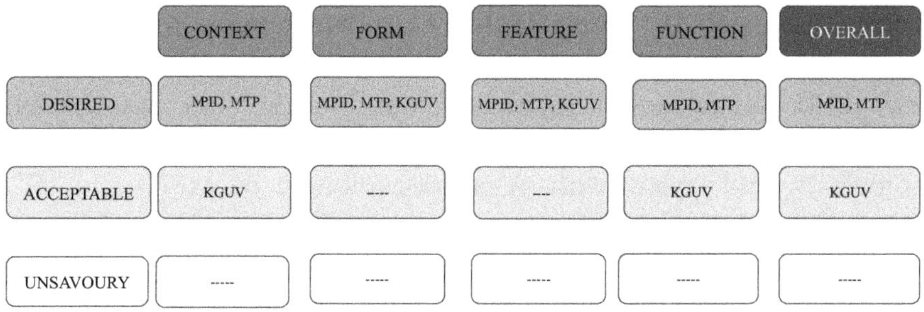

FIGURE 11.5 Performance matrix of the case innovation districts.

Note: MPID = Macquarie Park Innovation District; MTP = Monash Technology Precinct; KGUV = Kelvin Grove Urban Village.

TABLE 11.5 Typology Matrix of the Case Innovation Districts

Type	Name	Context	Function	Form	Feature	Overall	District
Type 1	Diverse, complex, and centrally located in a highly favorable context	Highly favorable	Diverse	Complex	Central	Desired	MPID MTP
Type 2	Diverse, complex, and centrally located in a favorable context	Favorable	Diverse	Complex	Central	Acceptable	KGUV

Note: Diverse = diversified (more than two) partnership or portfolio, Dual = diversified (two) partnership or portfolio, Single = specialized (single anchor/sector) investment. Complex = complex mixed use (work–learn–play–live), Semi-complex = mixed use (work–learn–play or live), Simple = single use (work or learn). Central = inner city or suburban, Remote = regional. Highly favorable = strong performance of hosting cities' systems (economic, societal, spatial, governance), Favorable = moderate performance of hosting cities' systems (economy, societal, spatial, governance), Unfavorable = weak performance of hosting cities' systems (economy, societal, spatial, governance). Desired = net scores above 60+, Acceptable = net scores above 30, Unsavory = net scores below 30.

Regarding the 'context' dimension, the investigation into the capital cities that host each of the case districts reveals that both Sydney's and Melbourne's performance are at the 'desired' level, with 'A' scores for the social and economic category compared to Brisbane's acceptable performance with 'B' scores (Table 11.4). Meanwhile, all three cities scored 'A' for the spatial and governance category, indicating their strength in this category. The investigation into the three case districts' 'forms' dimension reveals that KGUV is leading the category of 'complexity and layout' with 'A' scores, followed by MPID and MTP with a mixture of 'A' and 'B' scores. Furthermore, all three case districts equally achieved 'A' scores for the 'connectivity and space design' category, indicating their strength in this category. The investigation into the case districts' 'feature' dimension reveals that KGUV performed better than MPID and MTP in the 'centrality and amenity' category. The former district scored a mixture of 'A' and 'B' scores, whilst the latter two districts attained 'B' scores. On the other hand, all three districts performed well, with equal 'A' scores for the 'intelligence and concentration' category, indicating their strength in this category.

Finally, the investigation into the 'function' dimension of the case districts reveals that MPID and MTP are equally leading in 'specialization and diversity' with 'A' scores, whilst KGUV scored a mixture of 'A' and 'C' scores. In terms of the 'scale and support' category, MPID outperformed MTP with 'A' scores, whilst MTP scored a mixture of 'A' and 'B' and KGUV scored a mixture of 'A' and 'C' scores.

In sum, the discussion focuses on the 'A' and 'C' scores, as these are the subjects for calculating the net score. Based on the performance scores, MPID has consistent performance with 'A' scores (the desired level) for most of the indicators except for 'space design', 'social amenity', and 'locality setting', for which it scored 'B's (see Table 11.4). Note that it has no 'C' scores; hence, the 'A' score of 81% becomes an 81 net score. Likewise, MTP's performance is almost identical to MPID, except it has more 'B' scores than MPID. Its 'A' score of 75% becomes a 75 net score. KGUV, on the contrary, displays inconsistent performance as it exhibits 69% (n = 11) for 'A' scores, 19% (n = 3) 'B' scores, and 13% 'C' scores. The net score is calculated by subtracting 13% from 69%, resulting in a 56 net score. Hence, comparing the net scores, MPID's and MTP's performances are at the 'desired' level, whilst KGUV's performance falls into the 'acceptable' level.

The findings discussed above are summarized in the performance matrix presented in Figure 11.5.

Ranking of the case districts by performance scores shows that MPID is the top performer with net score of 81, edging out MTP on 75, both at the desired level, and Brisbane in third place with a net score of 56, which is an acceptable level.

Another important outcome of this study is to classify the case innovation districts into desired and acceptable typologies (see Table 11.5). Based on the final performance scores, MPID and MTP compose a typology named 'Type 1: Diverse, complex and centrally located in a highly favorable context' with the following characteristics.

Desired typologies: Suburban diversified (more than two) investment partnership dominated by multinational or large national firms. This type of innovation districts has:

- Complex mixed-use, semi-open space designs highly connected to external public spaces.
- Strong built environment and green and blue infrastructure.
- Strong human capital and skilled labor.
- Moderate social networking assets (social amenities).
- Technology-intensive firms.
- Macro (district-level) property management.

The second typology is named 'Type 2: Diverse, complex, and centrally located in a favorable context', which is represented by KGUV with the following characteristics.

Acceptable typologies: Inner city diversified (more than two) investment partnership dominated by small and medium enterprises/firms (SMEs). This type of innovation districts has:

- Complex mixed-use, open designs highly connected to external public spaces.
- Strong built environment and green and blue infrastructure.
- Strong human capital and skilled labor.
- Moderate social networking assets (social amenities).
- Business support services.

Macro (district-level) property management.

The results of the study show that both MPID's and MTP's performances are on par with each other at the 'desired' level, whilst

KGUV has an 'acceptable' performance level. Comparing their performance by dimensions, MPID and MTP exhibit consistent performances by sharing the desired performance cluster with KGUV for both the form and feature dimensions. However, in terms of the context and function dimensions, MPID and MTP are the two only districts with the desired performances.

Based on the net scores, the study ranked Sydney's MPID in first place with a score of 81, followed by Melbourne's MTP in second place with a score of 75. Brisbane's KGUV came in third place with a score of 56. The difference of six points between the MPID and MTP net scores indicates very high competition between these two districts and their hosting cities. MPID is the top scorer due to having stronger performance in 13 out of 16 indicators, whilst MTP has 12 out of 16, and KGUV is not far behind with 11 out of 16 strong performances. The difference in the scores between MPID and MTP is due to MPID being anchored by multinational enterprise—first-tier measure compared to MTP dominated by large national enterprise—second-tier measure. Meanwhile, KGUV's relatively lower score is because the district hosts more smaller- and medium-sized 'business support service' firms than large-to-multinational technology-intensive businesses.

In addition, KGUV displayed 'acceptable' performance for its context because the host city of Brisbane is lagging behind the other two cities, Melbourne and Sydney, which points toward the need for intervention to improve the areas of concern. This finding is consistent with previous related studies on smart cities (Yigitcanlar *et al.*, 2021; Tariq *et al.*, 2020), official government/industry reports on the world's most innovative cities (JLL, 2019; Bateman, 2022), and knowledge cities (Pratchett, 2017) that have ranked the performance of Sydney and Melbourne interchangeably between first and second place, whilst Brisbane continues to trail them in third place (BCC, 2018). Likewise, previous studies concerning MPID and MTP include Pancholi *et al.'s* (2019) research, which ranked the two districts as the top best practice innovation districts in Australia. On the other hand, KGUV is said to be Queensland's leading and best practice innovation district (Wardner & Hefferan, 2015; Esmaeilpoorarabi *et al.*, 2018).

In terms of context influence, the overall results of MPID, MTP, and KGUV is also influenced by the level of support from the respective states and city councils through their urban development policies and infrastructure development.

Based on the typology matrix developed in Table 11.5, two typologies are identified. MPID and MTP represent Type 1, namely 'Diverse, complex, and centrally located in a highly favorable context'. The main characteristics of the first typology include being initially developed and supported by more than two investment partners, having work–learn–play–live uses, designed as a semi-open space, and being in suburban areas. Innovation districts in this typology are dominated by multinational or large national firms in high-technology-intensive businesses with surrounding context (cities) having strong spatial, societal, economic, and governance systems. The second typology, Type 2, is 'Diverse, complex, and centrally located in a favorable context', represented by KGUV. Most of its characteristics are identical to typology 1's characteristics, except that they are designed as an open space, are in an inner-city area, and are dominated by small and medium business support service firms with the surrounding context having moderate-to-strong spatial, societal, economic, and governance systems.

It is important to note that the present study identified the innovation district typologies using only three case studies from the three largest cities of Australia. Therefore, we opine that the typology characteristics identified are only partial and should be expanded by future similar studies with more innovation district case studies to obtain a more representative typology characteristic. Hence, our future prospective study will investigate the performance of more innovation districts.

11.6 CONCLUSION

This study develops an innovation district typology matrix and evaluates its practicality with real-life innovation district data. Adopting the multidimensional innovation district performance framework to assess the performance of three eminent Australian innovation districts—i.e., Macquarie Park Innovation District (MPID), Monash Technology Precinct (MTP), and Kelvin Grove Urban Village (KGUV)—across the dimensions of form, feature, function, and context, this study confirms the suitability of the multidimensional performance framework to develop an innovation district typology matrix.

Based on the overall performance scores of the case innovation districts, MPID and MTP are placed in the desired performance category, whereas KGUV takes its place in the acceptable performance category. It is noteworthy that the top performers are suburban innovation districts, whilst the acceptable performer is an inner-city innovation district, indicating that innovation districts outside of inner-city locations can perform better than those located in inner cities. Furthermore, the study ranked MPID as the top performer, followed by MTP in second place and KGUV in third place. The six-point difference in the net scores between MPID and MTP indicates strong competition between them. Although KGUV has the potential to improve performance at the innovation district level, it will also rely on its host city of Brisbane to improve and boost its context and overall performance. The study also developed a typology matrix consisting of two distinctive typologies, where MPID and MTP represent Type 1: 'Diverse, complex, and centrally located in a highly favorable context' and KGUV represents Type 2: 'Diverse, complex, and centrally located in a favorable context'.

The findings of this study provide a number of practical implications. For example, identifying distinctive typologies for innovation districts is important for all stakeholders (i.e., urban planners, developers, managers, and local policymakers). Knowing and understanding the types and distinct characteristics of innovation districts can contribute to the decision making on where and what type of innovation district to develop, which industry type to invest in, and other areas that would most benefit from involvement. Consequently, these decisions and interventions can help innovation districts to be more successful.

Overall, the study sheds light on how to develop an innovation district typology matrix by carrying out a holistic assessment of the performance of innovation districts. The findings of this study inform our future research to implement the framework on existing innovation districts in South East Queensland (SEQ), Australia. They will also help in classifying the innovation districts into distinctive typologies for more precise assessment of their performances in their own typology groups.

The limitations of the study are as follows. First, the study took the context dimension as broad, and as it is not the primary focus of this study, not all of the context measures listed in Table 11.1 are assessed. Only measures with available data are selected—i.e., only diversity is audited and measured for the societal system. Furthermore, as the present study identified typologies using only three innovation districts, the typology characteristics identified are inevitably partial, meaning that future studies are needed to investigate a larger number of innovation districts to generate a deeper understanding of innovation district typologies. Addressing this issue will be the focus of our prospective research (Adu-McVie *et al.*, 2022).

ACKNOWLEDGMENTS

This chapter, with permission from the copyright holder, is a reproduced version of the following journal article: Adu-McVie, R., Yigitcanlar, T., Xia, B., & Erol, I. (2022). Innovation district typology classification via performance framework: insights from Sydney, Melbourne, and Brisbane. Buildings, 12(9), 1398.

REFERENCES

Australian Bureau of Statistics (2016). Australian Bureau of Statistics Data by Region. https://dbr.abs.gov.au/region.html?lyr=sa2&rgn=305031126.

Adu-McVie, R., Yigitcanlar, T., Erol, I., & Xia, B. (2021). Classifying innovation districts. Delphi validation of a multidimen-sional framework. *Land Use Policy*, 111, 105779. DOI:10.1016/j.landusepol.2021.105779

Adu-McVie, R., Yigitcanlar, T., Xia, B., & Erol, I. (2022). Innovation district typology classification via performance framework: insights from Sydney, Melbourne, and Brisbane. *Buildings*, 12(9), 1398. DOI:10.3390/buildings12091398

Advance Queensland. (2021). Advancing Regional Innovation Program. https://advance.qld.gov.au/entrepreneurs-and-startups-industry-investors-small-business-universities-and-researchers/advancing-regional-innovation-program.

Audretsch, B. D., & Belitski, M. (2022). The limits to open innovation and its impact on innovation performance. *Technovation*, 119, 102519. DOI:10.1016/j.technovation.2022.102519

Baehre, S., O'Dwyer, M., O'Malley, L., & Story, V. M. (2022). Customer mindset metrics: A systematic evaluation of the net promoter score (NPS) vs. alternative calculation methods. *J. Bus. Res.*, 149, 353–362. DOI:10.1016/j.jbusres.2022.04.048

Bateman, K. (2022). Cities and Urbanization. Which is the World's Most Innovative Cities in 2022? *World Economic Forum*. https://www.weforum.org/agenda/2022/02/innovative-global-cities-talent-property

Baum, S., Yigitcanlar, T., Horton, S., Velibeyoglu, K., & Gleeson, B. (2007). *The Role of Community and Lifestyle in the Making of a Knowledge City*. Griffith University.

BCC. (2018). Benchmarking Brisbane a Snapshot of Our Global Performance. https://www.brisbane.qld.gov.au/about-council/governance-and-strategy/business-in bris-bane/growing-brisbanes-economy/opportunity-brisbane/how-brisbane-compares

BCC. (2022). Brisbane's Key Economic Facts. https://www.brisbane.qld.gov.au/about-council/governance-and-strategy/business-in-brisbane/growing-brisbanes-economy/brisbanes-key-economic-facts.

BCC. (2019). Global Precincts. https://www.brisbane.qld.gov.au/about-council/governance-and-strategy/business-in-brisbane/growing-brisbanes-economy/opportunity-brisbane/opportunity-global-precincts

BCC. (2020). Opportunity: Kelvin Grove-Herston. https://www.brisbane.qld.gov.au/about-council/governance-and-strategy/business-in-brisbane/growing-brisbanes-economy/opportunity-brisbane/opportunity-inner-city/kelvin-grove-herston

Benefoh, D. T., Villamor, G. B., van Noordwijk, M., Borgemeister, C., Asante, W. A., & Asubonteng, K. O. (2018). Assessing land-use typologies and change intensities in a structurally complex Ghanaian cocoa landscape. *Appl. Geogr.*, 99, 109–119. DOI:10.1016/j.apgeog.2018.07.027

Boix, R., & Galletto, V. (2009). Innovation and Industrial Districts: A First Approach to the Measurement and Determinants of the I-District Effect. *Regional Studies*, 43, 1117–1133. DOI:10.1080/00343400801932342

Borsi, K., & Schulte, C. (2018). Universities and the City: From islands of knowledge to districts of innovation. *J. Arch.*, 23, 1143–1180. DOI:10.1080/13602365.2018.1514524

Business South Bank. (2017). Precinct. https://businesssouthbank.com.au/precinct

Cacar, K., & Aykol, S. (2020). Case study as a research method in hospitality and tourism research: A systematic literature review (1974–2020). *Cornell Hosp. Q.*, 62, 21–31.

Cohendet, P., Chenier, R., Simon, L., & Stojak, L. (2022). Centech, a world-class business incubator based in the Montréal innovation district, inspired by Barcelona 22@. *J. Evol. Stud. Bus.*, 7, 40–69.

Collier, D., Laporte, J., & Seawright, J. (2012). Putting typologies to work concept formation, analysis, and rigor. *Political Res. Q.*, 65, 217–232.

Cravo, T. A., & Piza, C. (2019). The impact of business-support services on firm performance: A meta-analysis. *Small Bus. Econ.*, 53, 753–770. DOI:10.1007/s11187-018-0065-x

Dahl, R. A. (1971). *Polyarchy: Participation and Opposition.* Yale University Press.

De Groot, R. S., Wilson, M. A., & Boumans, R. M. J. (2002). A typology for the classification, description and valuation of ecosystem functions, goods and services. *Ecol. Econ.*, 41, 393–408. DOI:10.1016/s0921-8009(02)00089-7

Díez-Vial, I., & Fern'andez-Olmos, M. (2015). Knowledge spill overs in science and technology parks: How can firms benefit most? *J. Technol. Transf.*, 40, 70–84.

DIIS. (2019). *Stocktake of Australian Innovation Precincts.* Australian Government, Canberra, Australia. http://www.data.gov.au/data/dataset/stocktake-of-australian-innovation-precincts

Durmaz, B., Platt, S., & Yigitcanlar, T. (2010). Creativity, culture tourism and place-making: Istanbul and London film industries. *Int. J. Cult. Tour. Hosp. Res.*, 4, 198–213.

Esmaeilpoorarabi, N., Yigitcanlar, T., & Guaralda, M. (2016). Towards an urban quality framework: Determining critical measures for different geographical scales to attract and retain talent in cities. *Int. J. Knowledge-Based Dev.*, 7, 290. DOI:10.1504/IJKBD.2016.078556

Esmaeilpoorarabi, N., Yigitcanlar, T., Guaralda, M., & Kamruzzaman, M. (2018). Does place quality matter for innovation districts? Determining the essential place characteristics from Brisbane's knowledge precincts. *Land Use Policy*, 79, 734–747. DOI:10.1016/j.landusepol.2018.09.016

Esmaeilpoorarabi, N., Yigitcanlar, T., Guaralda, M., & Kamruzzaman, M. (2018). Evaluating place quality in innovation districts: A Delphic hierarchy process approach. *Land Use Policy*, 76, 471–486. DOI:10.1016/j.landusepol.2018.02.027

Esmaeilpoorarabi, N., Yigitcanlar, T., Kamruzzaman, & Guaralda, M. (2020). How does the public engage with innovation districts? Societal impact assessment of Australian innovation districts. *Sustain. Cities Soc.*, 52, 101813. DOI:10.1016/j.scs.2019.101813

Forsyth, A. (2014). Alternative Forms of the High-Technology District: Corridors, Clumps, Cores, Campuses, Subdivisions, and Sites. *Environ. Plan. C Gov. Policy*, 32, 809–823. DOI:10.1068/c1267r

Gehrels, H., Meulen, S., Schasfoort, F., Bosch, P., Brolsma, R., Dinther, D., Geerling, G., Goossen, M., Jacobs, C., Jong, M., & et al. (2016). Designing Green and Blue Infrastructure to Support Healthy Urban Living. *Utrecht*. https://www.adaptivecircularcities.com/designing-green-and-blue-infrastructure-to-support-healthy-urban-living

Green Building Council Australia. (2021). Green Star Rating System. https://new.gbca.org.au/green-star/rating-system

Hansasooksin, S. T., & Tontisirin, N. (2021). Placemaking as an urban development strategy for making the Pattaya Innovation District. *Reg. Sci. Policy Pract.*, 13, 1930–1950. DOI:10.1111/rsp3.12400

Heaphy, L., & Wiig, A. (2020). The 21st century corporate town: The politics of planning innovation districts. *Telemat. Inform.*, 54, 101459. DOI:10.1016/j.tele.2020.101459

JLL. (2019). Australia's Got Talent: JLL Ranks World's Most Innovative Cities. https://www.jll.com.au/en/newsroom/australias-got-talent-jll-ranks-worlds-most-innovative-cities

Jones, A. L. (2017). Regenerating urban waterfronts-creating better futures- from commercial and leisure marketplaces to cultural quarters and innovation districts. *Plan. Pract. Res.*, 32(3), 333–344. DOI:10.1080/02697459.2016.1222146

Katz, B., & Wagner, J. (2014). *The Rise of Innovation Districts: A New Geography of Innovation in America*. Brookings Institution.

Kayanan, C. M. (2022). A critique of innovation districts: Entrepreneurial living and the burden of shouldering urban development. *Environ. Plan. A Econ. Space*, 54, 50–66.

Klein, R. (2014). How To Create A Building Typology? Typological matrix for mapping 19th century synagogues. *J. Fac. Civ. Eng.*, 30, 57–68. DOI:10.14415/zbornikgfs24.005

Koc, C. B., Osmond, P., & Peters, A. (2016). A Green Infrastructure Typology Matrix to Support Urban Microclimate Studies. *Procedia Eng.*, 169, 183–190. DOI:10.1016/j.proeng.2016.10.022

Komninos, N., Kakderi, C., Mora, L., Panori, A., & Sefertzi, E. (2022). Towards High Impact Smart Cities: A Universal Architecture Based on Connected Intelligence Spaces. *J. Knowl. Econ.*, 13, 1169–1197. DOI:10.1007/s13132-021-00767-0

Lawrence, S., Hogan, M., & Brown, E. G. (2019). *Planning for an Innovation District.* RTI Press.

Lee, K.-R. (2001). From Fragmentation to Integration: Development Process of Innovation Clusters in Korea. *Science Technology and Society*, 6, 305–327. DOI:10.1177/097172180100600202

Leon, N. (2008). Attract and connect: The 22@ Barcelona innovation district and the internationalization of Barcelona business. *Innovation (North Sydney)*, 10, 235–246. DOI:10.5172/imp.453.10.2-3.235

Macquarie Park Innovation District. (2022). About us. https://www.connectmpid.com.au/about-us

Markusen, A. (1996). Sticky Places in Slippery Space: A Typology of Industrial Districts. *Economic Geography*, 72(3), 293–313. DOI:10.2307/144402

Metaxiotis, K., Carrillo, F., & Yigitcanlar, T. (2010). *Knowledge-Based Development for Cities and Societies: Integrated Multi-Level Approaches.* Hershey, PA: Information Science Reference.

Monash Technology Precinct. (2022). About us. https://www.monashprecinct.com.au/about-us

Morais, P., & Camanho, A. (2011). Evaluation of performance of European cities with the aim to promote quality of life improvements. *Omega*, 39(4), 398–409. DOI:10/1016/j.omega,2010.09.003

Morawska, I., Anielska, K., Gądecki, J., & Afeltowicz, L. (2021). Changes in urban fabric–A cause or a result of an innovation district? *Journal of Urbanism*, ahead of print(ahead-of-print), 1–22. DOI:10.1080/17549175.2021.2005118

Morisson, A. (2020). A framework for defining innovation districts: Case study from 22@ Barcelona. In *Urban and Transit Planning*, 185–191. Cham, Switzerland: Springer.

Morisson, A., & Bevilacqua, C. (2019). Balancing gentrification in the knowledge economy: The case of Chattanooga's innovation district. *Urban Research & Practice.*, 12(4), 472–492. DOI:10/1080/17535069.2018.1472799

NSW Health. (2021). Healthy Built Environments. https://www.health.nsw.gov.au/urbanhealth/pages/default.aspx

NSW-IPC. (2018). *NSW Innovation Precincts: Lessons from International Experience*, 2018. NSW Government, Sydney, Australia.

Pancholi, S., Yigitcanlar, T., & Guaralda, M. (2019). Place making for innovation and knowledge-intensive activities: The Australian experience. *Technol. Forecast. Soc. Chang.*, 146, 616–625. DOI:10.1016/j.techfore.2017.09.014

Pancholi, S., Yigitcanlar, T., & Guaralda, M. (2015). Public space design of knowledge and innovation spaces: Learnings from Kelvin Grove Urban Village, Brisbane. *Journal of Open Innovation*, 1(1), 1–17. DOI:10.1186/s40852-015-0015-7

Pozas, M., & Gonzalez, F. J. (2018). Housing building typology definition in a historical area based on a case study: The Valley, Spain. *Cities*, 72, 1–7. DOI:10.1016/j.cities.2017.07.020

Pratchett, L., Hu, R., Walsh, M., & Tuli, S. (2017). The Knowledge City Index a Tale of 25 Cities in Australia. https://www.researchgate.net/publication/319472356_The_Knowledge_City_Index_A_tale_of_25_Cities_In_Australia_2017

Rashid, Y., Rashid, A., Warraich, M. A., Sabir, S. S., & Waseem, A. (2019). Case Study Method: A Step-by-Step Guide for Business Researchers. *Int. J. Qual. Methods*, 18, 1–13. DOI:10.1177/1609406919862424

Reichheld, F. F. (2003). The one number you need to grow. *Harvard Business Review*, 81(12), 46–54+124.

Retently. (2022). What Is a Good Net Promoter Score? https://www.retently.com/blog/good-net-promoter-score/

Rocks, B. (2016). Interval estimation for the "Net Promoter Score". *American Statistician.*, 70(4), 365–372. DOI:10.1080/00031305.2016.1158124

Roelandt, J. A. T., & Hertog, P. D. (1996, June). Assessing the knowledge distribution power of national innovation systems. In Proceedings of the OECD Conference on the New S&T Indicators for the Knowledge-Based Economy OECD, Paris.

Sarzynski, A., Galster, G., & Stack, L. (2014). Typologies of sprawl: Investigating United States metropolitan land use patterns. *Urban Geography*, 35(1), 48–70. DOI:10.1080/02723638.2013.826468

Schmahmann, L. (2019). Industry clusters and knowledge spill-overs-a comparative study of creative industry clusters in Sydney. *Planning News (Hawthorne, Vic.)*, 45(7), 10–12.

SGS. (2020). *From Alleys to Valleys: Creating Innovation Precincts through Inclusive Policy*. SGS Economics & Planning, Sydney, Australia.

Sisi, L. (2019). Streets place-making in Beijing's innovative districts: From the perspective of public assets and property rights. *China City Planning Review*, 28(3), 28–36.

Spirou, C. (2021). *Anchoring Innovation Districts: The Entrepreneurial University and Urban Change*. JHU Press.

Tariq, M. A. U. R., Faumatu, A., Hussein, M., Shahid, M. L. U. R., & Muttil, N. (2020). Smart City-Ranking of Major Australian Cities to Achieve a Smarter Future. *Sustainability*, 12(7), 2797. DOI:10.3390/su12072797.

Taylor, B. T., Fernando, P., Bauman, A. E., Williamson, A., Craig, J., & Redman, S. (2011). Measuring the Quality of Public Open Space Using Google Earth. *American Journal of Preventive Medicine*, 40(2), 105–112. DOI:10.1016/j.amepre.2010.10.024

Turkina, E., & Van Assche, A. (2018). Global connectedness and local innovation in industrial clusters. *Journal of International Business Studies*, 49(6), 706–728. DOI:10.1057/s41267-018-0153-9

Victoria Walks. (2021). Urban Design for Walking. https://www.victoriawalks.org.au/urban_design

Wardner, P., & Hefferan, M. (2015). Kelvin Grove Urban Village, Brisbane post implementation: Lessons for new urbanism. Australasian. *Journal of Regional Studies*, 21(3), 373–395.

Watson, C. (2020). Difference between the Mean & the Average. https://sciencing.com/difference-between-mean-average-6461324.html

Wong, P. K. (2022). An Evolutionary Analysis of the Development of the One North Innovation District in Singapore. *Journal of Evolutionary Studies in Business*, 7(2), 70–99. DOI:10.1344/jesb2022.2.j107

Wu, Y., Yang, Y., Xu, W., & Chen, Q. (2020). The Influence of Innovation Resources in Higher Education Institutions on the Development of Sci-Tech Parks' Enterprises in the Urban Innovative Districts at the Stage of Urbanization Transformation. *Land (Basel)*, 9(10), 1–36. DOI:10.3390/land9100396

Yang, T., & Wang, N. (2008). The Cultivation of Cluster's Sustainable Competence Based on Knowledge Management. *International Journal of Business and Management*, 3(7), 83–88. DOI:10.5539/ijbm.v3n7p83

Yigitcanlar, T., Adu-McVie, R., & Erol, I. (2020). How can contemporary innovation districts be classified? A systematic review of the literature. *Land Use Policy*, 95, 104595. DOI:10.1016/j.landusepol.2020.104595

Yigitcanlar, T., Corchado, J. M., Mehmood, R., Li, R. Y. M., Mossberger, K., & Desouza, K. (2021). Responsible Urban Innovation with Local Government Artificial Intelligence (AI): A Conceptual Framework and Research Agenda. *Journal of Open Innovation*, 7(1), 71. DOI:10.3390/joitmc7010071

Yigitcanlar, T., Dodson, J., Gleeson, B., & Sipe, N. (2007). Travel self-containment inmaster planned estates: analysis of rece nt Australian trends. *Urban Policy and Research*, 25(1), 129–149.

Yigitcanlar, T., & Inkinen, T. (2019). *Geographies of Disruption*. Springer International Publishing.

Yigitcanlar, T., Sabatini-Marques, J., da-Costa, E., Kamruzzaman, M., & Ioppolo, G. (2019). Stimulating technological innovation through incentives: Perceptions of Australian and Brazilian firms. *Technological Forecasting & Social Change*, 146, 403–412.

Yun, J. J., Zhao, X., Yigitcanlar, T., Lee, D., & Ahn, H. (2018). Architectural Design and Open Innovation Symbiosis: Insights from Research Campuses, Manufacturing Systems, and Innovation Districts. *Sustainability (Basel, Switzerland)*, 10(12), 4495. DOI:10.3390/su10124495

Zainal, Z. (2007). Case study as a research method. *Jurnal Kemanusiaan.*, 5(1), 1–6.

Zheng, J. (2011). Creative industry clusters and the entrepreneurial city of Shanghai. *Urban Studies (Edinburgh, Scotland)*, 48(16), 3561–3582. DOI:10.1177/0042098011399593

Zuo, J., Xia, B., Chen, Q., Pullen, S., & Skimore, M. (2016). Green building rating for office building-Lessons learnt. *Journal of Green Building*, 11(2), 131–146. DOI:10.3992/jgb.11.2.131.1

5

District Smartness

This part argues that there are limited studies investigating university and innovation district symbiosis, particularly in the context of place making. The part aims to investigate the socio-cultural role played by anchor universities in facilitating place making in innovation districts. The results revealed four key roles universities play, including the pivotal roles as facilitators of place making through societal integration in innovation districts and they act as platforms for collaboration and exchange of knowledge by adopting extroverted and collaborative approaches. Additionally, this part of the book explores how the smart campus is conceptualised and practiced as a smart district. The part reports on the process of conceptual framework development through the systematic literature review. This part identifies that society, economy, environment, and governance are the key smart campus domains to classify existing research. Nevertheless, the part also reveals that there is little evidence of a comprehensive real-world application of the smart campus towards addressing all four domains. This confirms the infancy of the current conceptualisation and practice of a smart campus as a smart precinct. The findings reported in this part offer a firm basis for comprehensive smart campus conceptualisation and provide directions for future research and development of smart campuses and smart districts.

DOI: 10.1201/9781032657431-16

CHAPTER 12

University and Innovation District Symbiosis

12.1 INTRODUCTION

In the knowledge and innovation economy era, cities and regions are investing in knowledge-based urban development precincts (KBUD) as a strategy to reinvigorate their local economy, society, and built environment (Yigitcanlar, 2014; Esmaeilpoorarabi *et al.*, 2020a). In this context, universities have acquired a renewed central position in society and development with the establishment of knowledge and creativity as the main engines for growth (Yigitcanlar & Sarimin, 2011; Morisson & Bevilacqua, 2019). Globally, KBUD has been increasingly positioned as an elixir for the contemporary and anticipated future growth challenges faced globally (Benneworth & Ratinho, 2014; Carrillo *et al.*, 2014). Cities around the world are investing into this kind of development, promoting innovation districts as spatial agglomerations of knowledge-intensive activities, universities, research institutes, and firms as well as engaging knowledge workers and local communities (Leon, 2008; Metaxiotis *et al.*, 2010).

Various terms are employed in associated literature to describe innovation districts, such as knowledge precincts, innovation precincts, science cities, science parks, technology parks, or knowledge hubs (Clark *et al.*, 2010; Pancholi *et al.*, 2017; Esmaeilpoorarabi *et al.*, 2020b). While Silicon Valley (USA) is one of the pioneers, other international best practice examples include 22@bcn (Barcelona), Arabianranta (Finland), One-North (Singapore), and Digital hub (Ireland). Innovation districts often adopt a quadruple helix model of governance—i.e., a partnership

DOI: 10.1201/9781032657431-17

between public, private, academia, and community—and most of them are connected to anchor universities (Youtie & Shapira, 2008; Brake & Kress, 2011; Edvardsson *et al.*, 2016).

During the last two decades, the coalescence of arts, culture, and creativity with technology as well as the acknowledgement of 'open innovation' has resulted in a multi-layered transformation in the structure of innovation districts—from their earlier introvert models to a more extrovert, mixed-use, vibrant, and connected environment (Chesbrough, 2003; Van Winden *et al.*, 2013). Nevertheless, innovation districts have been increasingly scrutinised for their contribution to the unintentional rise of specific socio-cultural conflicts and challenges in society (Espelt-Lleonart, 2012; Yigitcanlar *et al.*, 2020). These include—but are not limited to—gentrification, spatial segregation, social inequality due to marginalisation of certain groups, development of isolated societies, and lack of stakeholder coordination (Stehlin, 2016). As a result, the role of universities in this dynamic globalised space is continuously being redefined.

Despite the existence of extensive literature analysing the positioning of academia in the landscape of innovation and knowledge, a closer evaluation of these studies reveals that each of these has utilised a distinct lens—i.e., either institutional, economic, political, or formal networking lens. More importantly, universities' role in facilitating the place-making capacity or ability of innovation districts—acknowledged as a critical strategy to overcome these socio-cultural challenges and achieve the desired multidimensional developmental denouement from these locations—has not yet been thoroughly investigated (Yun *et al.*, 2018; Pancholi *et al.*, 2019).

This study aims to investigate the socio-cultural role played by anchor universities in facilitating place making in innovation districts. In this chapter, the particular interest of place making in innovation districts is to establish a cohesiveness within the district community and societal integration of the district with its surrounding communities at the local level. Henceforth, the two questions this study targets to answer are: (a) What the socio-cultural roles of anchor universities in innovation districts are in achieving a cohesive district community and societal integration with surrounding communities are, and (b) what the key policy implications for strengthening the role of universities as a critical place-making facilitator in the innovation districts are.

The chapter takes the Australian context for such an investigation. This selection is justified as follows. In their efforts to strengthen and establish a knowledge and innovation economy, Australian cities have increasingly invested in and branded themselves with the help of their innovation districts. Australian Technology Park (Sydney), Macquarie Park Innovation District (Sydney), Monash Employment and Innovation Cluster (Melbourne), Parkville Technology Precinct (Melbourne), Kelvin Grove Urban Village (Brisbane), and Diamantina Knowledge Precinct (Brisbane) are among the few examples including planned innovation districts as well as some organically formed ones (Yigitcanlar *et al.*, 2017). From those, two globally renowned cases from Australian cities are selected for investigation—namely Macquarie Park Innovation District (MPID) in Sydney and Kelvin Grove Urban Village (KGUV) in Brisbane. While these two cases are similar in terms of their underlying structure, where a university acts as an anchor, the differences in their spatio-temporal growth trajectory provide an interesting contrast—that became the reason to choose them.

The study adopts a qualitative approach grounded in a multi-dimensional framework for investigating university and innovation district symbiosis in the context of place making. The study generates insights into the redefined role of contemporary universities in innovation districts and reinforces their contribution as place-making facilitators in existing and emerging innovation districts.

12.2 LITERATURE BACKGROUND

12.2.1 Universities and the Knowledge and Innovation Economy

Several scholars have investigated the role of universities in the context of innovation districts and knowledge and innovation economy (Blakely & Hu, 2019). Spatially, the central location of the university—in most of the cases—as well as the considerable scale of its physical presence, acts as an anchor for the landscape of an innovation district (Pancholi *et al.*, 2019). While the university acts as the spatial nexus for the generation of creativity and innovation in the form of research that happens in its laboratories, it also plays a crucial role in placing any innovation district onto the global innovation landscape. Goldstein & Glaser (2012) discuss the role of the university in attracting talents and creative people. Studies have acknowledged the critical role of universities in the development of knowledge cities and the development of a location as an innovation hub

(Youtie & Shapira, 2008; Brake & Kress, 2011; Edvardsson *et al.*, 2016). In an exploration of the dynamics that come into play in an innovation landscape—owing to the required confluence of different sectors—the governance role of the university in improving the administrative capacity and effectiveness is critical (Mok, 2016).

The role of universities as triggers of innovation activities are widely acknowledged, where innovation is seen as the generator of positive economic and societal externalities (Abramo *et al.*, 2009; Aldieri *et al.*, 2018). Subsequently, in recent years, universities have started to adopt a culture of openness and external collaborations in order to initiate a spill-over effect (Aldieri *et al.*, 2019). Moreover, in the landscape of innovation districts, which involves multiple stakeholders, universities play a critical role in strengthening local/regional/national/international innovation networks (Casaleiro, 2011; Charles, 2011; Romein *et al.*, 2011; Den Heijer & Magdaniel, 2012). Henceforth, as an integral part of innovation districts, the role of universities—at metropolitan and regional levels—has extended from educational and research functions to the generation of innovation and creativity. The global rise of digitalisation has also resulted in bringing innovation in the process of how innovation districts are being shaped, function, and managed (Panori *et al.*, 2020).

The triple helix model—elaborated in more detail in the next section—propounds restructuring of the university from its traditional educational model to an entrepreneurial model as the source of regional economic development (Smith & Bagchi-Sen, 2010). The economic contribution of universities—by acting as a stimulator of economic activities related to innovation—is highlighted by the research undertaken by Goldstein & Glaser (2012). The contemporary economic model is also based on networks—rather than the earlier linear model of innovation—focusing on new and sustainable business/industry activities such as incubation of spin-offs from research in the form of incubators, accelerators, and technology or innovation districts (Arbolino *et al.*, 2018). Their contribution at the regional level is specifically singled out by other studies (Aquilani *et al.*, 2018). Focusing on the contribution of universities at the regional scale, Nord (2002) derives four critical roles of universities in the context of Nordic countries. These are: (a) providing access for higher education; (b) strengthening professional education and training; (c) contributing in regional economic growth; and (d) fostering new social and cultural development opportunities.

12.2.2 Innovation Districts and the Shifting Context of Anchor Universities

In order to understand the overall context of innovation districts, and how a shift of spatio-functional positioning of universities in these districts is happening, it is necessary to have a closer look at the evolution of innovation districts. The earliest typology of innovation districts—also the most documented one—was the agglomeration of firms, businesses, start-ups, research institutes with formal and operational links with the anchor university. This type, commonly known as science parks, were managed mainly by public companies with the government and university as crucial stakeholders (Zhang, 2005). With the increased prominence of technology as a strategic sector, the rationale to invest in science parks was to foster economic growth and enhance the economic impact of research (Phan *et al.*, 2005; Esmaeilpoorarabi *et al.*, 2020c). Stanford Research Park or Cambridge Science Park are some of the well-known initiatives undertaken by their respective universities to enhance knowledge spill-over and develop connections with industry (Van Winden *et al.*, 2013). In recent years, the prevalence of cities investing into their universities to attract knowledge-intensive global firms and as a source of regional economic diversification has increased—as seen in Singapore Science Park that hosts numerous multinational companies (Carvalho, 2009).

Nonetheless, with globalisation, the emergence of 'creative industries' as one of the pivotal growth sectors, and acceptance of the model of 'open innovation', there has been a shift in the geographies and sectoral structure of these innovation districts (Chesbrough, 2003; Durmaz *et al.*, 2010; Yigitcanlar *et al.*, 2016). With the integration of arts and creativity with local culture, these innovation districts are emerging as mixed-use urban districts to satisfy the specific requirement of the 'creative class' (Florida, 2005; Katz & Wagner, 2014). Socio-spatially, a clear transposition from the earlier 'work-learn' model to 'live-work-learn-play' urban developmental model is evident. These creative districts are defined by their socio-cultural diversity and vibrancy, network-rich and tolerant social environment, integrated culture and heritage, aim for socio-cultural denouements along with the economic outcomes (Pancholi *et al.*, 2015a; Esmaeilpoorarabi *et al.*, 2018a; Yigitcanlar & Inkinen, 2019).

In parallel to the recent spatio-physical and socio-cultural shift in the structure of innovation districts, there has also been an evident shift in their organisational structure. Etzkowitz & Leydesdorff (1995) highlight the 'triple-helix' model of partnership, in which innovation is generated by the interaction of three key spheres of academia, state, and industry. Nevertheless, in recent years, several innovative knowledge locations such as 22@Barcelona (Spain) and Arabianranta (Finland) have gone a step further by integrating the community into this helix and augmenting this to a quadruple-helix model (Van Winden *et al.*, 2013; Pancholi *et al.*, 2015b). In organisational terms, thus, innovation districts have witnessed a metamorphosis from their earlier triple-helix model to progressively adopting a quadruple-helix model of governance. The foundation of KBUD rests upon the confluence of four streams—public, private, academia, and, increasingly, community (Zhao, 2010). Organisationally, therefore, this multi-sectoral confluence, as seen in innovation districts, has strengthened foundations of these knowledge-nurturing locations. The role of universities in this context—acknowledged to be acting as a prime anchor for knowledge and innovation in a myriad of ways—has also been shifting, as discussed in the earlier section.

12.2.3 Place Making for Innovation Districts

It has been acknowledged that the success of KBUD rests upon the exchange of knowledge and creativity and the development of formal and informal networks at local and global levels. Nevertheless, studies have highlighted their failure—to a large extent—in doing so (Florida, 2005; Baum *et al.*, 2009). Due to the nature of the activities and functions they generate; innovation districts have been criticised for developing into islands of innovation within the cities in which they are located. They have been increasingly scrutinised for developing into locations/entities that are spatially and socially disconnected from their local surroundings—leading to gentrification as well as marginalisation of specific social and community groups (Toolis, 2017).

With an ever-increasing acknowledgement of the role of public participation, planners and developers are attempting to accommodate public consultation exercises in the process of developing these spaces. Nonetheless, owing to the conflicting mutual interests and past attachments of community, they have been facing socio-cultural friction in the process (Esmaeilpoorarabi *et al.*, 2018b; Pancholi *et al.*, 2018a). Besides, to keep up with the fast pace of the knowledge and innovation economy, it is also critical to facilitate a fast conception, execution, and delivery of these innovation districts. As a result of these various factors, the socio-cultural integration of these locations/districts in the local context emerges as one of the critical challenges faced by innovation districts.

Globally, place making has been adopted as a dynamic strategy to address the main challenges of these innovation districts (Pancholi *et al.*, 2019). Recent simultaneous studies have also established that place making needs to be defined multidimensionally through a lens that considers 'place' in its physical attributes as well as the socio-economic, political, and organisational processes that happen there, people's perceptions and experience of a locale as well as the context that shapes it (Pancholi *et al.*, 2015a). While the role of place making has been established in terms of attracting and retaining knowledge workers, recent studies have questioned the extent of its role in addressing societal integration issues. Studies conducted by Pancholi *et al.* (2017; 2018a) in MPID Sydney and Diamantina Knowledge Precinct (Brisbane) aimed at understanding place making in these innovation districts revealed how a space earlier defined by conflicts was transformed into a space defined by coordination. These studies also revealed the extension of the role of place-making strategies as per the unique requirements of these globalised spaces arising from their challenges, as mentioned earlier. Being placed into this globalised context, with an increased acknowledgement of place making, the role of academia—one of the key pillars over which quadruple-helix models of innovation districts rests—has transformed today to acquire more comprehensive roles.

12.3 METHODOLOGY

This study is an empirical investigation grounded in qualitative analysis of the literature and spatio-temporal investigation of the case studies of renown innovation districts from Australia. Adoption of a qualitative investigation is justified as follows: (a) qualitative research, also called soft science, is more likely to take place in a natural/physical setting investigating intangible aspects; (b) qualitative research is exploratory that involves the collection, analysis, and interpretation of data by observing human behaviour; (c) qualitative research provides a more realistic feel of the world that cannot be experienced in the numerical data (Boodhoo & Purmessur, 2009); and (d) qualitative research is a prominent method in place-making studies (Morisson & Bevilacqua, 2019).

A multidimensional conceptual framework of place making, developed by Pancholi *et al.* (2017), has guided the analysis (Figure 12.1). This framework identifies place in any innovation district to be defined

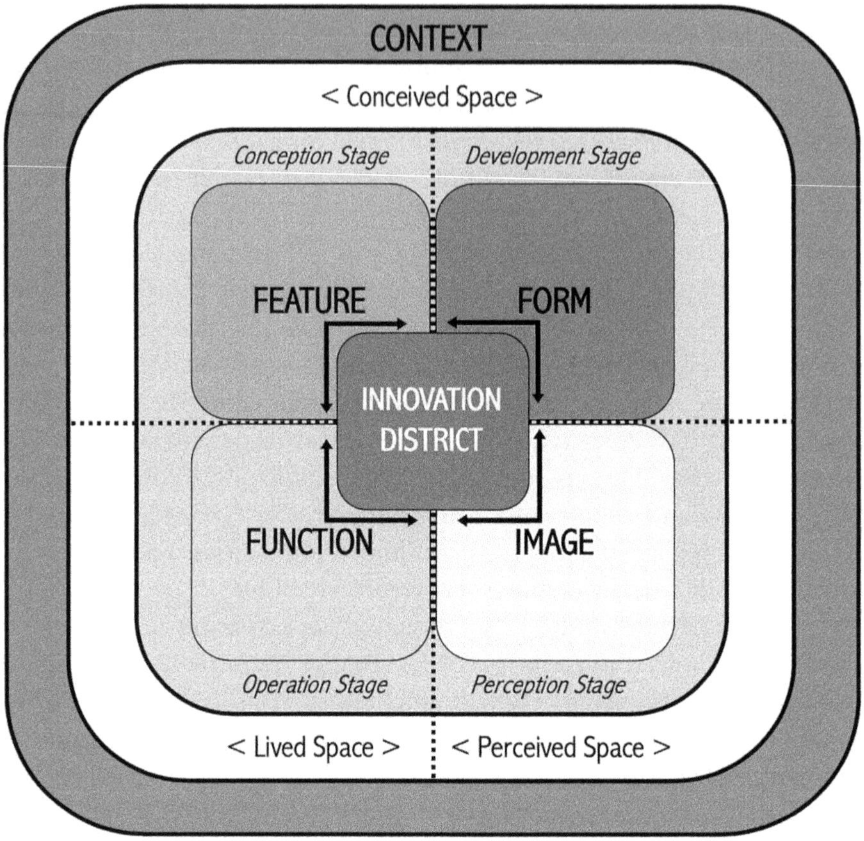

FIGURE 12.1 Conceptual framework of innovation district place making (derived from Pancholi *et al.*, 2017).

through its four core dimensions—i.e., (a) Feature, (b) Form, (c) Function, and (d) Image—each influenced by the surrounding context. The framework is, hence, being adopted, owing to its holistic approach in comprehending the role of universities in the landscape of innovation districts. This provides a comprehensive perspective through a multi-layered lens, examining our objective in four key aspects: (a) projecting innovation districts through the dimension of 'feature'; (b) conceiving them through the lens of 'form'; (c) exploring how they are lived in terms of 'function'; and (d) understanding how they are perceived, focusing on the aspect of 'image'. To understand the socio-cultural role of universities in its totality, it is imperative to explore it spatio-temporally as well as through the perception of its users and key stakeholders.

TABLE 12.1 Data and Research Methods

Place-Making Dimension	Data Source	Type of Data	Research Method
Feature	Government reports, design and policy documentations, scholarly literature, and innovation district websites	Secondary	Content analysis
Form	Aerial images and innovation district planning schemes	Secondary	Morphological analysis
Function	Key stakeholders, innovation district users, and local residents	Primary	Focus group and interviews Observations
Image	Key stakeholders, innovation district users, and local residents	Primary	Focus group and interviews Observations

Moreover, to ensure the validity and robustness of findings, different qualitative data, such as interviews, focus groups, and observations, were triangulated to investigate 'function' and 'image' (Farmer *et al.*, 2006). Spatio-physical data such as maps were utilised to analyse 'form'. Secondary documents, such as government design and policy documents, scholarly literature, past research studies, and websites of innovation districts were utilised for analysing 'feature' and setting up the context of the study. Table 12.1 provides a summary of the data and research methods.

This framework considers place making as a continuum, and space to be continuously shaped through different stages of a project—i.e., stages of conception or designing, development, management, functioning, and perception by users. Considering the specialised context of innovation districts, the first two dimensions of this framework, i.e., 'feature' refers to the conceived intangibles such as amenities, innovation infrastructure, and support; management; 'form' refers to conceived tangibles such as the physical and built environment. The third dimension, 'function', refers to lived space or socioeconomic activities and processes happening in the innovation district. The fourth dimension, 'image', refers to perceived space or space as shaped by the perception of users. This framework was developed, intersecting urban morphology and phenomenography as the dominion of enquiry. It builds on Lefebvre's theory, assuming that place is the production of concurrent tangible and intangible factors, as mediated by people's experience (Lefebvre &

Nicholson-Smith, 1991). In adopting this framework, the research study sets out the evolutionary context for each case and then specifically explores the socio-cultural role of the university in the spatio-temporal transformation of a place by looking at it through the lens of each core dimension—i.e., feature function, form, and image. Conclusions are drawn on the basis of inductive generalisations from both cases as well as understanding the unique patterns that shape the socio-cultural layer of each innovation district (Glaser & Strauss, 1967).

The case studies selected for this study include MPID (Sydney) and KGUV (Brisbane). The former is home to Macquarie University (MU) and the latter is home to the Queensland University of Technology (QUT). These cases are purposefully selected because of their mature state of development, their location within the broader urban fabric, as well as for their attributes of being rich in information, revelatory, and unique concerning each other (Baxter & Jack, 2008). To ensure the robustness of findings, these two cases were selected in order to represent distinct contexts in terms of spatial and economic scale, dominant sectors, and historical evolution and, hence, present unique conditions and challenges (Johansson, 2007). Spatially, in terms of area and economically (hosting over 180 large international and 200 small businesses), the scale of MPID is much larger than KGUV. Besides, they are unique concerning each other in terms of the sectors they host. While MPID has several technological, bio-tech companies, KGUV is primarily a 'creative industries' focused precinct. In addition to this, they also display different historical evolution. While MPID is a spontaneously formed innovation district with later interventions, KGUV is a deliberately planned district from its initial stages. For both, districts' participants were recruited and engaged in providing qualitative data on their experience. For KGUV, five formal interviews and two focus group experiments, involving nine participants, were conducted, with five and three residents, respectively. For MPID, a total of 12 interviews were undertaken.

12.4 RESULTS

12.4.1 Macquarie Park Innovation District

MPID forms one of the core axes of the global economic corridor of Sydney. It is 12 km from the centre of Sydney (Hawken & Han, 2017). Historically, the site had a series of land-use transformations, from the land of Aboriginals to a rural hamlet. In addition to traditional housing, there is an on-site housing estate, Ivanhoe Estate. The redevelopment resulted in a mixed-housing neighbourhood with 556 social housing dwellings (Pancholi *et al.*, 2018b). With amenities such as the Macquarie shopping centre, strong connectivity by road and rail, proximity to national parks, bushland, as well as being home to MU, this area has massive potential for becoming a community-oriented innovation district. In conjunction with the university, the government has also shifted its focus to make societal integration as key future strength and source of appeal for MPID (Pancholi *et al.*, 2018b). The contribution of the university—that acts as a critical anchor of the site—in place making through its four dimensions has been as follows.

Feature-wise, in the 1960s, the establishment of the university was a key spatio-temporal and politico-economic landmark in the development of the site as an innovation district. In line with the Stanford University model, the area was identified as an industrial growth area with the aim of lubricating a possible exchange of knowledge and innovation between university and companies with their agglomeration and clustering. Subsequently, ancillary developments such as housing, shopping centres, and infrastructure constructed—paving the way for a cluster formation. While today, the area boasts top-100 ranked global companies located on-site, the origin of cluster formation happens to be the university acting as the key anchor of the site. The university also gets featured as a critical resource in terms of available land—owing to its large site—and innovation-supportive infrastructure as stated by one of the interviewees, "*Universities are quite innovative in themselves. They have a lot of land, that they've got business parks inside their university*". Institutionally, the university has ventured into the creation of a networking group that aims to bring community, businesses, government, and academia together. In addition to this, other organisational infrastructures such as the establishment of groups focused on transport, i.e., 'Connect Macquarie'—a joint initiative by the NSW government, City of Ryde, and local businesses—has helped to establish a network between employees, community, and other users of the district.

Form-wise, while initially, MPID was criticised for development as an isolated island—with corporate offices behind boom gates, creating islands within an island. Recently proposed plans aimed to break the monotony of space. The innovation district is being planned to host a mix of housing and entertainment precincts on the site. In terms of scale, the size and location of the university on the site is significant yet

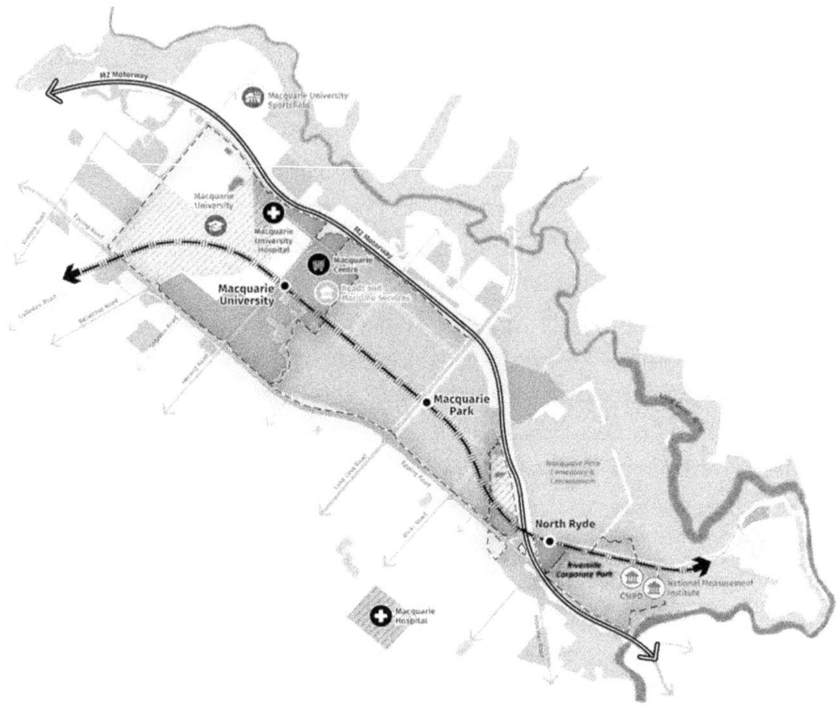

FIGURE 12.2 Layout of MPID (derived from MU, 2017).

confined to one end of the site (Figure 12.2). The central axis leads to MU; hence, holding the whole site together at one end. In the proposed government plans, the university plays a vital role by aligning their internal design and developmental plans with the proposed plans by the government for the development of MPID. The vision of MU, as outlined by one of the interviewees, depicts "*Our vision is for a borderless campus. So that we have innovation between the businesses, within the community and the university*", and further specifies, "*In the same breath is to create a transparency and accessibility through the university to the outer areas to get better connectivity*". To manifest this vision, a few key physical initiatives by the university include: (a) conversion from zoned land use to mixed-use land use; (b) commercial zones planned at the fringes; (c) visual display of creativity; and (d) providing more pedestrian-friendly campus by clearly identifiable entry statements and engaging spaces. This approach is appreciated by the interviewees from a community group, as evidenced by the comment, "*The university is doing a pretty good job in developing their public domain as well*". In addition to this, the university is open to the community with people having access to its amenities and infrastructure. One of the interviewed community members points out, "*We all use the university facilities for the swimming lesson, and for the gym*".

Function-wise, this research aimed to analyse the contribution of the university in strengthening the networks with the community. While it has been a challenge in the past to develop a collaborative environment amongst businesses and academia, the university has always been an epicentre for the dissemination of knowledge to the local community. In the absence of an adequately institutionalised common management at an innovation district level, the university itself is proactively running programs to engage the community. One of the interviewees at a key position in the university confirms, "*Programs such as 'path bisector'—an initiative for engaging students with community groups where it was a requirement for them to participate and work in conjunction with industry partner or community partner in order to get a degree—were run by the university*". More importantly, MU is considered by the community as a key socialising space where they not only get to congregate, play, and attend joint programs, events, public workshops, festivals, and seminars. Integrating with local community being one of the key agendas of the university, it helps in developing a well-informed community—which is more significant. Stated by one of the interviewees, "*University has a series of programs with the community, school programs, sports programs and other events that we invite them into campus, there is always something going on with the local community*". These initiatives, however, remain limited to specific groups due to the absence of an overall formalised group dedicated to organising such events or concerts to upscale the activities of community integration at the innovation district level.

Image-wise, the university is named after former NSW Governor Lachlan Macquarie. There are stories attached to the district that have shaped its past, which the community believes should help in the development of a sense of pride and belonging to this site. As one of the interviewees thoughtfully argues, "*What the university tries to take in actual naming of the university. So, it shows this is a place that has a history, that has a story, and these stories actually give us a sense of identity, a sense of connection*". Likewise, the existence of the university acts in the development of a sense of trust in the perception of the community. This is confirmed by the following statement, "*So you really need that air to collaborate, an environment that the university has*".

Another one points this out by distinguishing between university and corporate environment, "*If there's going to be any potential for collaboration, that's been right there in the university? Not here, this is corporate*". There is a growing consensus amongst the businesses and community regarding the university's role as collaborator, as highlighted by the following statement, to create a sense of coordination between stakeholders, "*The university is probably the place they are going to get through creative cooperation, rather than private creativity*".

12.4.2 Kelvin Grove Urban Village

KGUV is a joint initiative of the Queensland Government Department of Communities and QUT. With QUT as its key anchor, this master-planned community was developed as an exemplar model for compact, sustainable, and mixed-use urban development. It was developed with the aim of achieving a vibrant, healthy, diverse, and socially sustainable urban community that has access to a range of cultural, health, and educational resources (Pancholi *et al.*, 2015b). The masterplan acknowledged the importance of a strong organisation and good design principles, as well as consideration for the intangible aspects of community development and social inclusion (Byrne, 2004). The dominant sector here is creative industries—making its character dominated by integration of arts and culture.

Feature-wise, KGUV established itself as an innovative partnership of academia and public including Queensland government with QUT—along with the close involvement of the Brisbane City Council. It was a result of a comprehensive multi-disciplinary planning approach and a program of agency, industry, and community consultations. Hence, it emerges as a strong case of the triple-helix model of governance. This innovation district was branded and projected as a 'village' and 'diverse city fringe community'. To manifest this, the strategies for its design and development aimed at achieving place making, community and cultural development, economic and social sustainability outcomes. To ensure the development of a well-knit community, key strategies include appointing a community development manager, forming a community association, and using university services to form a community well-being program. Key stakeholders joined hands to establish 'Communify'—a community development and support group—with projects funded by the Queensland Government. It operates the Exchange Community Centre that runs a series of projects and sessions to build a sense of community—working particularly with disadvantaged and disenfranchised groups. Some of its initiatives include management of community gardens, homework club, knitting group, English classes, and conversations group work. University students are involved either as a part of their placements and work experience required by their courses or voluntarily in their homework club sessions. Organisation of a Saturday farmers market is also one of such initiatives where local farmers set out their stalls of local and organic products, as well as crafts and a diverse range of international foods. High rental prices emerge as key issues as evident by the following interviewee statements, "*I agree it is too expensive*" and "*Residential space in Kelvin Grove is highly unaffordable to either rent or buy*".

Form-wise, KGUV was developed, based on new urbanism principles (Klaebe, 2006); seen as an 'urban social laboratory', this project aimed to cater to diverse social groups and was developed as compact, mixed-use, and diverse built environment. Spatially, availability of wide, even pathways, active public spaces, and green spaces defined its form (Klaebe, 2006). The university has a dominant presence within KGUV, with its buildings dotted along the whole site (Figure 12.3). It offers arts,

FIGURE 12.3 Layout of KGUV (derived from QUT, 2004).

culture, entertainment areas, and cafes with blurred boundaries depicting no public-private distinction. QUT, therefore, has a key role in enhancing physical permeability as well as the integration of arts and culture. This is evident in the innovative design and sustainability principles of their buildings, the permeability of lower floors depicted through the public display of artwork as well as the respectful renovation of heritage buildings and local architecture (Ingrao *et al.*, 2018). KGUV has received several awards for urban design, such as the 2001 RAPI Queensland Award for Excellence in Urban Planning, the 2002 RAPI National Award for Urban Planning Achievement, and the 2003 Queensland Premier's Award for Sustainable Environment. Interviews and surveys revealed that people were mostly appreciative of the mixed-use environment that ensures availability of all amenities, food, and shopping at their doorstep. Nonetheless, interviewees simultaneously pointed out the need to have places of social interaction and community spaces, which incite "*More community interaction*" and "*Community space! Bringing people together where people can sit and enjoy*". Conversely, issues like lack of integration of art in public spaces, green spaces, clarity regarding the distinction between public and private, uneven topography, and climate-responsiveness of urban form are vital factors that emerged as a repellent for the user-group. Respondents considered the village to be "*Heartless*", vibrant, or lacking a personality, despite being a creative precinct. Interviewees expressed their preference to include more display of art and creativity in the innovation district in the form of public arts by mentioning that "*We need more art in KGUV*" and "*In the halls, on the street, everywhere, it's so boring here*".

Function-wise, the interviews revealed that KGUV displays attracting people beyond the district level. The key reasons for such connections are the existence of art and culture theatre La Boite, Saturday markets, QUT's open days, and other days of display of student works as well as the mixed-use environment developed in the precinct. At Communify sessions, an informal exchange of information happens, and it creates a space for a congregation of various cultures. Building a sense of community through the Communify organisation has been challenging due to a lack of information regarding its activities and limited funding for such organisation. Lack of vibrancy and nighttime activities in the precinct emerged as a critical negative factor in terms of having an impact on KGUV's perception as an urban village. Interviewee suggestions for 24-hour activation were common; for instance, "*I want some place for gathering late at night at weekday*", "*It's a very day-oriented village*", and "*There is not as many evening options with many shops closing in the evening*". Other interviewee's comments included, "*I would like to see more KGUV community events and a Friday night market*", as well as, "*At nights, it feels dead, more people need to be attracted to the cafes*". There is much untapped potential for the staff and students generating art and innovation in science to share with the local community for inspiration, entertainment, and learning. This was evident by comments from interviewees, such as, "*There could be more regular public displays forums performances and cross-cultural activities*", and "*Free events where design and music and dance and other students showcase their work to the public and run interactive workshops/activities where local residents can join in and learn free dance sessions*". While the existence of blurred boundaries was meant to spark an exchange of dialogue between academia and the public, the lack of its manifestation in an intangible layer—in terms of lack of clear communication of ideas and exchange of conversations with the community—leads to creating a sense of exclusion, despite not presenting any physical boundaries. This is reflected in statements such as "*Where are we allowed to go*", as pointed out by one of the interviewees in the focus group.

Image-wise, there were mixed responses. However, there was—to a certain degree—a kind of resentment in the community regarding planning authorities, as revealed through interviews. It was found that the perceived indecisiveness of authorities regarding any issue affects the sense of attachment of KGUV users. In addition to it, unresolved conflicts between authorities and stakeholders, as well as a lack of an effective follow-up on consultation processes by acting on the raised issues, leads to a loss of sense of trust. Lack of proper communication between different stakeholder groups was flagged as a central issue during the focus groups. An interviewee statement on that point was "*How do you find out what is going on in the village? Locals wanted to find out what was happening in the village, at QUT*". The focus group data reinforced the interest that the community expressed in regards to social learning spaces for the general population and the idea of hearing about what is going on within the university. There was also a lack of awareness of which QUT facilities were available to the broader public. Another issue emerging from interviews is the struggle to create a sense of identity and attachment for its users. The mobile nature of students and knowledge society and lack of a unique identity due to the existing social and ethnic mix are pointed out as critical reasons. One of the interviewee statements reveals this: "*It's more than that, it's the transient community and you don't have that establishing the connections and the rituals and habits and things like that*". Nevertheless, most of the respondents appreciated the multi-cultural and diverse environment of KGUV. This is evident through interviewee comments like "*Love the ethnic diversity*", and "*We are all pink inside*". The proximity to the university incites a sense of pride and belonging into the residents—especially the parents of young children. One of the interviewee statements expressed, "*My children perform at the markets and are living the urban village life along with QUT and QACI students. I like to think going into the village inspires them to think about their next level of study after high school*". Interviewees also expressed their desire to integrate more through arts as a medium, as evident in the following statement: "*I'd like to see more street arts/student performances outdoors as it is an ideal location*". In KGUV's case, hence, it can be said that the university played an important role as a facilitator in the shaping up of the innovation district from the stage of conception to its functioning, extending to the image formed in the minds of its users.

12.5 FINDINGS AND DISCUSSION

The study findings from the MPID and KGUV cases contribute to a better understanding of how universities can act as an anchor for flourishing knowledge and innovation landscapes—e.g., innovation districts. Nonetheless, despite a strong community-oriented branding, it remains questionable how to advance and how successful the engagement with residents and communities has been. Some participants have been highly critical of the actual integration between the university and its surrounding neighbourhoods. Exchange of knowledge and innovation needs to function two ways. Science, technology, research, and development activities happening behind the walls of the university should not be just displayed to people through events—e.g., seminars, workshops, or exhibitions—but the community also should be a part of these activities by participating and contributing, where/when appropriate, with their ideas. Additionally, the innovation district becoming an easily accessible space also brings the government and industry's ideas to the community. In that scenario, the university acts as a facilitator for the societal integration of innovation districts. Such societal integration is highly invaluable for the generation of a democratic, cohesive, and prosperous community.

TABLE 12.2 Comparative Case Study Findings

Place-Making Dimension	**Place-Making Role of the University**	**MPID**	**KGUV**
Feature	Leader	*****	***
	Identity builder	****	*****
Form	Catalyst	****	****
	Spatial laboratory	****	****
Function	Resource provider	****	***
	Knowledge disseminator	****	**
Image	Collaborator	*	*
	Communicator	***	***

Note: *****=very high, ****=high, ***=moderate, **=low, *=very low.

As seen in MPID, conversion of the university's spatial character from an introvert to extrovert model has led to a better societal integration outcome. Similarly, in KGUV, while physically, there are blurred boundaries, interviews have revealed a limited interaction and engagement at a functional level in terms of knowledge and resources of the university. This leads to the projection of an introverted image of the university within the local community. Interviews have, hence, revealed an aspiration by the community to have a more extrovert attitude from the university, and a more permeable innovation district environment. Table 12.2 presents the comparative case study findings. Additionally, the data triangulation and analysis lead to the following key insights on the role of universities as facilitators of place making in innovation districts.

12.6 CONCLUSION

12.6.1 Socio-Cultural Role of the University

The anchor universities in contemporary innovation districts can play a critical socio-cultural role as place making facilitator in achieving a cohesive community and their societal integration at the local level. This multi-layered role is defined by key attributes. These are discussed below.

Feature-wise, the university—in the first place—helps in developing a unique brand and identity of the innovation district that acts as an attraction factor for researchers, knowledge workers, companies—locally and internationally—as well as the local community. Due to their specialised knowledge, resources, and networks, while they become a crucial attraction for companies to co-locate, they also are perceived to be a strong anchor managing and facilitating the collaborations. In the context of the local community, the presence of a university features the area as a hub with easily accessible amenities, resources, and networks, including libraries, playgrounds, theatres, and art exhibition halls, enhancing its societal integration.

Form-wise, the university can provide a shared physical space for the exchange of knowledge as well as for people to congregate by the organisation of events. Its boundaries need to be physically blurred and integrated with surroundings by being open and permeable with the rest of the district. Universities also have the potential to add uniqueness to the built environment of innovation districts by being the spatio-physical location for the display of arts, new innovation, and integration of heritage and culture. Socio-culturally, this can result in the creation of a vibrant and innovative environment—contributing to developing a people-oriented and unique character. For society, it also acts as a physical platform to stay well informed of cutting-edge technology and creative arts.

Function-wise, universities act as the centre for exchange and dissemination of knowledge and innovation through their open days and public events, activities, seminars, and exhibitions. Hence, the university—with the help of its researchers and team of professors and students—plays a key role in developing formal and informal socio-economic connections beyond the boundaries of the precinct at local and global levels. At the district level, it acts as a linking point between community, government, and businesses. The university can assist companies in integrating internal and external communities as a 'living social laboratory' for the immediate implementation of research and innovation ideas. Techno-economically, this may lead to the development of products that are a result of an integration of local ideas—in this way, adding on to competitiveness and innovation quotient of the innovation district. Most importantly, as seen in KGUV and MPID, university's specialised role in innovation districts have been bringing the diverse ethnicities together and empowering marginalised groups by disseminating information and having their say.

Image-wise, universities through the joint research projects with key stakeholders, including authorities can become an appropriate catalyst for public participation in planning by assisting in effective communication of plans and projects, organising community workshops, and seminars. These tactics will develop a sense of ownership in the community by making the planning process transparent and inclusive. In addition to this, universities generate a sense of trust in the community; being the centre of higher education can become a key factor in developing an overall environment of trustworthiness within the innovation district, bringing various stakeholders together. By keeping society well-informed regarding the research happening beyond the walls, it also helps in developing a sense of pride and belonging for the residents regarding the innovation district as highlighted by the interviews. Moreover, universities also help in preserving heritage and enhancing local history. These end up being key aspects in the development of a sense of attachment with the site. Table 12.3 lists the overall study findings.

For an effective contribution to the success of an innovation district, universities need to identify and practice their multi-layered role for developing a cohesive community within innovation-scape, and its integration with its local community. In sum, we provide some strategic recommendations for universities to act as a place making facilitator in the development of a societally integrated innovation district. These are presented below.

Feature-wise: (a) As an initiator in institutionalisation and formation of networking groups between government, community, and businesses by providing resources and teams; and (b) as a brand image by contributing in identity formation of innovation districts at the local and international level.

Form-wise: (a) As a catalyst for developing innovation districts by aligning their development plans with proposed plans by government and being open, accessible, and creation of innovative built environment; and (b) as a laboratory by assisting in the exchange of knowledge achieved through providing the physical space and resources for the participation of the community in cutting-edge research, creativity, and innovation.

Function-wise: (a) As a collaborator in strengthening formal and informal networks within the innovation district and outside it by forming networking groups and actively engaging various stakeholders; and (b) as a facilitator in developing an inclusive society by empowering and integrating diverse ethnicities, marginalised people, and others through training, networking, and educating them.

Image-wise: (a) As a coordinator between stakeholders by assisting in devising innovative, participative, and transparent ways of communication of plans and proposals; and (b) as a communicator by being a trustworthy

TABLE 12.3 Overall Study Findings

Dimension	Broad Social Attribute	Specific Social Attribute	Implications
Feature	Management and networking	Lead and facilitate collaborations	University leading establishment and management of new institutional structure at precinct level for networking
	Identity and branding	Create and strengthen brand identity	University by virtue of its name, presence and identity
Form	Public space integration and permeability	Catalyse development and people-orientedness in design	University aligning its own plans with proposed development plans
	Uniqueness and engaging design	Provide precinct with a competitive edge and disseminate innovation	University designing an open space for display of arts and innovation and engagement of society in exhibited innovation
Function	Knowledge and resource sharing	Encourage community's engagement in innovation and exchange of ideas	University providing access to the community to its own space and resources and executing collaboration initiatives such as living laboratories
	Social inclusiveness and empowerment	Train marginalised and other sections	University training and educating the marginalised and other weaker sections
Image	Stakeholder relationship and trust	Develop a sense of ownership and trust within stakeholders	University assisting in devising innovative and participative planning approaches and transparent communication
	Sense of pride and belonging	Develop a sense of pride and belonging in the community	University communicating ideas of KBUD as well as history of the site

and well-equipped source of information—owing to its assets of researchers, professors, and other knowledgeable individuals—that enlightens community by communicating with them: (i) The idea of KBUD; (ii) plans and proposals related to the innovation district; (iii) research and innovation happening in the space; and (iv) stories related to the history of the site. In the process, they contribute to developing a sense of trust, place attachment, and pride of belonging to the site in the community.

12.6.2 Policy Implications

The study highlights some of the key policy implications for clearly defining as well as strengthening the socio-cultural role of universities as a place-making facilitator in the contemporary innovation districts. Universities' roles should not be limited to their physical presence as an anchor, in research activities, and as the core for innovation generation. These roles should be extended in socio-cultural networks and a key place-making facilitator for the innovation district. To project a trustworthy image amongst the local community, this refers to highlighting the socio-cultural activities of their universities in their branding and marketing promotion. In terms of design and planning, it is critical to align the plans of the development within innovation districts in accordance with the plans and development as envisaged by the university for its future growth. Seamless integration of space and function of the universities with the local community and other parts of innovation districts should be encouraged. To empower the societal integration role of universities, the management of a university should be involved in key decision-making processes, management of the space, and in developing internal networks within the innovation district by setting up university-laden networking organisations. In order to do so, financial and institutional support—wherever required—is to be facilitated. To become a contemporary innovation district, the real manifestation of innovation should not be limited simply to techno-economic output and innovative built environments. These innovation-generating locations should integrate and display innovation in their politico-institutional constitution and socio-cultural denouements as well.

On the one hand, as significant as it is to empower the role and position of universities within the innovation districts, on the other, it is also imperative to ensure through policies that the anchor universities adopt an open culture and be collaborative into their research, innovation, networking activities, and socio-cultural integration. For reaping the true benefits of KIUD, contemporary innovation districts should be defined as spaces of collaboration that bring together multiple stakeholders—rather than encouraging the dominance of any one stakeholder—whether it be public, private, or academia.

In sum, this research study reported in this chapter discloses the importance of university and innovation district symbiosis in the context of place making. One of the key derivatives from the research is that the co-location of diverse activities is not enough for the success of an innovation district. The proximity needs to extend itself to intangible layers of formal and informal connections, in order to be completely effective. The most critical aspects of the role of the university with the redefined requirements in a global knowledge and innovation economy; thus, are in strengthening formal and informal networks within the innovation district and being a socially inclusive and accessible platform to create, disseminate, and interchange ideas, innovation, and creativity. Besides, universities act as a facilitator in societal integration of these innovation districts—necessary for the generation of a democratic, cohesive, and prosperous community. In innovation districts, hence, the role of a university—specifically when they act as an anchor—is not just limited to the economic benefits linked to research and education; benefits also extend to politico-organisational and socio-cultural dimensions.

Lastly, the major contribution of this research study is in terms of filling up an important theoretical gap in terms of defining the socio-cultural role of universities as place-making facilitators in innovation districts and determining policy implications as per the insights generated from the selected case studies. Under the scope of the current research study based on qualitative data, the attributes are defined, and strategies are outlined. It is to be noted here that the objective of this study is to arrive at the most generalised results that remain uniform, irrespective of the context. Nevertheless, considering specialised socio-cultural context and unique challenges in varying locations and their internal dynamics, future research may arrive at the attributes and strategies, which will be specific to the selected cases. Furthermore, the results from this research study provide a base for prospective research studies that further extend the knowledge in this domain by quantifying these attributes and aim at developing an assessment base (Pancholi *et al.*, 2020).

ACKNOWLEDGEMENTS

This chapter, with permission from the copyright holder, is a reproduced version of the following journal article: Pancholi, S., Yigitcanlar, T., Guaralda, M., Mayere, S., Cadwell, G., & Medland, R., (2020). University and innovation district symbiosis in the context of place making: insights from Australian cities. Land Use Policy, 9(1), 105109.

REFERENCES

Abramo, G., D'Angelo, C. A., Di Costa, F., & Solazzi, M. (2009). University–industry collaboration in Italy: a bibliometric examination. *Technovation*, 29(6-7), 498–507. DOI:10.1016/j.technovation.2009.02.004

Aldieri, L., Kotsemir, M., & Vinci, C. P. (2018). The impact of research collaboration on academic performance: an empirical analysis for some European countries. *Socio-Economic Planning Sciences*, 62, 13–30. DOI:10.1016/j.seps.2017.11.001

Aldieri, L., Carlucci, F., Vinci, C. P., & Yigitcanlar, T. (2019). Environmental innovation, knowledge spillovers and policy implications: a systematic review of the economic effects literature. *Journal of Cleaner Production*, 239, 118051. DOI:10.1016/j.jclepro.2019.118051

Aquilani, B., Silvestri, C., Ioppolo, G., & Ruggieri, A. (2018). The challenging transition to bio-economies: towards a new framework integrating corporate sustainability and value co-creation. *Journal of Cleaner Production*, 172, 4001–4009. DOI:10.1016/j.jclepro.2017.05.109

Arbolino, R., De Simone, L., Carlucci, F., Yigitcanlar, T., & Ioppolo, G. (2018). Towards a sustainable industrial ecology: implementation of a novel approach in the performance evaluation of Italian regions. *Journal of Cleaner Production*, 178, 220–236. DOI:10.1016/j.jclepro.2017.12.036

Baxter, P., & Jack, S. (2008). Qualitative case study methodology: Study design and implementation for novice researchers. *The Qualitative Report*, 13(4), 544–559. https://nsuworks.nova.edu/tqr/vol13/iss4/2

Baum, S., O'Connor, K., & Yigitcanlar, T. (2009). The implications of creative industries for regional outcomes. *International Journal of Foresight and Innovation Policy*, 5(1-3), 44–64 DOI:10.1504/IJFIP.2009.023098

Benneworth, P., & Ratinho, T. (2014). Reframing the role of knowledge parks and science cities in knowledge-based urban development. *Environment and Planning C*, 32(5), 784–808. DOI:10.1177/0263774X14538862

Blakely, E. J., & Hu, R. (2019). *Crafting innovative places for Australia's knowledge economy*. Singapore: Springer. DOI:10.1007/978-981-10-8899-1

Boodhoo, R., & Purmessur, R. D. (2009). Justifications for qualitative research in organisations: a step forward. *The Journal of Online Education*. Accessed on 22 July 2020 from http://www.nyu.edu/classes/keefer/waoe/deeprosh2.pdf.

Brake, K., & Kress, C. (2011). Concepts of pride, brand, and communication: Architecture, urban design, and knowledge production: Berlin from 1810 to 2010. *International Journal of Knowledge-Based Development*, 2(2), 220–246. DOI:10.1504/IJKBD.2011.039428

Byrne, J. (2004). Origins, options and sustainability: Reflections on the Kelvin Grove urban village. *Australian Planner*, 41(4), 29–32. DOI:10.1080/07293682.2004.9960897

Carrillo, J., Yigitcanlar, T., Garcia, B., & Lönnqvist, A. (2014). *Knowledge and the city*. New York: Routledge.

Carvalho, L. (2009). Four challenges for a new science and technology park: AvePark in Guimaraes, Portugal. *Urban Research & Practice*, 2(1), 103–108. DOI:10.1080/17535060802661514

Casaleiro, P. (2011). Changing from a univer(s)city to a knowledge city: The case of Coimbra. *International Journal of Knowledge-Based Development*, 2(2), 166–184. DOI:10.1504/IJKBD.2011.040119

Charles, D. (2011). The role of universities in building knowledge cities in Australia. *Built Environment*, 37(3), 281–298. DOI:10.2148/benv.37.3.281

Chesbrough, H. W. (2003). *Open innovation: The new imperative for creating and profiting from technology*. Boston: Harvard Business Press.

Clark, J., Huang, H. I., & Walsh, J. P. (2010). A typology of 'innovation districts': what it means for regional resilience. *Cambridge Journal of Regions, Economy and Society*, 3(1), 121–137. DOI:10.1093/cjres/rsq001

Den Heijer, A., & Magdaniel, F. (2012). The university campus as a knowledge city: Exploring models and strategic choices. *International Journal of Knowledge-Based Development*, 3(3), 283–304. DOI:10.1504/IJKBD.2012.050172

Durmaz, B., Platt, S., & Yigitcanlar, T. (2010). Creativity, culture tourism and place-making: Istanbul and London film industries. *International Journal of Culture, Tourism and Hospitality Research*, 4(3), 198–213. DOI:10.1108/17506181011061564

Edvardsson, I. R., Yigitcanlar, T., & Pancholi, S. (2016). Knowledge city research and practice under the microscope: A review of empirical findings. *Knowledge Management Research and Practice*, 14(4), 537–564. DOI:10.1080/14778238.2016.1211759

Esmaeilpoorarabi, N., Yigitcanlar, T., Guaralda, M., & Kamruzzaman, M. (2018a). Does place quality matter for innovation districts? Determining the essential place characteristics from Brisbane's knowledge precincts. *Land Use Policy*, 79(1), 734–747. DOI:10.1016/j.landusepol.2018.08.019

Esmaeilpoorarabi, N., Yigitcanlar, T., Guaralda, M., & Kamruzzaman, M. (2018b). Evaluating place quality in innovation districts: a Delphic hierarchy process approach. *Land Use Policy*, 76(1), 471–486. DOI:10.1016/j.landusepol.2018.04.017

Esmaeilpoorarabi, N., Yigitcanlar, T., Kamruzzaman, M., & Guaralda, M. (2020a). How can an enhanced community engagement with innovation districts be established? Evidence from Sydney, Melbourne and Brisbane. *Cities*, 96(1), 102430. DOI:10.1016/j.cities.2019.102430

Esmaeilpoorarabi, N., Yigitcanlar, T., Kamruzzaman, M., & Guaralda, M. (2020b). How does the public engage with innovation districts? Societal impact assessment of Australian innovation districts. Sustainable *Cities and Society*, 52(1), 101813.

Esmaeilpoorarabi, N., Yigitcanlar, T., Kamruzzaman, M., & Guaralda, M. (2020c). Conceptual frameworks of innovation district place quality: an opinion paper. *Land Use Policy*, 90(1), 104166. DOI:10.1016/j.landusepol.2019.104166

Etzkowitz, H., & Leydesdorff, L. (1995). The triple-helix. *EASST Review*, 14(1), 14–19. DOI:10.18848/2154-8587/CGP/v02i03/44454

Espelt-Lleonart, P. (2012). Building Barcelona's 22@ Innovation District: a technical review. *International Journal of the Constructed Environment*, 2(3), 113–133.

Farmer, T., Robinson, K., Elliott, S. J., & Eyles, J. (2006). Developing and implementing a triangulation protocol for qualitative health research. *Qualitative Health Research*, 16(3), 377–394. DOI:10.1177/1049732305285708

Florida, R. (2005). *Cities and the creative class*. New York: Routledge.

Glaser, B. G., & Strauss, A. L. (1967). *The discovery of grounded theory*. Chicago: Aldine.

Goldstein, H. A., & Glaser, K. (2012). Research universities as actors in the governance of local and regional development. *The Journal of Technology Transfer*, 37(2), 158–174. DOI:10.1007/s10961-011-9226-9

Hawken, S., & Han, J. (2017). Innovation districts and urban heterogeneity: 3D mapping of industry mix in downtown Sydney. *Journal of Urban Design*, 22(5), 568–590. DOI:10.1080/13574809.2016.1235757

Ingrao, C., Messineo, A., Beltramo, R., Yigitcanlar, T., & Ioppolo, G. (2018). How can life cycle thinking support sustainability of buildings? Investigating life cycle assessment applications for energy efficiency and environmental performance. *Journal of Cleaner Production*, 201, 556–569. DOI:10.1016/j.jclepro.2018.08.120

Johansson, R. (2007). On case study methodology. *Open House International*, 32(3), 48. DOI:10.1016/S0169-2607(00)00053-8

Katz, B., & Wagner, J. (2014). *The rise of innovation districts: a new geography of innovation in America*. New York: Brookings Institution.

Klaebe, H. (2006). *Sharing stories: A social history of the Kelvin Grove Urban Village*. Brisbane: Focus Publishing.

Lefebvre, H., & Nicholson-Smith, D. (1991). *The production of space*. Oxford: Blackwell.

Leon, N. (2008). Attract and connect: The 22@ Barcelona innovation district and the internationalisation of Barcelona business. *Innovation*, 10(2-3), 235–246. DOI:10.5172/impp.2008.10.2-3.235

Metaxiotis, K., Carrillo, J., & Yigitcanlar, T. (2010). *Knowledge-based development for cities and societies: integrated multi-level approaches*. Hersey: IGI Global. DOI:10.4018/978-1-60566-928-1

Mok, K. H. (2016). Massifying and internationalising higher education, changing labour markets and social mobility: challenges for education and urban governance. *Journal of Higher Education Policy and Management*, 38(3), 233–241. DOI:10.1080/1360080X.2016.1151374

Morisson, A., & Bevilacqua, C. (2019). Balancing gentrification in the knowledge economy: the case of Chattanooga's innovation district. *Urban Research & Practice*, 12(4), 472–492.

MU (Macquarie University). (2017). *Submission to greater Sydney commission, draft north district plan*. Sydney: Macquarie University.

Nord, D. C. (2002). The role of universities in northern development. In *Higher education across the circumpolar north*, 175–194. Houldsmill: Palvegrave-Macmillan.

Pancholi, S., Yigitcanlar, T., & Guaralda, M. (2015a). Place making facilitators of knowledge and innovation spaces. *International Journal of Knowledge-Based Development*, 6(3), 215–240. DOI:10.1504/IJKBD.2015.070205

Pancholi, S., Yigitcanlar, T., & Guaralda, M. (2015b). Public space design of knowledge and innovation spaces: learnings from Kelvin Grove Urban Village, Brisbane. *Journal of Open Innovation: Technology, Market, and Complexity*, 1(1), 13. DOI:10.1186/s40852-015-0008-8

Pancholi, S., Yigitcanlar, T., & Guaralda, M. (2017). Governance that matters: identifying place making challenges of Melbourne's Monash Employment Cluster. *Journal of Place Management and Development*, 10(1), 73–87. DOI:10.1108/JPMD-08-2016-0058

Pancholi, S., Yigitcanlar, T., & Guaralda, M. (2018a). Attributes of successful place making in knowledge and innovation spaces: evidence from Brisbane's Diamantina Knowledge Precinct. *Journal of Urban Design*, 23(5), 693–711. DOI:10.1080/13574809.2018.1502816

Pancholi, S., Yigitcanlar, T., & Guaralda, M. (2018b). Societal integration that matters: place making experience of Macquarie Park Innovation District, Sydney. *City, Culture and Society*, 13, 13–21. DOI:10.1016/j.ccs.2018.06.002

Pancholi, S., Yigitcanlar, T., & Guaralda, M., (2019). Place making in knowledge and innovation spaces: the Australia experience. *Technological Forecasting and Social Change*, 146(1), 616–625. DOI:10.1016/j.techfore.2018.12.011

Pancholi, S., Yigitcanlar, T., Guaralda, M., Mayere, S., Cadwell, G., & Medland, R., (2020). University and innovation district symbiosis in the context of place making: insights from Australian cities. *Land Use Policy*, 9(1), 105109. DOI: 10.1016/j.landusepol.2020.105109

Panori, A., Kakderi, C., Komninos, N., Fellnhofer, K., Reid, A., & Mora, L. (2020). Smart systems of innovation for smart places: Challenges in deploying digital platforms for co-creation and data-intelligence. *Land Use Policy*, 104631, DOI:10.1016/j.landusepol.2020.104631.

Phan, P. H., Siegel, D. S., & Wright, M. (2005). Science parks and incubators: observations, synthesis and future research. *Journal of Business Venturing*, 20(2), 165–182. DOI:10.1016/j.jbusvent.2004.01.001

QUT (Queensland University of Technology). (2004). *Kelvin Grove Urban Village: integrated master plan.* Brisbane: QUT. DOI:10.1504/IJKBD.2011.041794

Romein, A., Fernández-Maldonado, A. M., & Trip, J. J. (2011). Delft blues: the long road from university town to knowledge city. *International Journal of Knowledge-Based Development*, 2(2), 148–165.

Smith, H. L., & Bagchi-Sen, S. (2010). Triple helix and regional development: a perspective from Oxfordshire in the UK. *Technology Analysis & Strategic Management*, 22(7), 805–818. DOI:10.1080/09537325.2010.509174

Stehlin, J. (2016). The post-industrial "shop floor": Emerging forms of gentrification in San Francisco's innovation economy. *Antipode*, 48(2), 474–493. DOI:10.1111/anti.12198

Toolis, E. E. (2017). Theorizing critical place making as a tool for reclaiming public space. *American Journal of Community Psychology*, 59(1–2), 184–199. DOI:10.1002/ajcp.12144

Van Winden, W., de Carvalho, L., Van Tuijl, E., Van Haaren, J., & Van den Berg, L. (2013). *Creating knowledge locations in cities*. London: Routledge.

Yigitcanlar, T., & Sarimin, M. (2011) The role of universities in building prosperous knowledge cities. *Built Environment* 37(3), 260–280. DOI:10.2148/benv.37.3.260

Yigitcanlar, T. (2014). Position paper: benchmarking the performance of global and emerging knowledge cities. *Expert Systems with Applications*, 41(12), 5549–5559. DOI:10.1016/j.eswa.2014.03.045

Yigitcanlar, T., Guaralda, M., Taboada, M., & Pancholi, S. (2016). Place making for knowledge generation and innovation: planning and branding Brisbane's knowledge community precincts. *Journal of Urban Technology*, 23(1), 115–146. DOI:10.1080/10630732.2015.1022049

Yigitcanlar, T., Edvardsson, I. R., Johannesson, H., Kamruzzaman, M., Ioppolo, G., & Pancholi, S. (2017). Knowledge-based development dynamics in less favoured regions: insights from Australian and Icelandic university towns. *European Planning Studies*, 25(12), 2272–2292. DOI:10.1080/09654313.2017.1385689

Yigitcanlar, T., & Inkinen, T. (2019). *Geographies of disruption: place making for innovation in the age of knowledge economy*. Switzerland: Springer.

Yigitcanlar, T., Adu-McVie, R., & Erol, I. (2020). How can contemporary innovation districts be classified? A systematic review of the literature. *Land Use Policy*, 95(1), 104595. DOI:10.1016/j.landusepol.2020.104595

Youtie, J., & Shapira, P. (2008). Building an innovation hub: A case study of the transformation of university roles in regional technological and economic development. *Research Policy*, 37(8), 1188–1204. DOI:10.1016/j.respol.2008.03.009

Yun, J. J., Zhao, X., Yigitcanlar, T., Lee, D., & Ahn, H. (2018). Architectural design and open innovation symbiosis: Insights from research campuses, manufacturing systems, and innovation districts. *Sustainability*, 10(12), 4495. DOI:10.3390/su10124495

Zhao, P. (2010). Building knowledge city in transformation era: knowledge-based urban development in Beijing in the context of globalisation and decentralisation. *Asia Pacific Viewpoint*, 51(1), 73–90. DOI:10.1111/j.1467-8373.2010.01405.x

Zhang, Y. (2005). The science park phenomenon: development, evolution and typology. *International Journal of Entrepreneurship and Innovation Management*, 5(1–2), 138–154. DOI:10.1504/IJEIM.2005.006656

CHAPTER 13

Smart Campus as an Innovation District

13.1 INTRODUCTION

Recent technological advancement in this era has far-reaching development phases in cloud computing, big data, Internet-of-Things (IoT), and artificial intelligence (AI) (Yigitcanlar *et al.*, 2019). Higher education institutions (HEI) have used these technologies in their campuses to reform teaching, learning, and research experiences of university students and staff, including alternative modes of attendance and collaboration (Zhou & Tang, 2022). Today, more and more university administrations are embracing digital campus transformations (Al-Shoqran & Shorman, 2021).

The digitally transformed university campus is also referred to as a "smart campus" (Hidayat & Sensuse, 2022; Horvath *et al.*, 2021). Smart campuses model the smart city ecosystem, and along with their traditional teaching, learning, and research functions, they also promote lifelong learning, research, innovation, and entrepreneurship, and offer various Living Lab programs (Huertas *et al.*, 2021). Smart campuses can be seen as a new form of innovation district.

Some scholars argue that technological progress, and other issues pertaining to the environment, economics, and social sector make it inevitable that all education processes will become "smart", whereby campuses will become learning and living environments that are integrated with advanced technological infrastructures, such as sensors, IoT, and artificial intelligence (AI) (Yang *et al.*, 2018; Villegas-Ch *et al.*, 2019). Some others see a campus as a place of convenience, safety, sustainability, leisure, and learning, enhanced by diverse and advanced systems and technologies (Gao, 2021). Nevertheless, there is a lack of explicit definition

DOI: 10.1201/9781032657431-18

of a smart campus beyond its heavy dependence on big data and digital technology (Hidayat & Sensuse, 2022; Ahmed *et al.*, 2020).

Existing definitions of a smart campus often consider the smart campus as being a miniature model of a smart city (Awuzie *et al.*, 2021). Other definitions describe it as a dynamic learning environment for fostering intelligence and smart solutions using people, physical infrastructure, and digital technologies (Martins *et al.*, 2021). It is an emerging development in the education sector and in the application of AI (Huang *et al.*, 2021), creating a cultural paradigm (Majumdar & Mandal, 2021) or a smartization process (that is a planned and organized process by which an organization adopts and implements smart technologies in their operations), where universities are opening their boundaries (Silva-da-Nóbrega *et al.*, 2022)

While many definitions are based on smart campus functions, commonly accepted conceptualizations of how these functions align to its idealized purposes are less understood. Table 13.1 summarizes the key definitions of a smart campus.

This study aims to develop a clearer understanding of how a smart campus is conceptualized and practiced by employing the PRISMA approach technique (Moher *et al.*, 2015; Shamseer *et al.*, 2015).

Following this introduction, Section 13.2 covers research design, search strategy, inclusion and exclusion criteria, and data analysis. Section 13.3 describes the results under general observations, and the four domains of smart campus. Section 13.4 presents and discusses the findings and its proposed conceptual framework, and Section 13.5 concludes the chapter.

13.2 METHODOLOGY

13.2.1 Research Design

This chapter draws on recent studies reported in the academic literature related to smart campus, where they attempt to comprehend the concept relative to their pursuits in a defined plan or protocol (Pittway, 2007). The systematic literature review approach was applied to identify and analyse papers relating to the concept of smart campus to gauge how this notion is conceptualized and practiced. The PRISMA approach with its four stages of identification, screening, eligibility, and inclusion was adopted to deal with the unknown stack of publications out there and to provide a structured and targeted search to obtain the key literatures for the construction of the conceptual framework.

TABLE 13.1 Definitions of a Smart Campus

Theme	Definition	Source
Smart city development focused	"Considered a small smart city that acts within the context of smart cities, which offer intelligent services and applications to their citizens to improve their quality of life."	Zaballos *et al.* (2020); Imbar *et al.* (2020, p. 6)
	"Acts within the context of smart cities, which offer intelligent services and applications to their citizens to improve their quality of life."	Imbar *et al.* (2020, p. 6)
Advanced technology infrastructure focused	"Refers to the hardware and software required to provide advanced intelligent context-aware services and applications to university students and staff."	Fernández-Caramés & Fraga-Lamas (2019, p. 3)
	"A data-oriented, networked, intelligent and collaborative teaching, management and scientific research system based on big data, internet-of-things, mobile internet and other advanced information technologies."	Cheng *et al.* (2022, p. 204)
	"A collaboration of technologies such as big data, cloud computing, IoT, internet and high-performance computing, virtualization, mobile network and social network, sensors and common communication interfaces."	Omotayo *et al.* (2021, p. 2)
	"Intelligent infrastructure where smart sensors and actuators collaborate to collect information and interact with the machines, tools, and users of a university campus."	Fraga-Lamas *et al.* (2019, p. 1)
	"Utilizes and integrates smart physical and digital spaces to establish responsive, intelligent, and improved services for creating productive, creative, and sustainable environment."	Min-Allah & Alrashed (2020, p. 3)
	"Not only about deploying smart platforms to effectively perform	Ikrissi & Mazri (2020, p. 255)

(*Continued*)

TABLE 13.1 *(Continued)* Definitions of a Smart Campus

Theme	Definition	Source
	campus-related services but it is a broad concept that includes many electronic and physical objects that communicate and interact with each other."	
Enhanced education experience focused	"An educational environment that is penetrated with enabling technologies for smart services to enhance educational performance while meeting stakeholders' interests, with broad interactions with other interdisciplinary domains in the smart city context."	Dong *et al.* (2020, p. 4)
	"Is the high-end form of education systems."	Zhang (2022, p. 16145)
	"To integrate information technology into teaching and education to provide a teaching environment with network, data, integration, and intelligence."	Chen & Liu (2021, p. 450)

The PRISMA study included four primary phases as shown in Figure 13.1. An online search, using the Scopus database, was utilized for the research. The search was confined to journal articles published between 2017 and 2022 to capture recent studies, considering the rapid phase of transformation in digital technology development. Only articles published in English were considered to avoid difficulties in translation and interpretation.

The first phase of identification in planning involved developing the research aim, research question, a list of keywords, and the criteria for the inclusion and exclusion of articles. The aim of the research was framed to generate insights into the application of the four domains in making smart campuses. The criteria for inclusion included peer-reviewed journal articles in English, available online, and relevant to the research aim. The search was conducted using the query string of ((("smart" OR "intelligent") AND ("campus" OR "universit*" OR "college" OR "higher education institution" OR "HEI"))) to search the titles and abstracts of available articles. The first stage of the search gathered 3,601 journal articles. Further exclusions were applied on

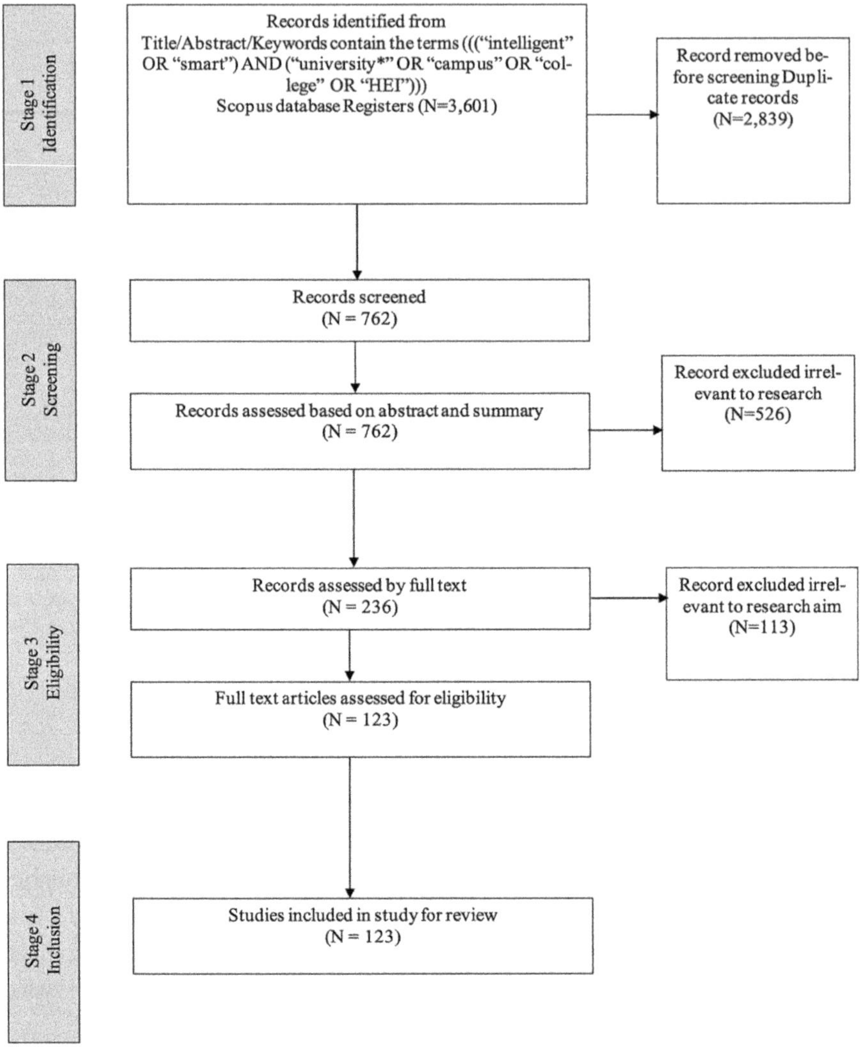

FIGURE 13.1 PRISMA review flow chart.

relevant subject areas, peer-review status, and online full-text availability, resulting in the selection of 762 articles.

The second phase of screening involved an assessment of the abstract and summaries of the articles, where 526 articles were removed, leaving 236 articles. The exclusion focused on the relevance of the articles to the objective of the paper on the four domains in the making of a smart campus. The third phase of eligibility evaluated full-texts of the articles where irrelevant articles were excluded, and 123 articles were retained,

TABLE 13.2 Exclusion and Inclusion Criteria

Primary Inclusionary Criteria	Primary Exclusionary Criteria	Secondary Inclusionary Criteria	Secondary Exclusionary Criteria
Academic journal articles	Duplicate records	Relevant to the smart campus notion	Irrelevant to the smart campus notion
Peer-reviewed	Books and chapters	Relevant to research objective	Irrelevant to research objective
Full-text available online	Industry reports		
Published in English	Government reports		

suggesting discussions related to the four smart campus domains. Finally, the fourth stage of inclusion focused on the reading of the full-text of the 123 articles, which met the research objectives. These articles provided sufficient grounding for the four smart campus domains. Figure 13.1 presents the PRISMA review flow chart, and the inclusion and exclusion criteria are listed in Table 13.2.

Relevant publications to the smart campus notion are those that cover the application of advanced AI in university campuses, focusing on capturing the recent developments in AI, IoT, cloud computing, and big data. In addition, relevant publications to the research objective included those that contribute to the research aim to consolidate a clearer understanding of how the smart campus is conceptualized and practiced. They should address the array of smart campus definitions, demonstrate the latest technological development in smart campuses, and finally, discuss the four domains for the conceptual framework of smart campuses. On the other hand, any endeavor that does not suggest the application of AI, IoT, cloud computing, and big data for university campuses are deemed irrelevant to the smart campus notion. Additionally, any endeavor that does not discuss smart campus definitions, smart technological development for campuses, and the four domains are also deemed irrelevant to the research objective.

Out of 123 selected and reviewed papers, 81 relate to the smart campus domains (or categories) elaborated in the next section. The list of these papers is provided in Appendix 13.A. These categorization criteria, or the analysis domains, are elaborated in the following section. The selected 123 articles are read and, following the categorization, they are analyzed manually for their key aspects and contributions to address the research question of this study.

13.2.2 Category Formulation

Providing the review results in a structured way requires categorizing the literature on smart campuses under widely accepted domains. For that purpose, the study adopted the four domains of economy, society, environment, and governance, with an overarching (digital) technology and (big) data perspective. These domains (illustrated in Figure 13.2) are widely used in urban research (Yigitcanlar, 2021), and align with common understandings of sustainable development at the microlevel for an inclusive framework (Omotayo *et al.*, 2021), which is generally a key rationale for the "smart" concept (Sujata *et al.*, 2016; Song *et al.*, 2017).

The economy domain is concerned with business or capital investment activities that are responsible for the efficient provision of services for the operation of organizations and citizens. Universities are expected to promote entrepreneurial activities, while serving their role in academic research (Sciarelli *et al.*, 2021).

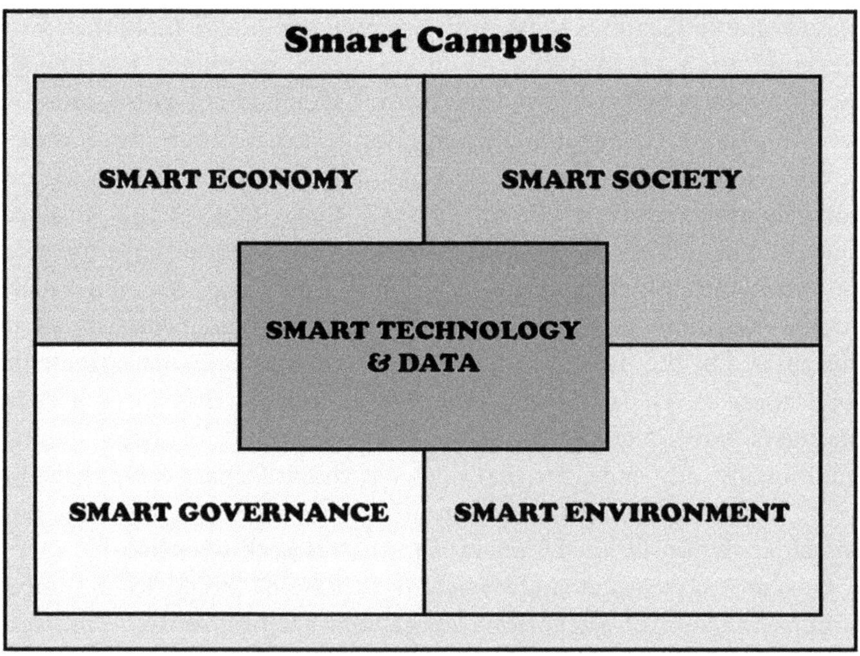

FIGURE 13.2 Smart campus domains.

The societal domain covers the social issues relating to the life of campus users (Yigitcanlar & Dur, 2013). It includes matters relating to safety, health, and equitable access to resources, opportunities, and services, with full allowance for cultural and spiritual needs. Normative social aspects also relate to deliberative processes that seek to empower campus citizens in the management and operation of the campus. As such, citizens are meaningfully engaged in any implementation of smart campus approaches (Zhang *et al.*, 2020).

The environmental domain is concerned with improving both the natural and built environments. It seeks to maintain species and habitats and ensure long-term sustainability, which also mean the sound integration of built and natural environments (Yigitcanlar, 2009). The environmental domain recognizes that the way natural resources are used and considered can influence the quality of campus life (Baum *et al.*, 2007; Yigitcanlar *et al.*, 2017).

The governance domain is the enabler to achieve the smart campus as it involves strategic decision making and implementation with the engagement of stakeholders. The emergence of the "third mission" on universities induced the role of collaborative governance, shifting from public administration and political science in public policymaking to reach across the society through an inclusive approach to adopt specific public policies (Kern & Smutko, 2021).

These four domains are supported by an overarching (digital) technology and (big) data perspective—that could be seen as the fifth category that has high interaction with the other four—in smart campuses (Villegas-Ch *et al.*, 2019; Omotayo *et al.*, 2021; Xu *et al.*, 2022). For instance, big data handles the huge and heterogeneous information collected, stored, and sorted for the operations of the campus (Hicks *et al.*, 2021; Cheong & Nyaupane, 2022). Digital technologies, such as sensors, IoT, AI, 5G, cloud computing, and blockchain, sustain big data and provide the means to facilitate smart campus functions (Alhazmi, 2021; Cao, 2022). The results of the analysis under these domains are provided in the next section.

13.3 RESULTS

13.3.1 General Observations

There is a body of evidence raised in recent research claiming that smart campus is currently being applied in universities. The application is mainly in smart architecture and smart applications but is limited in both functional span and in the number of universities. The research studies are primarily focused on the technological developments, including AI, IoT, cloud computing, big data, sensors, and blockchain and not the human aspect or the end user. Researchers have, however, acknowledged this gap where studies are scant in relation to human issues and its significance for the smart campus. The need for collaborative approaches involving the campus citizens has been raised in most recent research studies. This is significant throughout all the stages of a smart campus from conceptualization to implementation. User integration has been crucial. It is demonstrated in recent research studies that a smart campus is a tangible paradigm for capturing the surging development of information technology, AI, IoT, and cloud computing. It provides a means or framework to embrace the rapid technological advancements toward progressive solutions to improve quality, efficiency, waste reduction, cost reduction, and overall sustainability.

A total of 123 relevant articles and conference papers were included for the PRISMA review, where 81 (66%) were discussed on the research on at least one of the four domains of a smart campus, while 42 (34%) were discussed in the other sections of the chapter in the introduction and the materials and methods section. A total of 60 articles (74%) out of the 81 articles discussed the concepts of the domains, while 21 (26%) discussed prototypes and case studies on the four domains. This indicated that more research was conducted on the concepts than on prototypes and case studies. Hence, the development of the conceptual framework is highly conceptual, which fulfills the aim of the study.

The number of domains covered by the 81 articles is shown in Appendix 13.A. As shown, it is demonstrated that 34 (42%) articles covered research on at least one domain, 23 (28%) on two domains, 20 (25%) on three domains, and 4 (5%) on all four domains. Therefore, it is evident that more than two-thirds of the research focused on a single domain, about a third on two, about a quarter on three, and very few on all the four domains. It is also demonstrated that smart campus research has so far been dominated by the focus on a single domain which was evident by the numerous discrete studies being conducted. This has a significant bearing on the development of the conceptual framework at this stage in not reaching its full development to cover all four domains. The desired outcome of focus on all four domains is, however, very small right now.

The reporting on each of the four domains is shown in Appendix 13.A. As shown, about 69% (n = 56) covered the society domain, 48% (n = 39) covered the environment domain, 42% (n = 34) covered the economy domain, and 36% percent (n = 29) covered the governance domain. It should be noted that the coverage ratio for each of the domains is calculated from the total number of 81 articles focused on each domain separately to compare the degree of attention devoted to each of them. Hence, the total percentage does not necessarily add up to 100%. The study demonstrated that the society domain was given the most coverage, followed by environment, economy, and the least coverage was that of governance.

13.3.2 Economy Domain of Smart Campus

The economy domain of the smart campus entails the entrepreneurial capacity of the campus (Fraga-Lamas *et al.*, 2019; Yigitcanlar *et al.*, 2016). This is reviewed further under four aspects: business opportunity, improved efficiency, innovation ecosystem, and utility cost saving. A total 34 articles discussed the economy domain, where 27 of them discussed the conceptual phase and 7 were focused on prototyping and case study investigations.

13.3.2.1 Business Services

A business services opportunity is created within the economy domain of smart campus using smart technology. The global agenda of Industry 4.0 has forced universities to transform to meet its requirements by modernizing existing programs, facilities, and infrastructure (Mian *et al.*, 2020). A knowledge economy is created through the growing global smart education market (Dong *et al.*, 2020) for income generation (Deraman *et al.*, 2021). Rich and diverse business systems are developed (Chen & Liu, 2021), spurring new organizational and business models (Kuzior & Sira, 2022). This created a connection between universities and industry (Fernández-Morante *et al.*, 2022) for economic benefits by maximizing funds and providing the best service to campus stakeholders with open business opportunities for web developers (Soegoto & Rinaldy, 2020). Dedicated smart systems are developed, including the “Blockchain Business Model for Higher Education” (BBM-HE) created to transform business model for tokenized education (Son-Turan, 2022). In addition, the control and energy management system to boost and sell energy to the grid (Zainuddin *et al.*, 2021), the smart waste management system which recycles solid wastes for circular economy (Longo *et al.*, 2021), the

BIM-based smart system in the saving of building stock for investment (Wehbe & Shahrour, 2021), and finally, a smart parking platform providing profits for participants through electric vehicles (Silva *et al.*, 2019) to lower costs, improve user experience and maintain a certain level of revenue (Sutjaritthan, 2021) are also developed. It is, therefore, demonstrated that there are very few attempts in the development of smart technology for business services.

13.3.2.2 Improved Efficiency

Improved efficiency benefits are created with smart technologies where smart campus provides advanced education systems using AI for information and communication to enhance efficiency (Zhang *et al.*, 2022). There are opportunities for better experiences to improve campus efficiency (Adamkó, 2018) through Internet of Things, and community-based digital technology to enhance resource utilization in a university (Debauche *et al.*, 2020). In addition, there is financial stability and entrepreneurial leadership (Alrashed, 2020), reduction in production costs (Al-Shoqran & Shorman, 2021), effective management of energy, and other resources in a university (Eltamaly *et al.*, 2021; Amaral *et al.*, 2021). Moreover, the emergence of technology-enabled efficient resource control in university facilitates, improving information dissemination, and generating the organization income (Deraman *et al.*, 2021). Finally, a smart campus involves the development of a trading platform for optimizing resource allocation (Dong *et al.*, 2020), including blockchain-based applications, to foster efficient production at low-cost in new business approaches (Kuzior & Sira, 2022). Though it is demonstrated that there are a few attempts in smart technology development for improving efficiency, it is, however, worth noting that efforts have begun.

13.3.2.3 Innovation Ecosystem

An innovative ecosystem comprises two main areas of sustainability and smart campus solutions. Sustainability innovations explored campuses as urban living labs (ULL) (Martínez-Bello *et al.*, 2021) for prototyping an open architecture for sustainable evolution (Adamkó, 2018). These included the following developments: a platform for promoting sustainable campus (Acuña *et al.*, 2021), smart green campus (Abdulmouti *et al.*, 2022) to reduce the carbon footprint using solar photovoltaic (PV) power (Teah *et al.*, 2019), mobility for promoting decarbonization and credit trading (Cirrincione *et al.*, 2022), and finally, a smart waste bin

(Longo *et al.*, 2021). On the other hand, smart campus solutions uncovered significant resource efficiency gains, offering better experiences (Sutjarittham, 2021), which included smart grid innovations for integrating power generation and infrastructures through digital technology networks (Alrashed, 2020). In addition, pertinent innovations, including a hybrid renewable energy system (Eltamaly *et al.*, 2021; Nasir *et al.*, 2021), and several energy management systems (EMS) for distributed energy resources (DERs) (Nasir *et al.*, 2021), were developed.

Furthermore, a blockchain-based energy trading platform (Silva *et al.*, 2019) increasing sustainable energy by green energy suppliers (Bin *et al.*, 2022), integrating automated distribution systems and energy storage technologies (Javed *et al.*, 2021), and a Hybrid Firefly Lion Algorithm (FLA) (Ullah *et al.*, 2021) for optimizing energy management were also developed. Other innovations included a data governance platform (Chen & Liu, 2021), the (R)evoCampus (Debauche *et al.*, 2020), OnCampus mobile platform (Dong *et al.*, 2016), and a Blockchain Business Model for Higher Education (BBM-HE) (Son-Turan, 2022). Additionally, smart parking innovations, such as an IoT Raspberry Pi-based parking management system (IoT-PiPMS) (Waheb *et al.*, 2021), and a cloud-based automated parking system (Yee *et al.*, 2020), were developed. Finally, the human-centered learning-oriented smart campus (HCLSC) framework (Dong *et al.*, 2020) for smart campus development was developed. Therefore, it is evident that there are already smart technology developments promoting an innovation ecosystem in universities and such developments will continue in the future.

13.3.2.4 Utility Cost Savings

Utility cost savings is an economic benefit created by smart campus using smart technology in electrical energy reduction through the development of energy management systems, smart buildings, and smart parking. A smart campus facilitates electrical energy reduction (Yee *et al.*, 2020) through the development of energy management systems minimizing grid electricity cost (Nasir *et al.*, 2021; Bin *et al.*, 2022; Ullah *et al.*, 2021). These were achieved through the application of smart technology in green energy production (Abdulmouti *et al.*, 2022) where the hybrid renewable power distribution is supported through IoT (Eltamaly *et al.*, 2021). In addition, there was a reduction of operation cost, including electricity (Debauche *et al.*, 2020), water, and energy in a self-sufficient university building (Cortese *et al.*, 2022). Finally, a parking management system was developed to reduce management costs and save energy consumption (Waheb *et al.*, 2021). Although the impact is not widespread at present, there has been a reduction in utility costs due to the use of smart technology and there are opportunities for further cost savings with smart technologies.

13.3.3 Society Domain of Smart Campus

The society domain of the smart campus addresses the crosscutting issues that affect the life of people that included strategies for community engagement, quality of campus life, social responsibility, and versatile learning. A total of 56 articles were found discussing the domain where 46 were conceptual and 10 were prototypes.

13.3.3.1 Community Engagement

Community engagement is a benefit offered by smart campus, where the university community becomes an IoT integrated environment (Yang *et al.*, 2018), creating a digital campus using advanced information technologies in IoT, big data, network, and AI (Wan & Yuan, 2022). The university campus becomes smarter by the inclusion and advancements of technology (Sneesl *et al.*, 2022), increasing automation (Aldowah *et al.*, 2017) where technology is adapted creating a connected and responsive campus (Deraman *et al.*, 2021). In addition, a collective university experience is created with the focus on human citizens provided by the three elements of a smart campus, comprising: (a) the influence by campus citizens, (b) networking of operations and businesses, and (c) emphasis on infrastructure and services development (Awuzie *et al.*, 2021). Moreover, higher education institutions (HEIs) provide technological solutions for a dynamic ecosystem created from students and academics in a digital community (Curralo *et al.*, 2022). The smart solutions created include sensing, adaptation, inferring, and community contentment (Adamkó, 2018). Finally, there is connectivity with everything forming an ecosystem of robust communications networks (Deraman *et al.*, 2021). However, these initiatives are limited to the concept of smart technology sustaining community engagement in universities. There are undoubtedly practical applications in universities that have not been documented.

13.3.3.2 Quality of Campus Life

A smart campus is a smaller version of a smart city (Zhang *et al.*, 2022) that has the potential of impacting the quality of life of its citizens (Pérez *et al.*, 2021). In addition, a smart campus is human-centered and user-driven for value creation to improve the total university experience (Sutjarittham, 2021). IoT is a revolutionary technology that has upgraded human lives through the smartization of services (Swati & Dimple, 2021). There is an integration of new technologies focusing on enhancing the life of campus citizens (Villegas-Ch *et al.*, 2019; Esmaeilpoorarabi *et al.*, 2018). It has smart features that improve the quality and support the overall experience of its citizens, providing value-added services in developing a much better and more effective environment achieved through technological changes using smart devices through the internet (Adamkó, 2018).

There is a wide application of internet-on-community services and management (Awuzie *et al.*, 2021) and smart living in smart transport, smart canteen, and smart navigation (Zhang *et al.*, 2022). Moreover, expert systems are provided to improve the educational environment and quality (Al-Shoqran & Shorman, 2021), such as a dedicated smart campus solution architecture called RevoCampus developed to enhance room comfort control for a smart campus (Debauche *et al.*, 2020). For ease of parking, a parking management system was developed to assist matching of individual drivers to vacant parking places (Waheb *et al.*, 2021; Yee *et al.*, 2020). Finally, there was a development of social systems to assist life on campus through a smart platform that forms social circles based on interests mining (Dong *et al.*, 2016). These are but a few recent developments in smart campus improving the quality of campus life, which demonstrates the beginning of a large potential yet to be developed.

13.3.3.3 Social Responsibility

A smart campus is an emerging development in IoT technology (Sari *et al.*, 2017) that provides the overall influence on campus citizens, including the development of digital systems to enhance services in a university (Adamkó, 2018). In addition, a dynamic ecosystem is created by campus citizens in a digital community (Curralo *et al.*, 2022), forming a robust communications network (Demartini *et al.*, 2020). Moreover, a community of human-centered and user-driven value creation to improve the total university experience is created (Awuzie *et al.*, 2021), facilitated by a Wi-Fi hotspot-based mobile application to construct friendship networks (Zheng *et al.*, 2018). In addition, a platform was developed to enable formation of teams for IoT applications appropriate in a university context of academic groups (Acuña *et al.*, 2021). Furthermore, external stakeholders are now able to participate in defining higher education institutions' external interests (Kangas & Aarrevaara, 2020). Moreover, a new management system is developed to replace the present system, which has become obsolete (Zhang, 2022). On the other hand, robust systems technologies were developed, including a novel DMS (disaster management system) called the Named Data Networking Disaster Management (NDN-DM) (Ali *et al.*, 2020) and a surveillance system for enabling a safer space (Anagnostopolos *et al.*, 2021). There is evidence on the emphasis in the social factor promoting stakeholders' participation that complements the enhancement of social responsibility.

13.3.3.4 Versatile Learning

The rapid change in technology is transforming learning and teaching, evolving from basic e-learning to smart campuses and s-learning (Saiyeda, 2020). It is allowing education to be delivered anywhere and anytime through distance learning, growing, and creating new programs dedicated for implementation (Silvestru *et al.*, 2022). In addition, learning opportunities are now accessible to all segments of society for geographical boundaries are no longer a barrier (Abeer, 2021). Moreover, digital transformation has created new learning using digital technology to provide high-quality resources to a wider audience and improved access to information that is physically inaccessible (Agbehadji *et al.*, 2021). Online learning is now encouraging students to become more participative and active learners (Al-Mekhlafi *et al.*, 2022). Furthermore, continuous innovation in technology has created a high-end form of education system with remote, personalized, and ubiquitous features (Zhang *et al.*, 2022).

Learning efficiency can be promoted using wireless communication and virtual reality (VR) technology, reforming college education and entering a stage of rapid development (Li, 2022). Moreover, a smart classroom is implemented through mobile learning for modernizing and internationalizing education networks (Omirzak *et al.*, 2022) as new learning approaches are developed through the integration of IT (Tan & Du, 2022). Furthermore, the intelligent classroom stimulates students' learning process and comprehension quality (Qi & Huang, 2022) as a smart learning environment (Shi *et al.*, 2022). However, online learning has shortcomings, including monotonous and uneven teaching quality, absence of interaction, inadequate teaching audit, and the lack of students' motivation in learning (Xue *et al.*, 2021). This might be due to that teaching and learning are innately human processes that simply work best with direct and interpersonal instruction and relationships. It has been demonstrated that the development of smart technology for learning has advanced and will continue to progress, thus increasing the versatility in learning.

13.3.4 Environment Domain of a Smart Campus

The environment domain of the smart campus is concerned with the management of the natural environment and sustenance of resources (Fraga-Lamas *et al.*, 2019). The environment domain is reviewed under four aspects: environmentally friendly services, renewable energy, sustainable development, and zero waste, which are discussed below. A total of 39 articles were found to discuss the environment domain, where 29 focused on the conceptual phase, while 10 focused on prototyping and case study.

13.3.4.1 Environmentally Friendly Services

Environmentally friendly services are created on a smart campus through smart technology by integrating technologies into a centralized environment for generating a sustainable relationship between people and the natural environment (Villegas-Ch *et al.*, 2019). Moreover, environmental sustainability has been enhanced by reducing travel to and from campuses, minimizing the use of facilities, and curtailing energy demand. In addition, there is reduction of paper waste and tree harvests (Yigitcanlar *et al.*, 2007; Falola *et al.*, 2022). Finally, a wireless sensor network, IoT cyber-physical system, was developed in the University of Malaga in Spain to mitigate environmental pollution, including noise, by assisting campus managers in scheduling classes to avoid peaks and clusters (Toutouh & Alba, 2021). The online learning promoting remote learning and reducing on-campus learning has created environmentally friendly services.

13.3.4.2 Renewable Energy

Renewable energy is created by incorporating green energy into smart system applications to minimize the environmental impact through a hybrid renewable energy system operated by an IoT-based architecture in a university campus (Eltamaly *et al.*, 2021). Green IT (information technology) assists in reducing electrical power consumption (Zainuddin *et al.*, 2021) in a global move by universities towards energy-saving and carbon-emission mitigation thereby improving sustainability and stakeholders' participation (Waheb *et al.*, 2021). An energy management system, involving development of smart grid and microgrid project integration in sustainable energy generation, was pursued at the University of Genoa's Savona Campus. Its simulation was successful on the optimal management of heating, ventilation, and air-conditioning systems and plant (Bianco *et al.*, 2020). In addition, a proposed energy management system was developed to integrate renewable energy resources with a campus microgrid for cost reduction (Javed *et al.*, 2021). Finally, energy management and efficiency are pursued in major universities demonstrated by the development of a first ICT-centered methodology and software toolchain (Falola *et al.*, 2022). Sustainable transport, relying on smart energy, is yet another key area for smart campuses (Yigitcanlar *et al.*, 2008). The aspiration of integrating renewable energy resources, by using smart technology to promote green energy, however, is still at its infancy stage.

13.3.4.3 Sustainable Development

Industry 4.0 and sustainable development goals (Mian *et al.*, 2020; Deraman *et al.*, 2021) were among the global agendas that triggered transformations in higher education institutions (HEIs) for influencing future generations to deal with global issues (Son-Turan, 2022). Therefore, HEIs have begun focusing on sustainability management throughout the globe (Adenle *et al.*, 2020), facilitated effectively by advanced technology for sustainable development (Villegas-Ch *et al.*, 2020). Environmental specific factors were among the issues considered in the pursuit of IoT-based smart campus (Sneesl *et al.*, 2022). Smart campuses provide urban living labs (ULLs) (Martínez-Bello *et al.*, 2021) and a testbed (Ceccarini *et al.*, 2021) designed to facilitate experimentation on sustainability solutions in energy production and water utilization and cleanliness, meeting the global SDGs (Zhang *et al.*, 2022).

Moreover, digital technology is used in treating water, environmentally friendly facilities, and mobility and waste management (Alrashed, 2020), driven by smart solution systems (Adamkó, 2018) through the long-range wide-area network infrastructure (Fraga-Lamas *et al.*, 2019) and IoT platforms (Acuña *et al.*, 2021). Additionally, integrative approaches to sustainability (Amaral *et al.*, 2021) promoting open and integrative participation of stakeholders (Silva-da-Nóbrega *et al.*, 2022) were adopted by HEIs. Finally, environmental sustainability is boosted by e-learning because emissions go down due to a reduction in travel to campuses, as virtual classes can be attended from anywhere in the world (Falola *et al.*, 2022). There has been attempts to use smart technology to enhance the pursuit of sustainable development on campuses and it is expected to increase together with the rapid development of digital technology.

13.3.4.4 Zero Waste

Zero waste is about the reduction of waste on a smart campus to improve the state of the natural environment addressed through waste management and green campus. Waste management systems developed included the following three prototypes: firstly, the smart waste bin for effective waste collection (Agbehadji *et al.*, 2022); secondly, the Smartbottle Ecosystem developed as a means to eliminate single-use plastic water bottles to mitigate waste, where the bottle communicates with a smart refilling station through IoT and ICT for refill (Curralo *et al.*, 2022); and finally, the third is a smart waste management system for a university campus operated by 5G to mitigate health and environmental problems. It is based on the premise of recycling materials and reducing emissions from landfills and industrial sites to rectify environmental issues of water/air pollution and littering (Longo *et al.*, 2021). Green campus development, on the other hand, is demonstrated by two prototypes. Firstly, the reduction of carbon footprint of a university using solar photo volt system to save 31% of electrical energy and decrease the annual emissions from campus electricity use by 27% (Teah *et al.*, 2019). The second prototype is on smart agriculture through IoT to improve soil quality, water management, and environmental sustainability (Tabuenca *et al.*, 2020). The few prototypes demonstrated the beginning of tangible research on zero waste and is expected to expand, coinciding with the rapid development of digital technology.

13.3.5 Governance Domain of Smart Campus

The governance domain of the smart campus enables users to be involved in decision making and in the use of public services (Fraga-Lamas *et al.*, 2019). This is elaborated further into cybersecurity, data governance, decision making, and service management, which facilitate the governance domain. A total of 29 articles were found on the governance domain, where 19 were conceptual and 10 were prototypes and case studies.

13.3.5.1 Cybersecurity

Cybersecurity has emerged swiftly, along with the surge in IoT, prompting an urgent challenge for all institutions and business organizations and is growing (Alzahrani & Bamhdi, 2022). Smart campus development involves significant informatization, emphasizing the need for effective cybersecurity (Chen *et al.*, 2022). Cybersecurity is demanded literally in networks due to their vulnerability to attacks and is thus becoming a focus of recent research (Aljabri *et al.*, 2021). Smart systems require machine learning (ML) and deep learning (DL) intelligence to detect attacks in IoT networks (Alzahrani & Alsaade, 2022). However, intrusion detection systems (IDS) using traditional machine-learning algorithms in online systems are becoming weak as they are shallow learning and cannot counter new attacks (Kanna & Santhi, 2022). Shallow intelligent algorithms cannot process the voluminous amount of data now generated. Hence, deep learning algorithms are required and are gaining focus on recent research to address the emerging problem of cybersecurity in cloud computing architectures (Jauro *et al.*, 2020). Finally, blockchain frameworks also provide effective counter cyber-attack measures in quantum tools for large-scale quantum computers with cryptographic mechanisms (El-Latif *et al.*, 2021). Cybersecurity is developed at a fast phase with deep learning algorithm technologies to deal with the increasing trend in hacking.

13.3.5.2 Data Governance

Data governance emphasizes the significant role that data provides in facilitating the governance domain on a smart campus, including a scientific information system framework required on a smart campus (Cao *et al.*, 2022) to create a system for managing information for data detection through the integration of IoT technology and intelligent computing model (Lei *et al.*, 2022). A main facet for governance in public organizations is information technology (Telino *et al.*, 2020) and, as such, a smart campus requires a data governance platform to manage and control education data resources objectively (Chen & Liu, 2021). In addition, there is a need for effective systems to deal with the increasing challenges from student privacy, complex data, and the digital gap through smart campus innovation and social data governance (Cheong & Nyaupane, 2022).

Hence, universities are encouraged to fully implement measures to manage a large amount of information through big data technology for decision making and development. The research, therefore, suggests a focus on development of student management systems using big data technology in smart campus (Zhang, 2022). Finally, big data management is effectively managed by blockchain technology, which is found to be vital for creating advanced solutions where university leaders are embracing as a robust solution (Kuzior & Sira, 2022). Data governance facilitates enhance management of information through smart technology. Universities comprised operational facets that serve a wide spectrum of stakeholders beyond its traditional role in teaching and research.

13.3.5.3 Decision Making

The robust decision making created by a smart campus facilitates its governance domain through smart decision making based on objective principles (Dong *et al.*, 2016). University administrators, industry, and policymakers need to know the crucial issues of a smart campus comprising specific technologies, organizational traits, environmental values, and the end-user requirements to improve the adoption and utilization of an IoT-based smart campus effectively (Sneesl *et al.*, 2022). Moreover, a university transition to smart campus in adopting smart technology is oriented to the improvement of processes outlining evolution paths and prioritize improvement measures, forming a road map that facilitates decision making (Rico-Bautista *et al.*, 2022).

In addition, the making of a smart campus emulates a smart city for improvement in administration, management, and decision making (Fortes *et al.*, 2019). As such, innovations in decision making adopted the approach in making university campuses like smart cities having more centralized body in smart governance (Alrashed, 2020) for sustainability management and implementation (Amaral *et al.*, 2021) through an integrated management system to facilitate decision making (Al-Shoqran & Shorman, 2021). In addition, a mobile platform called "On Campus" was developed to address accurate context awareness, resource allocation, and smart decision making (Dong *et al.*, 2016). Finally, a smart platform for delivering smart university projects through an IT conceptual framework was developed to facilitate good governance for a university strategic plan (Pérez *et al.*, 2021). These developments have demonstrated the emergence of smart technology for decision making in universities.

13.3.5.4 Service Management

Innovation in service management pursued an open extensible architecture to support service integrations and create a dynamic digital ecosystem (Adamkó, 2018). As such, artificial intelligence (AI) is used to enhance all educational processes and students' achievements in servicing the education function in a university (Al-Shoqran & Shorman, 2021). In addition, the emergence of a technology-enabled university is helping the management control resource utilization and improve information dissemination (Deraman *et al.*, 2021), thereby increasing efficiency and facilitating better space utilization (Waheb *et al.*, 2021). An architectural framework was developed for the integration of technologies to improve administrative and academic management, where its versatility creates solutions for implementing the various services in the university (Villegas-Ch *et al.*, 2019). Smart parking development addressed the problem of unauthorized visitors on the campus by enhancing the security service through the coordination of vehicle identification (Yee *et al.*, 2020) and a novel optimal allocation framework developed to allow a campus manager to re-domain a car park for revenue generation through electric vehicles (Sutjarittham, 2021).

Moreover, the management of a smart campus is enhanced by an ICT-centered methodology and software toolchain for monitoring and controlling infrastructure on energy management (Silva *et al.*, 2021). Finally, a smart fire evacuation system was developed, based on building information modeling (BIM) to develop a spatial building modeling plan for fire evacuation (Wehbe & Shahrour, 2021). These are a few recent developments in smart technology assistance in service management that demonstrates a potential for more opportunities in the near future.

13.4 FINDINGS AND DISCUSSION

This study investigated the making of a smart campus through a framework of the four domains of society, economy, environment, and governance, where each is aligned closely with the central smart campus concepts of smart technology and big data. The domains were investigated further to determine the extent of their development, where four subthemes for each were discovered to create the conceptual framework shown in Figure 13.3. Moreover, a profile of the spread of the recent research on the domains is shown in Appendix 13.A. The study also gauged the level of development between the conceptual and prototype initiatives conducted so far, also shown in Appendix 13.A.

The study review results revealed a variation in the amount of research conducted on each domain. The society domain had more research followed by the environment, the economy, and lastly, the governance

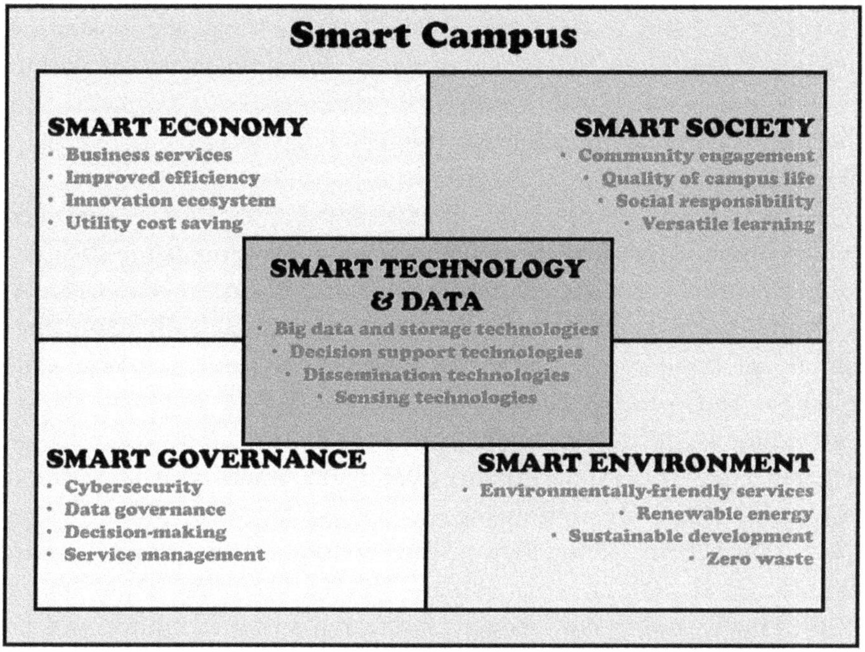

FIGURE 13.3 Conceptual framework of a smart campus.

domain. These are based solely on the count of recent publications located in the search and did not take into account any other basis for differences in the outcome. Moreover, the count was completed on the content analysis of the reviewed literature, where a spread on the coverage of each article on the domains was revealed, demonstrating that the majority were devoted to a single domain, followed by the coverage of two, then three, and lastly, all four domains. It was also revealed that more than half of the research was of conceptual nature and only a few were prototypes, indicating the level of research on smart campus to be more conceptual at this present stage.

The results revealed the economy domain with business opportunities provided by smart campus to be at its infancy stage where there is no evidence of expanded businesses established on campuses at present. Current research has only provided an overview of the business setting, such as meeting the global agenda of Industry 4.0 to modernize university operations that aim to create rich and diverse business systems (Mian *et al.*, 2020). Moreover, few discrete studies were completed on technology to sustain business, such as blockchain technology (Son-Turan, 2022) to control the sales of energy (Zainuddin *et al.*, 2021), a circular economy created by recycling waste (Longo *et al.*, 2021), building stock sustenance (Wehbe & Shahrour, 2021), and revenue generation using car parking spaces (Silva *et al.*, 2019). Research, to date, has only highlighted the possibility of improving efficiency through smart technology (Kuzior & Sira, 2022).

However, there is no evidence of widespread application. Again, there has only been a few discrete smart technology developments, such as a trading platform for efficient resource control (Dong *et al.*, 2016). An innovation ecosystem has so far been limited to sustainability and smart campus solutions where an urban living lab (ULL) (Martínez-Bello *et al.*, 2021) has been propagated predominantly to test for sustainability and green campus. Moreover, there has been limited studies focused on resource efficiency (Sutjarittham, 2021) and smart grid innovations (Alrashed, 2020) for optimizing energy management (Nasir *et al.*, 2021), blockchain technology (Silva *et al.*, 2019), smart parking (Waheb *et al.*, 2021), and human-centered learning (Dong *et al.*, 2020). Finally, utility cost reduction research so far is limited and has only begun with electrical energy reduction (Bin *et al.*, 2022) and cost reductions through smart buildings (Martínez-Bello *et al.*, 2021) and smart parking (Yee *et al.*, 2020).

On the other hand, the results on the society domain revealed a rise in smart technology development with few discrete studies being conducted. Community engagement development addressed connectivity with building an ecosystem of integrated environments for a digital campus (Yang *et al.*, 2018). The campus becomes connected, responsive, and smarter when it comprises a collection of people, amenities, and assets in an atmosphere of advanced digital infrastructures and services. A group and community concept for social-level fulfillment is created (Sarimin & Yigitcanlar, 2012). The quality of campus life is influenced by the smart city concept, which then impacts the citizens' quality of life (Swati & Dimple, 2021). The total university experience is human-centered and user-driven for value creation to upgrade human lives through technological changes using smart devices (Sutjarittham, 2021). Smart living is provided through expert systems, such as smart architecture (Zhang *et al.*, 2022) and a smart platform forming social circles (Dong *et al.*, 2016).

Social responsibility is about the overall influence on campus citizens (Adamkó, 2018) influenced by the paradigm of the IoT (Sari *et al.*, 2017), where friendship networks are facilitated by a mobile application (Zheng *et al.*, 2018), and work groups created by a platform (Acuña *et al.*, 2021). Disaster management (Ali *et al.*, 2020), and surveillance systems (Anagnostopolos *et al.*, 2021) were also created through smart technology development. In addition, versatile learning is characterized by its predominant benefit of a borderless delivery providing learning opportunity to all societies' segments (Abeer, 2021). Moreover, a high-end form of education system with remote, personalized, and ubiquitous features is created (Zhang, 2022), including smart classrooms for modernized learning integrating IT (information technology) (Omirzak *et al.*, 2022). However, online learning has certain disadvantages, such as monotonous delivery, lack of sharing knowledge, and insufficient education evaluation standards, which leads to a decrease in motivational drive for students (Xue *et al.*, 2021).

The study review results on the smart environment domain revealed the environmentally friendly services being created by generating a sustainable environment to harmonize the integration between people and nature (Villegas-Ch *et al.*, 2019). The use of smart technology mitigated the use of systems responsible for polluting the environment (Falola *et al.*, 2022). In addition, renewable energy was found to integrate green energy and smart system applications which mitigated ill-environmental impacts (Eltamaly *et al.*, 2021). Moreover, the hybrid renewable energy systems in Green IT created savings in energy and reduced carbon emissions (Yigitcanlar, 2021; Zainuddin *et al.*, 2021) through energy management systems (Javed *et al.*, 2021). Sustainable development is a global agenda (Son-Turan, 2022) addressed by smart solutions through integrative approaches in favor of humanity and sustainability (Villegas-Ch *et al.*, 2020). Finally, zero waste was found to involve the application of smart technology in waste management (Curralo *et al.*, 2022; Agbehadji *et al.*, 2022) and contributes to a green campus (Teah *et al.*, 2019; Tabuenca *et al.*, 2020).

Finally, the review on the governance domain revealed that cybersecurity development is reaching a heightened level of complex dynamics (Alzahrani & Bamhdi, 2022) to counter advanced attacks through deep learning algorithms (Jauro *et al.*, 2020) and blockchain technology (El-Latif *et al.*, 2021). These technologies provide a robust security system needed to protect the data governance function for the information framework in a university (Cao *et al.*, 2022). In addition, big data information is required for decision making and development (Zhang, 2022), which is effectively managed by blockchain technology (Kuzior & Sira, 2022). Moreover, there is an open extensible architecture to support service integration and creation (Adamkó, 2018). Finally, smart technology is helping the management control resource utilization and information dissemination (Deraman *et al.*, 2021) for service provisions in learning and other functions of a university (Villegas-Ch *et al.*, 2019; Wehbe & Shahrour, 2021; Silva *et al.*, 2021).

13.5 CONCLUSION

This review study used four major domains of the smart campus—namely society, economy, environment, and governance—to classify existing academic knowledge on smart campuses. These domains are each aligned to the central smart campus concepts of digital technology and big data. The analysis revealed little evidence of a comprehensive real-world application of the smart campus towards addressing all four domains. This finding indicates the infancy of the current conceptualization and practice. The study contributes to the development of a new conceptual foundation and research directions for the smart campus notion and informs its practice through a conceptual framework, and lastly advocates further empirical studies on smart campus for discovering the recipes for successful practice. In other words, the study findings reported in this chapter offer a firm basis for comprehensive smart campus conceptualization and provide directions for future research and development of smart campuses.

This study has the following limitations that should be noted: (a) The study did not engage any bibliometric or visualization software in the data analysis, which may result in some possible minor human errors. (b) The study did not undertake bibliometric (Donthu *et al.*, 2021), scientometric (Faisal *et al.*, 2021), or meta (Wirtz & Müller, 2021) analyses as the field is not advanced enough to expand the review to that level. The study also did not employ an analysis or visualization software and instead conducted the tasks manually. (c) The study adopted the generic domains of economy, society, environment, governance, and technology and data, and identified key indicators from the literature reviewed under these domains, where some other contextual domains might have existed. (d) The study focused on academic literature and only peer-reviewed journal articles; this might have resulted in neglecting some of the key developments reported in other academic or gray literature outlets. (e) The review was conducted manually, which might result in a reviewer bias. (f) There was no investigation on the technology and data domain of the framework. (g) The measurability of the 16 elements of the four domains is yet to be undertaken to add to the robustness of the framework. Lastly, our prospective research will focus on addressing these limitations and further expanding knowledge in these areas. Along with this, our prospective study will focus on investigating the functioning of smart campus as a new form of innovation district (Polin *et al.*, 2023).

ACKNOWLEDGMENTS

This chapter, with permission from the copyright holder, is a reproduced version of the following journal article: Polin, K., Yigitcanlar, T., Limb, M., & Washington, T. (2023). The making of smart campus: a review and conceptual framework. Buildings, 13(4), 891.

REFERENCES

Abdulmouti, H., Skaf, Z., & Alblooshi, S. (2022). Smart Green Campus: The Campus of Tomorrow. In Proceedings of the 2022 Advances in Science and Engineering Technology International Conferences (ASET), 1–8. IEEE. DOI:10.1109/ASET53988.2022.9735087

Abeer, A. (2021). Studies in Computational Intelligence. In *The Fourth Industrial Revolution: Implementation of Artificial Intelligence for Growing Business Success Studies in Computational Intelligence*, edited by Hamdan, A., 935, 295–310. Springer, Cham, Switzerland.

Acuña, L. C., Narváez, R., Salas, C., Magre, L. A., & González, M. J. (2021, October). Smart UTB: An IoT Platform for Smart Campus. In Proceedings of the 8th Workshop on Engineering Applications, WEA, Medellín, Colombia, 239–249. Springer, Cham, Switzerland.

Adamkó, A. (2018). Building Smart University Using Innovative Technology and Architecture. In *Smart Universities, Smart Innovation, Systems and Technologies*, edited byUskov, V. L., Bakken, J. P., Howlett, R. J., & Jain, L. C. Springer, Cham, Switzerland. DOI:10.1007/978-3-319-59454-5

Adenle, Y. A., Chan, E. H. W., Sun, Y., & Chau, C. K. (2020). Modifiable Campus-Wide Appraisal Model (MOCAM) for Sustainability in Higher Education Institutions. *Sustainability*, 12(17), 6821. DOI:10.3390/SU121 76821

Agbehadji, I. E., Abayomi, A., Bui, K. N., Millham, R. C., & Freeman, E. (2022). Nature-Inspired Search Method and Custom Waste Object Detection and Classification Model for SmartWaste Bin. *Sensors*, 22(16), 6176. DOI:10. 3390/s22166176

Agbehadji, I. E., Millham, R. C., Awuzie, B. O., & Ngowi, A. B. (2021). Stakeholder's Perspective of Digital Technologies and Platforms Towards Smart Campus Transition: Challenges and Prospects. In Proceedings of the

First International Conference, ICIIA 2021 Revised Selected Papers Informatics and Intelligent Applications Communications in Computer and Information Science, 197–213. Springer Nature.

Ahmed, V., Alnaaj, K. A., & Saboor, S. (2020). An Investigation into Stakeholders' Perception of Smart Campus Criteria: The American University of Sharjah as a Case Study. *Sustainability*, 12(12), 5187. DOI:10.3390/su12125187

Aldowah, H., Rehman, S., Ghazal, S., & Umar, I. N. (2017). Internet of Things in Higher Education: A Study on Future Learning. *Journal of Physics: Conference Series*, 892(1), 12017. DOI:10.1088/17426596/892/1/01017

Alhazmi, F. N. (2021). Cloud Computing Big Data Adoption Impacts in Teaching and Learning in Higher Education: A Systematic Review. In *Fostering Communication and Learning with Underutilized Technologies in Higher Education*, Ali, M. B., & Wood-Harper, A. T., 53–66. IGI Global: Hershey, PA, USA.

Ali, Z., Shah, M. A., Almogren, A., Din, I. U., Maple, C., & Khattak, H. A. (2020). Named Data Networking for Efficient IoT-based Disaster Management in a Smart Campus. *Sustainability*, 12(8), 3088. DOI: 10.33 90/SU12083088

Aljabri, M., Aljameel, S. S., Mohammad, R. M. A., Almotiri, S. H., Mirza, S., Anis, F. M., Aboulnour, M., Alomari, D. M., Alhamed, D. H., & Altamimi, H. S. (2021). Intelligent Techniques for Detecting Network Attacks: Review and Research Directions. *Sensors*, 21(21), 7070. DOI:10.3390/s21217070

Al-Mekhlafi, A. A., Othman, I., Kineber, A. F., Mousa, A. A., Ahmad, M. A., & Zamil, A. M. A. (2022). Modeling the Impact of Massive Open Online Courses (MOOC) Implementation Factors on Continuance Intention of Students: PLS-SEM Approach. *Sustainability*, 14(9), 5342. DOI:10.3390/su14095342

Alrashed, S. (2020). Key Performance Indicators for Smart Campus and Microgrid. *Sustainable Cities & Society*, 60, 102264. DOI:10.1016/j.scs.2020.102264

Al-Shoqran, M., & Shorman, S. (2021). A Review on Smart Universities and Artificial Intelligence. In *The Fourth Industrial Revolution: Implementation of Artificial Intelligence for Growing Business Success Studies in Computational Intelligence*, edited by Hamdan, A., 281–294. Springer Nature.

Alzahrani, M. S., & Alsaade, F. W. (2022). Intelligence Approaches in Developing Cyberattack Detection System. *Computational Intelligence and Neuroscience*, 2022, 4705325-16. DOI:10.1155/2022/4705325

Alzahrani, M. Y., & Bamhdi, A. M. (2022). Hybrid Deep-Learning Model to Detect Botnet Attacks over Internet of Things Environments. *Soft Computing (Berlin, Germany)*, 26(16), 7721–7735. DOI:10.1007/s00500-022-06750-4

Amaral, A. R., Rodrigues, E., Gaspar, A. R., & Gomes, A. (2021). Lessons from Unsuccessful Energy and Buildings Sustainability Actions in University

Campus Operations. *Journal of Cleaner Production*, 297, 1–12. DOI:10.1016/j.jclepro.2021.126665

Anagnostopolos, T., Kostakos, P., Zaslavsky, A., Kantzavelou, I., Tsotsolasi, N., Salmon, I., Morley, J., & Harle, R. (2021). Challenges and Solutions of Surveillance Systems in IoT-Enabled Smart Campus: A Survey. *IEEE Access*, 9, 131926–131954. DOI:10.1109/ACCESS.2021.3114447

Awuzie, B., Ngowi, A. B., Omotayo, T., Obi, L., & Akotia, J. (2021). Facilitating Successful Smart Campus Transitions: A Systems Thinking-SWOT Analysis Approach. *Applied Sciences*, 11(5), 1–21. DOI:10.3390/app11052044

Baum, S., Yigitcanlar, T., Horton, S., Velibeyoglu, K., & Gleeson, B. (2007). *The Role of Community and Lifestyle in the Making of a Knowledge City.* Griffith University, Brisbane, Australia.

Bianco, G., Bracco, S., Delfino, F., Gambelli, L., Robba, M., & Rossi, M. A. (2020). Building Energy Management System Based on an Equivalent Electric Circuit Model. *Energies*, 13(7), 1689. DOI:10.3390/en13071689

Bin, L., Shahzad, M., Javed, H., Muqeet, H. A., Akhter, M. N., Liaqat, R., & Hussain, M. M. (2022). Scheduling and Sizing of Campus Microgrid Considering Demand Response and Economic Analysis. *Sensors*, 22(16), 6150. DOI:10.3390/s22166150

Cao, H., He, H., & Tian, J. (2022). A Scientific Research Information System via Intelligent Blockchain Technology for the Applications in University Management. *Mobile Information Systems*, 2022, 7512692. DOI: 10.1155/2022/7512692

Cao, R. (2022, July). Promoting Digital Campus to Smart Campus Based on Artificial Intelligence. In Proceedings of the International Conference on Computer Engineering and Artificial Intelligence (ICCEAI), 730–734. IEEE. DOI:10.1109/ICCEAI55464.2022.00001

Ceccarini, C., Mirri, S., Salomoni, P., & Prandi, C. (2021). On exploiting Data Visualization and IoT for Increasing Sustainability and Safety in a Smart Campus. *Mobile Networks and Applications*, 26(5), 2066–2075. DOI:10.1007/s11036-021-01742-4

Chen, Z., & Liu, Y. (2021, October). Research and Construction of University Data Governance Platform Based on Smart Campus Environment. In 2021 Proceedings of the 3rd International Conference on Artificial Intelligence and Advanced Manufacture (AIAM), 450–455. ACM. DOI:10.1145/AIAM3495018.3495097

Chen, Z., Zhou, M., Feng, L., & Li, B. (2022). Statistical Analysis of threatening IP in Universities Based Automated Script. In Proceeding of the 2nd International Conference on Control and Intelligent Robotics (ICCIR), 825–828. ACM. DOI:10.1145/ICCIR3548608.3559315

Cheng, Y., Wei, J., Tan, X., Tan, Y., & Lei, Y. (2022. January). Research on Key Technologies of Data-Oriented Intelligent Campus in 5G Environment. In Proceedings of the 2nd International Conference on Consumer Electronics and Computer Engineering (ICCECE), 203–208. IEEE. DOI:10.1109/ICCECE54139.2022.9712690

Cheong, P. H., & Nyaupane, P. (2022). Smart Campus Communication, Internet of Things, and Data Governance: Understanding Student Tensions and Imaginaries. *Big Data & Society*, 9(1), 1–13. DOI:10.1177/20539517221092656

Cirrincione, L., Dio, S. D., Peri, G., Scaccianoce, G., Schillaci, D., & Rizzo, G. (2022). A Win-Win Scheme for Improving the Environmental Sustainability of University Commuters' Mobility and Getting Environmental Credits. *Energies*, 15(12), 396. DOI:10.3390/en15020396

Cortese, T. T. P., Almeida, J. F. S. de, Batista, G. Q., Storopoli, J. E., Liu, A., & Yigitcanlar, T. (2022). Understanding Sustainable Energy in the Context of Smart Cities: A PRISMA Review. *Energies*, 15(7), 2382.

Curralo, A. F., Lopes, S. I., Mendes, J., & Curado, A. (2022). Joining Sustainable Design and Internet of Things Technologies on Campus: The IPVC Smartbottle Practical Case. *Sustainability*, 14(10), 5922. DOI:10.3390/su14105922

Debauche, O., Abdelouahid, R. A., Mahmoudi, S., Moussaoui, Y., Marzak, A., & Manneback, P. (2020, September). RevoCampus: A Distributed Open Source and Low-cost Smart Campus. In Proceedings of the 3rd International Conference on Advanced Communication Technologies and Networking (CommNet), 1–10. IEEE. DOI:10.1109/CommNet49926.2020.9199640

Demartini, C. G., Benussi, L., & Gatteschi, V. (2020). Education and Digital Transformation: The "Riconnessioni" Project. *IEEE Access* 8, 186233–186256. DOI:10.1109/ACCESS.2020.3018189

Deraman, N. A., Buja, A. G., Samah, K. A. F. A., Jono, M. N. H. H., & Isa, M. A. M. (2021). I-Campus: Towards the Information Integration for Uitm Cawangan Melaka Implementation of Smart Campus. *Turkish Journal of Computer and Mathematics Education*, 12(5), 1699–1709. DOI:10.17762/turcomat.v12i5.2164

Dong, X., Kong, X., Zhang, F., Chen, Z., & Kang, J. (2016). On Campus: A Mobile Platform Towards a Smart Campus. *SpringerPlus*, 5(1), 1–9. DOI:10.1186/s40064-016-2608-4

Dong, Z. Y., Zhang, Y., Yip, C., Swift, S., & Beswick, K. (2020). Smart campus: Definition, Framework, Technologies, and Services. *IET Smart Cities*, 2(1), 43–54. DOI:10.1049/iet-smc.2019.0072

Donthu, N., Kumar, S., Mukherjee, D., Pandey, N., & Lim, W. M. (2021). How to conduct a bibliometric analysis: An overview and guidelines. *Journal of Business Research*, 133, 285–296. DOI:10.1016/j.jbusres.2021.04.070

El-Latif, A. A. A., Abd-El-Atty, B., Mehmood, I., Muhammad, K., Venegas-Andraca, S. E., & Peng, J. (2021). Quantum-Inspired Blockchain-Based Cybersecurity: Securing Smart Edge Utilities in IoT-Based Smart Cities. *Information Processing & Management*, 58(4), 102549. DOI:10.1016/j.ipm.2021.102549

Eltamaly, A. M., Alotaibi, M. A., Alolah, A. I., & Ahmed, M. A. (2021). IoT-Based Hybrid Renewable Energy System for Smart Campus. *Sustainability*, 13(15), 8555. DOI:10.3390/su13158555

Esmaeilpoorarabi, N., Yigitcanlar, T., Guaralda, M., & Kamruzzaman, Md. (2018). Evaluating place quality in innovation districts: A Delphic hierarchy process approach. *Land Use Policy*, 76, 471–486.

Faisal, A., Yigitcanlar, T., Kamruzzaman, M., & Paz, A. (2021). Mapping Two Decades of Autonomous Vehicle Research: A Systematic Scientometric Analysis. *Journal of Urban Technology*, 28(3-4), 45–74. DOI:10.1080/10630732.2020.1780868

Falola, H. O., Ogueyungbo, O. O., Adeniji, A. A., & Adesina, E. (2022). Exploring Sustainable E-Learning Platforms for Improved Universities' Faculty Engagement in the New World of Work. *Sustainability*, 14(7), 3850. DOI:10.3390/su14073850

Fernández-Caramés, T. M., & Fraga-Lamas, P. (2019). Towards Next Generation Teaching, Learning, and Context-Aware Applications for Higher Education: A Review on Blockchain, IoT, Fog and Edge Computing Enabled Smart Campuses and Universities. *Applied Sciences*, 9(21), 4479. DOI:10.3390/app9214479

Fernández-Morante, C., Cebreiro-López, B., Rodríguez-Malmierca, M., & Casal-Otero, L. (2022). Adaptive Learning Supported by Learning Analytics for Student Teachers' Personalized Training during in-School Practices. *Sustainability*, 14(1), 124. DOI:10.3390/su14010124

Fortes, S., Santoyo-Ramón, J. A., Palacios, D., Baena, E., Mora-García, R., Medina, M., Mora, P., & Barco, R. (2019). The Campus as a Smart City: University of Málaga Environmental, Learning, and Research Approaches. *Sensors*, 19(6), 1349. DOI:10.3390/s19061349

Fraga-Lamas, P., Celaya-Echarri, M., Lopez-Iturri, P., Castedo, L., Azpilicueta, L., Aguirre, E., Suárez-Albela, M., Falcone, F., & Fernández-Caramés, T. M. (2019). Design and Experimental Validation of a LoRaWAN Fog Computing Based Architecture for IoT Enabled Smart Campus Applications. *Sensors*, 19(15), 3287. DOI:10.3390/s19153287

Gao, M. (2021). Smart Campus Teaching System Based on ZigBee Wireless Sensor Network. *Alexandria Engineering Journal*, 61(4), 2625–2635. DOI:10.1016/j.aej.2021.09.001

Hicks, S., Huang, C., Lee, E., Patel, K., & Vesonder, G. (2021). Smart Campus for Better Study Spaces. In Proceedings of the IEEE 12th Annual Ubiquitous Computing, Electronics & Mobile Communication Conference (UEMCON), 325–327. IEEE. DOI:10.1109/UEMCON53757.2021.9666667

Hidayat, D. S., & Sensuse, D. I. (2022). Knowledge Management Model for Smart Campus in Indonesia. *Data*, 7(1), 7. DOI:10.3390/data7010007

Horvath, D., Csordas, T., Ásvanyi, K., Faludi, J., Cosovan, A., Simay, A. E., & Komar, Z. (2021). Will Interfaces Take Over the Physical Workplace in Higher Education? A Pessimistic View of the Future. *Journal of Corporate Real Estate*, 24(2), 108–123. DOI:10.1108/JCRE-10-2020-0052

Huang, J., Saleh, S., & Liu, Y. (2021). A Review on Artificial Intelligence in Education. *Academic Journal Interdisciplinary Studies*, 10(3), 206–217.

Huertas, J. I., Mahlknecht, J., Lozoya-Santos, J. J., Uribe, S., López-Guajardo, E. A., & Ramirez-Mendoza, R. A. (2021). Campus City Project: Challenge Living Lab for Smart Cities. *Applied Sciences*, 11(23), 11085. DOI:10.3390/app112311085

Ikrissi, G., & Mazri, T. (December 7-8, 2020). A Study of Smart Campus Environment and its Security Attacks. In Proceedings of the 5th International Conference on Smart City Applications, The International Archives of the Photogrammetry, Remote Sensing and Spatial Information Sciences–ISPRS Archives, 44(4), 255–261. DOI:10.5194/isprs-archives-XLIV-4-W3-2020-255-2020

Imbar, R. V., Supangkat, S. H., & Langi, A. Z. R. (November 19-20, 2020). Smart Campus Model: A Literature Review. In Proceedings of the 2020 International Conference on ICT for Smart Society (ICISS), 1–7. IEEE, New York, NY, USA. DOI:10.1109/ICISS50791.2020.9307570

Jauro, F., Chiroma, H., Gital, A. Y., Almutairi, M., Abdulhamid, S. M., & Abawajy, J. H. (2020). Deep Learning Architectures in Emerging Cloud Computing Architectures: Recent Development, Challenges and Next Research Trend. *Applied Soft Computing*, 96, 106582. DOI:10.1016/j.asoc.2020.106582

Javed, H., Muqeet, H. A., Shehzad, M., Jamil, M., Khan, A. A., & Guerrero, J. M. (2021). Optimal Energy Management of a Campus Microgrid Considering Financial and Economic Analysis with Demand Response Strategies. *Energies*, 14(24), 8501. DOI:10.3390/en14248501

Kangas, R., & Aarrevaara, T. (2020). Higher Education Institutions as Knowledge Brokers in Smart Specialisation. *Sustainability*, 12(7), 3044. DOI:10.3390/su12073044

Kanna, P. R., & Santhi, P. (2022). Hybrid Intrusion Detection using MapReduce based Black Widow Optimized Convolutional Long Short-Term Memory Neural Networks. *Expert Systems with. Applications*, 194, 116545. DOI:10.1016/j.eswa.2022.116545

Kern, M. A., & Smutko, S. (2021). Collaborative governance: The Role of University Centers, Institutes, and Programs. *Conflict Resolution Quarterly*, 39(1), 29–50. DOI:10.1002/crq.21314

Kuzior, A., & Sira, M. (2022). A Bibliometric Analysis of Blockchain Technology Research Using VOSviewer. *Sustainability*, 14(13), 8206. DOI:10.3390/su14138206

Lei, Q., Li, Y., & Yan, S. (2022). Design and Optimization of University Management Information System Based on Internet of Things and Intelligent Computing Model. *Journal of Sensors*, 2022, 1–11. DOI:10.1155/2022/1049535

Li, G. (2022). Methods of College Education Reform under the Background of Wireless Communication and VR. *Wireless Communications and Mobile Computing*, 2022, 1–6. DOI:10.1155/2022/2589533

Longo, E., Sahin, F. A., Redondi, A. E. C., Bolzan, P., Bianchini, M., & Maffei, S. (2021). A 5G-Enabled Smart Waste Management System for University Campus. *Sensors*, 21(24), 8278. DOI:10.3390/s21248278

Majumdar, S., & Mandal, M. (2021). Transforming the Environment of Education by Internet of Things: A Review. In *Recent Advances in Technology Acceptance Models and Theories Studies in Systems, Decision and Control*, edited byAl-Emran, M., & Shaalan, K., 335, 411–418. Springer Nature, Cham, Switzerland. DOI:10.1007/978-3-030-64987-6-29

Martínez-Bello, N., Cruz-Prieto, M. J., Güemes-Castorena, D., & Mendoza-Domínguez, A. (2021). A Methodology for Designing Smart Urban Living Labs from the University for the Cities of the Future. *Sensors*, 21(20), 6712. DOI:10.3390/s21206712

Martins, P., Lopes, S. I., da Cruz, A. M. R., & Curado, A. (2021). Towards a Smart & Sustainable Campus: An Application-Oriented Architecture to Streamline Digitization and Strengthen Sustainability in Academia. *Sustainability*, 13(6), 3189. DOI:10.3390/su13063189

Meddeb, O., Maraoui, M., & Zrigui, M. (2021). Personalized Smart Learning Recommendation System for Arabic Users in Smart Campus. *International Journal of Web-Based Learning and Teaching Technologies*, 16(6), 1–21. DOI:10.4018/IJWLTT.20211101.oa9

Mian, S. H., Salah, B., Ameen, W., Moiduddin, K., & Alkhalefah, H. (2020). Adapting Universities for Sustainability Education in Industry 4.0: Channel of Challenges and Opportunities. *Sustainability*, 12(15), 6100. DOI:10.3390/su12156100

Min-Allah, N., & Alrashed, S. (2020). Smart Campus—A Sketch. *Sustainable Cities and Societies*, 59, 102231. DOI:10.1016/j.scs.2020.102231

Moher, D., Shamseer, L., Clarke, M., Ghersi, D., Liberati, A., Petticrew, M., Shekelle, P., Stewart, L. A., & PRISMA-P Group. (2015). Preferred Reporting Items for Systematic Review and Meta-Analysis Protocols (PRISMA-P) 2015 Statement. *Systematic Reviews*, 4(1), 1. DOI:10.1186/20464053-4-1

Nasir, T., Raza, S., Abrar, M., Muqeet, H. A., Jamil, H., Qayyum, F., Cheikhrouhou, O., Alassery, F., & Hamam, H. (2021). Optimal Scheduling of Campus Microgrid Considering the Electric Vehicle Integration in Smart Grid. *Sensors*, 21(21), 7133. DOI:10.3390/s21217133

Omirzak, I., Alzhanov, A., Kartashova, O., & Ananishnev, V. (2022). Integrating Mobile Technologies in a Smart Classroom to Improve the Quality of the Educational Process: Synergy of Technological and Pedagogical Tools. *World Journal on Educational. Technology. Current Issues*, 14(3), 560–578.

Omotayo, T., Awuzie, B., Ajayi, S., Moghayedi, A., & Oyeyipo, O. (2021). A Systems Thinking Model for Transitioning Smart Campuses to Cities. *Frontiers in Built Environment*, 7, 755424. DOI:10.3389/fbuil.2021.755424

Omotayo, T., Moghayedi, A., Awuzie, B., & Ajayi, S. (2021). Infrastructure Elements for Smart Campuses: A Bibliometric Analysis. *Sustainability*, 13(14), 7960. DOI:10.3390/su13147960

Pérez, F. M., Martínez, J. V. B., & Fonseca, I. L. (2021). Modelling and Implementing Smart Universities: An IT Conceptual Framework. *Sustainability*, 13(6), 3397. DOI:10.3390/su13063397

Pittway, L. (2007). Systematic Literature Reviews. In *The SAGE Dictionary of Qualitative Management Research*, edited by Holt, R., & Thorpe, R., 216–218. SAGE, London, UK.

Polin, K., Yigitcanlar, T., Limb, M., & Washington, T. (2023). The making of smart campus: a review and conceptual framework. *Buildings*, 13(4), 891. DOI:10.3390/buildings13040891

Qi, Y., & Huang, Y. (2022). A Chinese Intelligent Teaching Platform for Colleges Based on Cloud Computing. *Mobile Information Systems*, 2022, 3487248. DOI:10.1155/2022/3487248

Rico-Bautista, D., Guerrero, C. D., Collazos, C. A., Maestre-Gongora, G., Sánchez-Velásquez, M. C., Medina-Cárdenas, Y., Parra-Sánchez, D. T., Barreto, A. C., & Swaminathan, J. (2022). Key Technology Adoption Indicators for Smart Universities: A Preliminary Proposal. In *Intelligent Sustainable Systems, Lecture Notes in Networks and Systems*, 333, 651–663. Springer, Berlin/Heidelberg, Germany.

Saiyeda, A. (2020). From E-Learning to S-Learning: A Review. In Proceedings of the 2nd International Conference on ICT for Digital, Smart, and Sustainable Development, ICIDSSD 2020, edited by Afshar Alam, M., Biwas, R., Ahmed, J., & Siddiqui, F., 27–28. Jamia Hamdard.

Sari, M. W., Ciptadi, P. W., & Hardyanto, R. H. (2017). Study of Smart Campus Development Using Internet of Things Technology. *IOP Conference Series: Materials Science and Engineering*, 190(1), 12032. DOI:10.1088/1757-899X/190/1/012032

Sarimin, M., & Yigitcanlar, T. (2012). Towards a comprehensive and integrated knowledge-based urban development model: status quo and directions. *International Journal of Knowledge-Based Development*, 3(2), 175–192.

Sciarelli, M., Landi, G. C., Turriziani, L., & Tani, M. (2021). Academic Entrepreneurship: Founding and Governance Determinants in University Spin off Ventures. *The Journal of Technology Transfer*, 46(4), 1083–1107. DOI:10.1007/s10961-020-09798-2

Shamseer, L., Moher, D., Clarke, M., Ghersi, D., Liberati, A., Petticrew, M., Petticrew, M., Shekelle, P., Stewart, L. A., & PRISMA-P Group. (2015). Preferred Reporting Items for Systematic Review and Meta-Analysis Protocols (PRISMA-P) 2015: elaboration and explanation. *BMJ (Online)*, 349(jan02 1), 7647. DOI:10.1136/bmj.g7647

Shi, Y., Peng, F., & Sun, F. A. (2022). Blended Learning Model Based on Smart Learning Environment to Improve College Students' Information Literacy. *IEE Access*, 10, 89485–89498. DOI:10.1109/ACCESS.2022.3201105

Silva, F. C., Ahmed, M. A., Martínez, J. M., & Kim, Y. (2019). Design and Implementation of a Blockchain-Based Energy Trading Platform for Electric Vehicles in Smart Campus Parking Lots. *Energies*, 12(24), 4814. DOI:10.3390/en12244814

Silva, L. C. P., Meira, P. C. M., Cypriano, J. G. I., Azzini, H. A. D., & Santos, A. Q. (2021). Software Toolchain to Enhance the Management and Integration of a Sustainable Campus Model. *Energy Informatics*, 4(Suppl 2), 1–11. DOI:10.1186/s42162-021-00154-8

Silva-da-Nóbrega, P. I., Chim-Miki, A. F., & Castillo-Palacio, M. (2022). A Smart Campus Framework: Challenges and Opportunities for Education Based on the Sustainable Development Goals. *Sustainability*, 14(15), 9640. DOI:10.3390/su14159640

Silvestru, C., Firulescu, A., Iordoc, D., Icociu, V., Stoica, M., Platon, O., & Orzan, A. (2022). Smart Academic and Professional Education. *Sustainability*, 14(11), 6408. DOI:10.3390/su14116408

Sneesl, R., Jusoh, Y. Y., Jabar, M. A., Abdullah, S., & Bukar, U. A. (2022). Factors Affecting the Adoption of IoT-Based Smart Campus: An Investigation Using Analytical Hierarchical Process (AHP). *Sustainability*, 14(14), 8359. DOI:10.3390/su14148359

Soegoto, E. S., & Rinaldy, M. F. (2020). Smart Campus with A Learning Management System. *IOP Conference Series: Materials Science and Engineering*, 879(1), 12003. DOI:10.1088/1757899X/879/1/012003

Song, H., Srinivasan, R., Sookoor, T., & Jeschke, S. (2017). Smart Cities: Foundations, Principles, and Applications. In edited by Song, H., Srinivasan, R., Sookar, T., & Jeschke, S. John Wiley & Sons. DOI:10.1002/9781119226444

Son-Turan, S. (2022). Fostering Equality in Education: The Blockchain Business Model for Higher Education (BBM-HE). *Sustainability*, 14(5), 2955. DOI:10.3390/su14052955

Sujata, J., Saksham, S., Tanvi, G., Shreya. (2016). Developing Smart Cities: An Integrated Framework. In Proceedings of the 6th International Conference on Advances on Computing & Communications, ICACC 2016, Procedia Computer Science, 93, 902–909. DOI:10.1016/j.procs.2016.07.258

Sutjarittham, T. (2021). Modelling and Optimisation of Resource Usage in an IoT Enabled Smart Campus. [Doctoral dissertation, UNSW]. DOI:10.48550/arxiv.2111.04085

Swati, J., & Dimple, C. (2021). A Smart Education Model for Future Learning and Teaching Using IoT Smart Innovation, Systems and Technologies. In *Information and Communication Technology for Intelligent Systems Proceedings of ICTIS 2020*, 2, 67–75. Springer.

Tabuenca, B., García-Alcántara, V., Gilarranz-Casado, C., & Barrado-Aguirre, S. (2020). Fostering Environmental Awareness with Smart IoT Planters in Campuses. *Sensors*, 20(8), 2227. DOI:10.3390/s20082227

Tan, L., & Du, F. (2022). Integrating Entrepreneurship and Innovation Education into Higher Vocational Education Teaching Methods Based on Big Data Analysis. *Wireless Communications and Mobile Computing*, 2022, 1–9. DOI:10.1155/2022/4616446

Teah, H. S., Yang, Q., Onuki, M., & Teah, H. Y. (2019). Incorporating External Effects into Project Sustainability Assessments: The Case of a Green Campus Initiative Based on a Solar PV System. *Sustainability*, 11(20), 5786. DOI:10.3390/su11205786

Telino, V., Massa, R., Mota, I., Gomes, A., & Moreira, F. (2020). A Methodology for Creating a Macro Action Plan to Improve IT Use and Its Governance in Organizations. *Information*, 11(9), 1–7. DOI:10.3390/info11090427

Toutouh, J., & Alba, E. A. (2022). Low Cost IoT Cyber-Physical System for Vehicle and Pedestrian Tracking in a Smart Campus. *Sensors*, 22(17), 6585. DOI:10.3390/s22176585

Ullah, H., Khan, M., Hussain, I., Ullah, I., Uthansakul, P., & Khan, N. (2021). An Optimal Energy Management System for University Campus Using the Hybrid Firefly Lion Algorithm (FLA). *Energies*, 14(19), 6028. DOI:10.3390/en14196028

Villegas-Ch, W., Molina-Enriquez, J., Chicaiza-Tamayo, C., Ortiz-Garcés, I., & Luján-Mora, S. (2019). Application of a Big Data Framework for Data Monitoring on a Smart Campus. *Sustainability*, 11(20), 5552. DOI:10.3390/su11205552

Villegas-Ch, W., Palacios-Pacheco, X., Luján-Mora, S. (2019). Application of a Smart City Model to a Traditional University Campus with a Big Data Architecture: A Sustainable Smart Campus. *Sustainability*, 11(10), 2857. DOI:10.3390/su11102857

Villegas-Ch, W., Palacios-Pacheco, X., & Román-Cañizares, M. (2020). Integration of IoT and Blockchain in the Processes of a University Campus. *Sustainability*, 12(12), 4970. DOI:10.3390/su12124970

Waheb, A. J., Chong, W. W., Nur, A. A. M. A., & Nur, A. H. (2021). An IoT Raspberry Pi-Based Parking Management System for Smart Campus. *Internet of Things*, 14, 2542–6605. DOI:10.1016/i.iot.2021.100387

Wan, J., & Yuan, H. (2022). Application of Intelligent Algorithm Big Data Analysis in Smart Campus Construction. In *Cyber Security Intelligence and Analytics, Lecture Notes on Data Engineering and Communications Technologie (CSIA 2022)*, 125, 824–828. Springer, Berlin/Heidelberg, Germany.

Wehbe, R., & Shahrour, I. (2021). A BIM-Based Smart System for Fire Evacuation. *Future Internet*, 13(9), 221. DOI:10.3390/fi13090221

Wirtz, B., & Müller, W. (2021). A Meta-Analysis of Smart City Research and its Future Research Implications. *International Public Management Review*, 21(2), 18–39.

Xu, D., Gao, Y., Tu, T., & Xiao, X. (2022). A Big Data Integration Platform for Ideological and Political Education for Smart Campuses. *Security and Communication Networks*, 2022, 1–12. DOI:10.1155/2022/5973920

Xue, L., Fu, R., Lin, D., Kuok, K., Huang, C., Su, J., & Hong, W. (2021). Exploring the Innovative Blockchain-Based Application of Online Learning System in University. In *Advances in Web-Based Learning–ICWL 2021*, 20, 90–101. Springer Nature, Cham, Switzerland.

Yang, A., Li, S., Ren, C. H., Liu, H., Han, Y., & Liu, L. (2018). Situational Awareness System in the Smart Campus. *IEEE Access*, 6, 63976–63986. DOI:10.1109/ACCESS.2018.2877428

Yee, O. C., Yaakob, N., Elshaikh, M., & Azahar, F. (2020). A Cloud-Based Automated Parking System for Smart Campus. *IOP Conference Series:*

Materials Science and Engineering, 767(1), 12049. DOI:10.1088/1757-899X/767/1/012049

Yigitcanlar, T. (2009). Planning for Smart Urban Ecosystems: Information Technology Applications for Capacity Building in Environmental Decision Making. *Theoretical and Empirical Researches in Urban Management*, 4(3 (12)), 5–21.

Yigitcanlar, T. (2021). Smart City Beyond Efficiency: Technology–Policy–Community at Play for Sustainable Urban Futures. *Housing Policy Debate*, 31(1), 88–92. DOI:10.1080/10511482.2020.1846885

Yigitcanlar, T., Dodson, J., Gleeson, B., & Sipe, N. (2007). Travel Self-Containment in Master Planned Estates: Analysis of Recent Australian Trends. *Urban Policy Res.*, 25, 129–149.

Yigitcanlar, T., & Dur, F. (2013). Making space and place for knowledge communities: lessons for Australian practice. *Australasian Journal of Regional Studies*, 19(1), 36–63.

Yigitcanlar, T., Edvardsson, I. R., Johannesson, H., Kamruzzaman, M., Ioppolo, G., & Pancholi, S. (2017). Knowledge-Based Development Dynamics in Less Favoured Regions: Insights from Australian and Icelandic University Towns. *European Planning Studies*, 25(12), 2272–2292. DOI:10.1080/09654313.2017.1358699

Yigitcanlar, T., Fabian, L., & Coiacetto, E. (2008). Challenges to Urban Transport Sustainability and Smart Transport in a Tourist City: The Gold Coast, Australia. *Open Transportation Journal*, 2, 29–46.

Yigitcanlar, T., Guaralda, M., Taboada, M., & Pancholi, S. (2016). Place Making for Knowledge Generation and Innovation: Planning and Branding Brisbane's Knowledge Community Precincts. *The Journal of Urban Technology*, 23(1), 115–146. DOI:10.1080/10630732.2015.1090198

Yigitcanlar, T., Sabatini-Marques, J., da-Costa, E. M., Kamruzzaman, M., & Ioppolo, G. (2019). Stimulating Technological Innovation Through Incentives: Perceptions of Australian and Brazilian Firms. *Technological Forecasting and Social Change*, 146, 403–412. DOI:10.1016/j.techfore.2017.05.039

Zaballos, A., Briones, A., Massa, A., Centelles, P., & Caballero, V. (2020). A Smart Campus' Digital Twin for Sustainable Comfort Monitoring. *Sustainability*, 12(21), 1–33. DOI:10.3390/su12219196

Zainuddin, A. A., Bhattacharjee, S., Kalliat, S., Shrestha, S., Sivaraman, S., Khalique, M. M., Naidu, D. J., & Manokaran, P. (2021). Trends and Challenges of Internet-of-Things in the Educational Domain. *MJSAT*, 1, 81–88.

Zhang, Y. (2022). Challenges and Strategies of Student Management in Universities in the Context of Big Data. *Mobile Information Systems*, 2022, 1–10. DOI:10.1155/2022/3537468

Zhang, Y., Dong, Z. Y., Yip, C., & Swift, S. (2020). Smart Campus: A User Case Study in Hong Kong. *IET Smart Cities*, 2(3), 146–154. DOI:10.1049/iet-smc.2020.0047

Zhang, Y., Yip, C., Lu, E., & Dong, Z. Y. (2022). A Systematic Review on Technologies and Applications in Smart Campus: A Human-Centered Case Study. *IEEE Access*, 10, 6134–16149. DOI:10.1109/ACCESS.2022.3148735

Zheng, W., Pan, H., & Pengi, Y. (2018). Explore the Ubiquitous Learning on Campus: A Friendship-Based Knowledge Diffusion Approach. *IEEE Access*, 6, 56238–56245. DOI:10.1109/ACCESS.2018.2872785

Zhou, L., & Tang, Q. (2022). Construction of a Six-Pronge Intelligent Physical Education Classroom Model in Colleges and Universities. *Scientific Programming*, 2022, 9003864. DOI:10.1155/2022/9003864

APPENDIX 13.A REVIEWED ARTICLES, PAPER TYPES, AND DOMAINS

No.	Reference	Title	Paper Type		Domain Covered			
			Concept	Prototype	Economy	Society	Governance	Environment
1	Abdulmouti *et al.* (2022)	Smart Green Campus: The Campus of Tomorrow	√		√	√		√
2	Abeer (2021)	Studies in Computational Intelligence	√			√		√
3	Acuña *et al.* (2021)	Smart UTB: An IoT Platform for Smart Campus	√		√			√
4	Adamkó (2018)	Building Smart University Using Innovative Technology and Architecture	√		√	√	√	
5	Adenle *et al.* (2020)	Modifiable Campus-Wide Appraisal Model for Sustainability		√				√
6	Agbehadji *et al.* (2021)	Stakeholder's Perspective of Digital Technologies and Platforms Towards Smart Campus Transition	√			√		
7	Agbehadji *et al.* (2022)	Nature-Inspired Search Method and Custom Waste Object Detection and Classification Model		√				√
8	Aldowah *et al.* (2017)	Internet-of-Things in Higher Education: A Study on Future Learning	√			√		

(Continued)

No.	Reference	Title	Paper Type		Domain Covered			
			Concept	Prototype	Economy	Society	Governance	Environment
9	Ali *et al.* (2020)	Named Data Networking for Efficient IoT-based Disaster Management	√			√		
10	Eltamaly *et al.* (2021)	IoT-Based Hybrid Renewable Energy System	√		√			√
11	Aljabri *et al.* (2021)	Intelligent Techniques for Detecting Network Attacks: Review and Research Directions	√				√	
12	Al-Mekhlafi *et al.* (2022)	Modeling the Impact of Massive Open Online Courses Implementation Factors on Continuance Intention of Students	√			√		
13	Alrashed (2020)	Smart Campus and Microgrid	√		√		√	√
14	Al-Shoqran & Shorman (2021)	Review on Smart Universities and Artificial Intelligence	√		√	√		
15	Alzahrani & Alsaade (2022)	Computational Intelligence Approaches in Developing Cyberattack Detection System		√			√	
16	Alzahrani & Bamhdi (2022)	Hybrid deep-learning model to detect botnet attacks over internet-of-things environments		√			√	
17	Amaral *et al.* (2021)	Lessons from unsuccessful energy and buildings sustainability actions	√		√		√	√
18	Anagnostopolos *et al.* (2021)		√			√		

		Challenges and Solutions of Surveillance Systems in IoT-Enabled Smart Campus						
19	Awuzie *et al.* (2021)	Facilitating Successful Smart Campus Transitions: A Systems Thinking-SWOT Analysis Approach	√			√		√
20	Bianco *et al.* (2020)	A Building Energy Management System Based on an Equivalent Electric Circuit Model		√		√		√
21	Bin *et al.* (2022)	Scheduling and Sizing of Campus Microgrid	√		√			
22	Cao *et al.* (2022)	Promoting Digital Campus to Smart Campus Based on Artificial Intelligence	√			√		
23	Ceccarini *et al.* (2021)	On exploiting Data Visualization and IoT for Increasing Sustainability and Safety	√			√		√
24	Chen & Liu (2021)	Research and Construction of University Data Governance	√		√	√	√	
25	Chen *et al.* (2022)	Statistical Analysis of threatening IP in Universities Based Automated Script	√				√	
26	Cheong & Nyaupane (2022)	Smart campus communication, Internet-of-Things, and data governance: Understanding student tensions and imaginaries	√			√	√	
27	Cirrincione *et al.* (2022)	A Win-Win Scheme for Improving the Environmental Sustainability of	√		√	√		√

(Continued)

No.	Reference	Title	Paper Type		Domain Covered			
			Concept	Prototype	Economy	Society	Governance	Environment
		University Commuters' Mobility and Getting Environmental Credits						
28	Curralo *et al.* (2022)	Joining Sustainable Design and Internet-of-Things Technologies on Campus: The IPVC Smartbottle		√		√		√
29	Debauche *et al.* (2020)	Improving the Environmental Sustainability	√		√	√		√
30	Demartini *et al.* (2020)	Education and Digital Transformation		√		√		
31	Deraman *et al.* (2021)	I-Campus: Towards the Information Integration	√		√	√	√	√
32	Dong *et al.* (2016)	On Campus: a mobile platform towards a smart campus		√	√	√	√	√
33	Dong *et al.* (2020)	Smart campus: definition, framework, technologies, and services	√		√	√		√
34	El-Latif *et al.* (2021)	Quantum-Inspired Blockchain-Based Cybersecurity: Securing Smart Edge Utilities in IoT-Based Smart Cities		√			√	
35	Nasir *et al.* (2021)	Optimal Scheduling of Campus Microgrid Considering the Electric Vehicle Integration	√		√	√		
36	Falola *et al.* (2022)	Exploring Sustainable E-Learning Platforms for Improved Universities'	√			√		√

		Faculty Engagement in the New World of Work						
37	Fernández-Morante *et al.* (2022)	Adaptive Learning Supported by Learning Analytics for Student Teachers' Personalized Training during in-School Practices	√		√	√		√
38	Fortes *et al.* (2019)	The Campus as a Smart City: University of Málaga Environmental, Learning, and Research Approaches		√		√		√
39	Fraga-Lamas *et al.* (2019)	Design and Experimental Validation of a LoRaWAN Fog Computing Based Architecture for IoT Enabled Smart Campus	√		√	√		√
40	Jauro *et al.* (2020)	Deep learning architectures in emerging cloud computing architectures	√				√	
41	Javed *et al.* (2021)	Optimal Energy Management of a Campus Microgrid		√	√			√
42	Kangas & Aarrevaara (2020)	Higher Education Institutions as Knowledge Brokers in Smart Specialization	√			√		
43	Kanna & Santhi (2022)	Hybrid Intrusion Detection using MapReduce based Black Widow Optimized Convolutional Long Short-Term Memory		√			√	
44	Kuzior & Sira (2022)		√		√		√	

(*Continued*)

No.	Reference	Title	Paper Type		Domain Covered			
			Concept	Prototype	Economy	Society	Governance	Environment
		A Bibliometric Analysis of Blockchain Technology Research Using VOSviewer						
45	Lei *et al.* (2022)	Design and Optimization of University Management Information System Based on Internet-of-Things and Intelligent Computing Model	√				√	
46	Li (2022)	Methods of College Education Reform under the Background of Wireless Communication and VR Wireless	√			√		
47	Longo *et al.* (2021)	A 5G-Enabled Smart Waste Management System		√	√	√	√	√
48	Martínez-Bello *et al.* (2021)	A Methodology for Designing Smart Urban Living Labs	√		√	√		√
49	Mian *et al.* (2020)	Adapting Universities for Sustainability Education in Industry 4.0	√		√			√
50	Nasir *et al.* (2021)	Optimal Scheduling of Campus Microgrid Considering the Electric Vehicle Integration	√		√	√		
51	Omirzak *et al.* (2022)	Integrating mobile technologies in a smart classroom to improve the quality of the educational process:	√			√		

		Synergy of technological and pedagogical tools						
52	Pérez *et al.* (2021)	Modeling and Implementing Smart Universities: An IT Conceptual Framework		√		√	√	
53	Sari *et al.* (2017)	Study of Smart Campus Development Using Internet-of- Things	√			√		
54	Shi *et al.* (2022)	A Blended Learning Model Based on Smart Learning Environment to Improve College Students' Information Literacy	√			√		√
55	Silva *et al.* (2019)	Design and Implementation of a Blockchain-Based Energy Trading Platform for Electric Vehicles		√	√			
56	Silva *et al.* (2021)	Software toolchain to enhance the management and integration of a sustainable campus		√	√		√	√
57	Silva-da-Nóbrega *et al.* (2022)	A Smart Campus Framework: Challenges and Opportunities for Education Based on the Sustainable Development Goals	√			√	√	√
58	Silvestru *et al.* (2022)	Smart Academic and Professional Education	√			√		
59	Sneesl *et al.* (2022)	Factors Affecting the Adoption of IoT-Based Smart Campus: An	√			√	√	√

(Continued)

No.	Reference	Title	Paper Type		Domain Covered			
			Concept	Prototype	Economy	Society	Governance	Environment
		Investigation Using Analytical Hierarchical Process (AHP)						
60	Soegoto & Rinaldy (2020)	Smart Campus with A Learning Management System	√		√	√		
61	Son-Turan (2022)	Fostering Equality in Education: The Blockchain Business Model for Higher Education (BBM-HE)	√		√	√		√
62	Sutjarittham (2021)	Modeling and Optimization of Resource Usage in an IoT Enabled Smart Campus		√	√	√	√	
63	Swati & Dimple (2021)	A Smart Education Model for Future Learning and Teaching Using IoT Smart Innovation, Systems and Technologies	√			√		
64	Tabuenca *et al.* (2020)	Fostering Environmental Awareness with Smart IoT Planters		√		√		√
65	Tan & Du (2022)	Integrating Entrepreneurship and Innovation Education into Higher Vocational Education Teaching Methods Based on Big Data Analysis	√			√		
66	Teah *et al.* (2019)		√		√		√	√

		Incorporating External Effects into Project Sustainability Assessments: The Case of a Green Campus Initiative Based on a Solar PV System						
67	Telino *et al.* (2020)	A Methodology for Creating a Macro Action Plan to Improve IT Use and Its Governance in Organizations	√				√	
68	Toutouh & Alba (2022)	A Low-Cost IoT Cyber-Physical System for Vehicle and Pedestrian Tracking	√			√		√
69	Nasir *et al.* (2021)	An Optimal Energy Management System for University Campus Using the Hybrid Firefly Lion Algorithm (FLA)	√		√	√		
70	Villegas-Ch *et al.* (2019)	Application of a Big Data Framework for Data Monitoring	√			√	√	√
71	Villegas-Ch *et al.* (2020)	Integration of IoT and Blockchain to in the Processes of a University Campus	√			√		√
72	Waheb *et al.* (2021)	An IoT Raspberry Pi-based parking management system	√		√	√	√	√
73	Wan & Yuan (2022)	Application of Intelligent Algorithm Big Data Analysis in Smart Campus Construction	√			√		
74	Wehbe & Shahrour (2021)	A BIM-Based Smart System for Fire Evacuation		√	√		√	
75	Xue *et al.* (2021)	Exploring the Innovative Blockchain-Based Application of Online Learning System in University	√			√		

(*Continued*)

No.	Reference	Title	Paper Type		Domain Covered			
			Concept	Prototype	Economy	Society	Governance	Environment
76	Yang *et al.* (2018)	Situational Awareness System	√			√		
77	Yee *et al.* (2020)	A cloud-based automated parking system for smart campus	√		√	√	√	
78	Zainuddin *et al.* (2021)	Trends and Challenges of Internet-of-Things in the Educational Domain	√		√			√
79	Zhang *et al.* (2022)	A Systematic Review on Technologies and Applications in Smart Campus: A Human-Centered Case Study	√			√		√
80	Zhang (2022)	Challenges and Strategies of Student Management in Universities in the Context of Big Data Mobile Information Systems	√			√	√	√
81	Zheng *et al.* (2018)	Explore the Ubiquitous Learning on Campus: A Friendship-Based Knowledge Diffusion Approach		√		√		
Total			60	21	34	58	29	39

Afterword

FROM TECHNOPOLES TO INNOVATION DISTRICTS?

I started my PhD almost two decades ago with the late Professor Sir Peter Hall on innovation milieu, which, at that time, largely referred to the 'campus-park-suburb'-style science and technology parks. As documented in our edited book, *Making 21st Century Knowledge Complexes* (Miao *et al.*, 2015), the contemporary technopole concept encompassed synergy-building, inclusive governance arrangements, and new evaluation mechanisms as distinct features. What was left untouched, however, was the physical features and spatial evolution of these complexes, although some contributors in our volume then had noticed the transformation from science parks to science cities and a broader knowledge-based urbanism.

Now the cities once again triumph over their suburban counterparts in attracting and agglomerating talent and capital that are crucial for knowledge and innovation economy. *Innovation districts* in particular, are praised as the response to not only the distinct nature of creative, smart, and innovative work but also the changing preferences of society. The attraction of these localities, as this literature goes, comes not only from their modern built environments such as denser, mix-used, public-transport-accessible, and 24-7 neighbourhoods, but also their all-around, human-scale social atmospheres that are authentic, 'hip', tolerant, diverse, and full of street-level activities and interactions. These features, in turn, are taught in classrooms and coded into planning strategies, adopted by both public and private initiatives with the hope of revitalising urban centres or capitalising grassroots developments that already show signs of enterprise and creative activity clustering.

Yet as the current book rightly points out, there is a lack of nuanced understanding on how to plan, deliver, and manage innovation districts that are competitive in both productivity and quality of living. It is not unusual that we see the planning manipulated innovation districts run up against their limits in terms of inequality, social exclusion, and environmental deterioration. Some of the industry-parks-turned-innovation-districts are still operating on car-oriented spatial scales, land-use separations and building setbacks, with limited public spaces, interaction venues, and social programs that bring the whole community together.

THE NATURE OF INNOVATION DISTRICTS?

Planning *per se* is not to be blamed for some of the negative consequences of innovation districts and other economic infrastructures. Rather, a deeper understanding of the multidimensional nature and diverse histories of innovation spaces across a variety of policy makers has been lacking, as rightly emphasised in the current book.

A first question that emerges, therefore, is whether innovation districts are part of the bigger family of innovation infrastructures, or whether they are primarily policy tools for urban regeneration, although these two positions are not completely mutually exclusive. Innovation infrastructures are the assemblage of hard, soft, and culture that facilitates to support innovation-related activities. It responds to the need to better understand the location, physical, and social needs of innovation activities that remain in flux. Studies of innovation infrastructures should be centred on the industries, their compositions, economic fundamentals, and interconnections, and will be constantly dealing with the dilemma between flexibility and stability. Tools for urban regeneration, on the other hand, are place-centric. The design and delivery of a particular tool should take into account the local context, history, demographic features, social dynamics, and economic baselines among others. It is also here that creative place making becomes most relevant, as it contributes to the development of a unique character, fosters a sense of place, and facilitates the generation and dissemination of knowledge and innovation, as documented clearly in this volume. Current policy designs and planning for innovation districts, however, are often juggled between these two positionings without recognising their different rationales and focal points, leading to ineffective and/or diluted policy interventions.

The second question following from the above relates to the feasibility or possibility of transforming all 'old-style' industry parks into innovation districts. A partial answer to this question would likely be provided by a revision and extension of extant innovation district taxonomies. Arguably, it is the science and technology park authorities in Asia, especially China, that have been the most eager to change. Famous cases such as Boston Seaport, Barcelona 22@, London Silicon Roundabout, and New York DUMBO are studied as 'bibles' for innovation district-led rebranding, and numerous groups of planning experts gather together to 'brainstorm' and 'blue print' innovation district policies. Yet whether this enthusiasm is based on solid evidence and scientific evaluation, or simply represents the worst copying or sprawling of policy, is still unclear. On the one hand, the transformation of suburb science parks towards urban innovation districts involves not only a focus on changes in the physical amenities and public transportation but also often the industrial composition. But in many places, industrial/economic planning and spatial planning are dealt with by different departments who hardly talk to each other. On the other, it is all too easy to talk about 'mixed land use'. But in reality, the hugely different land revenues and tax streams associated with different land uses make the negotiation between different jurisdictions a hostile debate instead of a harmonious, win-win cooperation.

WHERE NEXT?

The current volume by Tan and colleagues has already covered some of the trickiest questions facing the planning of innovation districts, especially regarding place making, classification, evaluation, and the engagement of universities. Building around these, we could argue that establishing a district speciality, attuning to inclusiveness, and strengthening sustainability are among the key tasks planners and policymakers should pay attention to. Each task has its own economic, social, and environmental dimensions. A district speciality, for example, ideally encompasses an industrial speciality that fits the unique physical and social heritage of a locality, and maximises and leverages the skills of existing residents while attracting newcomers. To this end, and echoing the central theme of the current book, the successful development of innovation districts requires aligned interests, collaboration, and knowledge exchange among stakeholders of all sorts. This perhaps lies the biggest promise of innovation district planning, and here in this volume, we have important insights that contribute to an understanding of this promise.

Julie Miao
The University of Melbourne

REFERENCE

Miao, J., Benneworth, P., & Phelps, N. (2015). *Making 21st century knowledge complexes: technopoles of the world revisited.* Routledge.

Index

For Product Safety Concerns and Information please contact our EU
representative GPSR@taylorandfrancis.com
Taylor & Francis Verlag GmbH, Kaufingerstraße 24, 80331 München, Germany

www.ingramcontent.com/pod-product-compliance
Ingram Content Group UK Ltd.
Pitfield, Milton Keynes, MK11 3LW, UK
UKHW020025200926
475702UK00014B/2058

* 9 7 8 1 0 3 2 6 5 7 0 1 1 *